METHODS IN MOLECULAR BIOLOGY™

Series Editor
John M. Walker
School of Life Sciences
University of Hertfordshire
Hatfield, Hertfordshire, AL10 9AB, UK

For other titles published in this series, go to
www.springer.com/series/7651

Plant Systems Biology

Edited by

Dr. Dmitry A. Belostotsky

School of Biological Sciences,
University of Missouri, Kansas City, MO, USA

Humana Press

Editor
Dmitry A. Belostotsky
School of Biological Sciences
University of Missouri
Kansas City, MO 64110
USA
belostotskyd@umkc.edu

ISSN 1064-3745 e-ISSN 1940-6029
ISBN 978-1-60327-562-0 e-ISBN 978-1-60327-563-7
DOI 10.1007/978-1-60327-563-7
Springer Dordrecht Heidelberg London New York

Library of Congress Control Number: 2009921811

Printed on acid-free paper

Springer is part of Springer Science+Business Media (www.springer.com)

Dedicated in memory of Dmitry Belostotsky, who passed away just before the publication of the volume. Special thanks are extended to Julia Chekanova for her gracious help during the book's final weeks.

Foreword

Systems biology has been called many things by many people. Rather than making another attempt at an all-encompassing definition, it may be better to take an historical perspective. Back at the dawn of time there was molecular biology, whose goal was to identify individual genes. With a gene in hand, one then searched upstream and downstream for other genes that acted on it or that it targeted. This led to the description of linear pathways with little arrows between each of the genes. Then came genomics with its high-throughput technologies to determine the expression of all genes, proteins, metabolites, etc. The output was usually a long list of cellular components ordered by expression level or some other metric. These were parsed for meaning based on where something was found on the list.

What systems biology has brought that is new and different is an emphasis on finding the connections among the parts. From these connections, the hope is that new properties will be identified that were not apparent from just staring at the list of parts. These are called "emergent properties." But systems biology does not stop there. After the connections are found and networks begin to emerge, the next step is to characterize the dynamic properties of these networks. Accomplishing this requires perturbing the system and then determining how the system responds. In biology, perturbations can take the form of external stimuli such as sunlight or withholding a nutrient. They can also be at the level of mutations that alter gene function or expression.

A distinguishing feature of systems biology is the integration of quantitative analytical and modeling approaches. In the days of molecular biology, the view of quantitative analysis was, "If you have to use statistics, it means you need to do another experiment." With the advent of genomics, most scientists realized that they needed help to make sense of the masses of data. Nevertheless, the general approach was that of a "hand-off" – the experimental biologist would find someone with quantitative expertise to "analyze my data." When the analysis was completed it would be handed back to the biologist and that was the end of the interaction. The complexity of dynamic systems has convinced most biologists that the human brain needs mathematical formalisms to make any sense of the processes being studied. This means that systems biology is by and large practiced by collaborative teams, which comprise experimentalists and theorists, with equal weighting between them.

In this book you will find chapters that describe how to identify cellular components as well as the interactions among these components. You will also find chapters that describe methods for perturbing biological systems such as the use of small molecules in chemical genomics. Fittingly, a large portion of the book is devoted to quantitative approaches to analyze and model the interactions, emergent properties, and dynamics of the networks identified.

This work focuses on systems biology applied to plants. For many of the approaches described here there is no distinction between plants and animals. However, it is appropriate to focus an entire book on plant systems biology as plants have been in the vanguard of this field. The sequencing of the *Arabidopsis* genome opened the way for a host of new and innovative approaches to understanding plant biology. From live imaging of protein dynamics in floral meristems to the ability to follow chromosome dynamics in individual cells, plant biologists are among the pioneers in this area. No matter how you choose to define systems biology, it is likely to play an increasingly important role in elucidating the mysteries of plant biology.

Philip N. Benfey

Preface

Plant Systems Biology: Shooting a Moving Target

Plant systems biology is a fairly new art form. Unsurprisingly, its practitioners come in a variety of different flavors, and accordingly, there exist a great many conflicting definitions of what this art form really is (although this probably is true of any art form). Researchers have been entering this field from all walks of scientific life – there are classical plant physiologists by training, wet bench gene expression biologists like myself, cell biologists, mathematicians, statisticians, bioinformaticians, software engineers, and other more esoteric types ranging all the way to astrophysicists, etc.

The eclectic nature of this proverbial melting pot is also reflected in the content of this volume, which contains sections covering topics from systems biology of plant gene expression to analysis of networks, pathways, specific statistical issues and novel computational tools, imaging-based tools as well as chemical genetic, metabolomic, and integrative methods that cannot be easily pigeonholed.

While the definition of what plant systems biology really is may still be evolving, its key leading figures have clearly emerged and who they are is largely beyond dispute. Indeed, it is quite obvious who is driving the field forward and paving the way for others who follow in their wake and broaden the path. It is for that reason that the foreword to this volume is written by Philip Benfey, whose pioneering studies in the field of systems biology of gene expression have received wide recognition far beyond the plant community.

While the natural evolution of the field has been rapid and successful, it has become quite obvious that the time has come for setting up dedicated training programs in order to sustain this remarkable progress. This is already happening of course, in the form of IGERT and other training grants, iPlant initiative, etc., but additional modalities are needed. It is also the hope of the editor that this volume will make a contribution to achieving this goal as well.

In closing, I would like to acknowledge the contributions of the members of my own group over the years, and particularly that of Julia Chekanova, as well as the expert editorial assistance of Teresa Crew, without whom this volume would never have seen the light of day. Gene expression studies in my lab have been supported by grants from NSF, USDA, BARD, and NIH.

Dmitry A. Belostotsky

Contents

Contributors

RÉKA ALBERT • *Physics Department, Penn State University, University Park, PA, USA*
DAVID B. ALLISON • *Section on Statistical Genetics, Department of Biostatistics, University of Alabama at Birmingham, Birmingham, AL, USA*
SARAH M. ASSMANN • *Biology Department, Penn State University, University Park, PA, USA*
JULIA BAILEY-SERRES • *Center for Plant Cell Biology and Department of Botany and Plant Sciences, University of California, Riverside, CA, USA*
ALICE BARKAN • *Institute of Molecular Biology, University of Oregon, Eugene, OR, USA*
DMITRY A. BELOSTOTSKY • *School of Biological Sciences, University of Missouri Kansas City, Kansas City, MO, USA*
PHILIP N. BENFEY • *Department of Biology, Duke University, Durham, NC, USA; IGSP Center for Systems Biology, Duke University, Durham, NC, USA*
ANNICK BLEYS • *Department of Plant Systems Biology, Flanders Institute for Biotechnology (VIB), Gent, Belgium; Department of Molecular Genetics, Ghent University, Gent, Belgium*
SIOBHAN M. BRADY • *Department of Biology, Duke University, Durham, NC, USA; IGSP Center for Systems Biology, Duke University, Durham, NC, USA*
BHAVNA CHAUDHURI • *Department of Plant Biology, Carnegie Institution for Science, Stanford, CA, USA*
JOSÉ R. DINNENY • *Department of Biology, Duke University, Durham, NC, USA; IGSP Center for Systems Biology, Duke University, Durham, NC, USA*
NATALIA DUDAREVA • *Department of Horticulture and Landscape Architecture, Purdue University, West Lafayette, IN, USA*
SERGEI FILICHKIN • *Department of Botany and Plant Pathology and Center for Genome Research and Biocomputing, Oregon State University, Corvallis, OR, USA*
SAMUEL FOX • *Department of Botany and Plant Pathology and Center for Genome Research and Biocomputing, Oregon State University, Corvallis, OR, USA*
WOLF B. FROMMER • *Department of Plant Biology, Carnegie Institution for Science, Stanford, CA, USA*
GARY L. GADBURY • *Department of Statistics, Kansas State University, Manhattan, KS, USA*
KAREN A. GARRETT • *Department of Plant Pathology, Kansas State University, Manhattan, KS, USA*
BRIAN D. GREGORY • *Plant Biology Laboratory, The Salk Institute for Biological Studies, La Jolla, CA, USA*
ERICH GROTEWOLD • *Department of Plant Cellular & Molecular Biology and Plant Biotechnology Center, The Ohio State University, Columbus, OH, USA*
GLENN R. HICKS • *Institute for Integrative Genome Biology and Center for Plant Cell Biology, Department of Botany and Plant Sciences, University of California, Riverside, CA, USA*

PIERRE HILSON • *Department of Plant Systems Biology, Flanders Institute for Biotechnology (VIB), Gent, Belgium; Department of Molecular Genetics, Ghent University, Gent, Belgium*

EDWARD L. HUTTLIN • *Department of Biochemistry, University of Wisconsin-Madison, Madison, WI, USA*

PIYADA JUNTAWONG • *Center for Plant Cell Biology and Department of Botany and Plant Sciences, University of California, Riverside, CA, USA*

MANSOUR KARIMI • *Department of Plant Systems Biology, Flanders Institute for Biotechnology (VIB), Gent, Belgium; Department of Molecular Genetics, Ghent University, Gent, Belgium*

DANIEL J. KLIEBENSTEIN • *Department of Plant Sciences, University of California, Davis, CA, USA*

JEREMY D. KOCH • *3051 Cassel PI, Davis, CA 95616*

IAN F. KORF • *Genome Center and Section of Molecular and Cellular Biology, University of California, Davis, CA, USA*

ERIC LAM • *Biotechnology Center for Agriculture and the Environment, Rutgers University, New Brunswick, NJ, USA*

ANN LORAINE • *Bioinformatics Research Center , University of North Carolina-Charlotte, Kannapolis, NC, USA*

CHONGYUAN LUO • *Biotechnology Center for Agriculture and the Environment, Rutgers University, New Brunswick, NJ, USA*

TODD C. MOCKLER • *Department of Botany and Plant Pathology and Center for Genome Research and Biocomputing, Oregon State University, Corvallis, OR, USA*

KENGO MOROHASHI • *Department of Plant Cellular & Molecular Biology and Plant Biotechnology Center, The Ohio State University, Columbus, OH, USA*

ANGELIKA MUSTROPH • *Center for Plant Cell Biology and Department of Botany and Plant Sciences, University of California, Riverside, CA, USA*

TOTTE NIITTYLAE • *Department of Plant Biology, Carnegie Institution for Science, Stanford, CA, USA*

LORENA NORAMBUENA • *Plant Molecular Biology Laboratory, Department of Biology, Faculty of Sciences, University of Chile, Santiago, Chile*

DAVID A. ORLANDO • *Department of Biology, Duke University, Durham, NC,USA; IGSP Center for Systems Biology, Duke University, Durham, NC, USA*

ANTHONY V. QUALLEY • *Department of Horticulture and Landscape Architecture, Purdue University, West Lafayette, IN, USA*

NATASHA V. RAIKHEL • *Institute for Integrative Genome Biology and Center for Plant Cell Biology, Department of Botany and Plant Sciences, University of California, Riverside, CA, USA*

G. VENUGOPALA REDDY • *Department of Botany and Plant Sciences, University of California, Riverside, CA, USA*

ALAN B. ROSE • *Section of Molecular and Cellular Biology, University of California, Davis, CA, USA*

A. ROY-CHOWDHURY • *Department of Electrical Engineering, University of California, Riverside, CA, USA*

UWE SAUER • *Institute for Molecular Systems Biology, ETH Zürich, Zürich, Switzerland*

EDGAR P. SPALDING • *Department of Botany, University of Wisconsin, Madison, WI, USA*

MICHAEL R. SUSSMAN • *Department of Biochemistry, UW Biotechnology Center, University of Wisconsin-Madison, Madison, WI, USA*
DANA J. WOHLBACH • *Department of Genetics, University of Wisconsin-Madison, Madison, WI, USA*
ZIDIAN XIE • *Department of Plant Cellular & Molecular Biology and Plant Biotechnology Center, The Ohio State University, Columbus, OH, USA*
CHEN-HSIANG YEANG • *Institute of Statistical Science, Academia Sinica, Nankang, Taipei, Taiwan, R.O.C.*

Section I

Systems Biology of Plant Gene Expression

Chapter 1

Gene-Specific and Genome-Wide ChIP Approaches to Study Plant Transcriptional Networks

Kengo Morohashi, Zidian Xie, and Erich Grotewold

Abstract

Chromatin immunoprecipitation (ChIP) provides a versatile tool to investigate the in vivo location of DNA-binding proteins on genomic DNA. ChIP approaches are gaining significance in plants, in cases when entire genome sequences are available (e.g., *Arabidopsis*), for which several high-density oligo arrays have been or are being developed. Nevertheless, plant ChIP and ChIP-chip still present some technical challenges. Here, we describe general methods for ChIP and ChIP-chip, which have been successfully applied to maize and *Arabidopsis*.

Key words: Regulatory network, chromatin immunoprecipitation, transcription factor, histone.

1. Introduction

Protein–DNA interactions are central for life, for example as part of normal chromatin assembly and in the recognition of specific *cis*-regulatory elements (CRE) by transcription factors (TFs). CREs provide the blueprints for the integration of cellular signals on the DNA, with the proper gene expression response furnished by the tethering of sets of TFs to specific DNA motifs and their interactions with the basal transcription machinery. Understanding which of the thousands of TFs expressed by plant genomes recognize which CREs and establishing how combinations of TFs on specific promoters contribute to the regulation of gene expression pose significant challenges in elucidating the architecture of plant transcriptional regulatory networks (1).

Dmitry A. Belostotsky (ed.), *Plant Systems Biology, vol. 553*

DOI 10.1007/978-1-60327-563-7_1 Springerprotocols.com

Methods to identify protein–DNA interactions include experimental and computational approaches, or combinations thereof (1). Experimental approaches involve investigating the formation of protein–DNA complexes for example by electrophoretic mobility shift assays (EMSA) or by exploring the specific DNA sequence recognized by a TF on a given fragment of DNA using chemical or nuclease footprinting techniques. These techniques, however, involve in vitro protein–DNA interactions and their application depends on the availability of a DNA fragment containing the regulatory sequences. Two main approaches are currently available to identify and/or validate the direct in vivo targets of a TF. The first one involves expressing a fusion of the TF to GR (GR corresponds to the hormone-binding domain of the glucocorticoid receptor) and identifying the mRNAs induced/repressed in the presence of the GR ligand (dexamethasone, DEX), in the presence of an inhibitor of translation (e.g., cycloheximide, CHX) (2–6). The second one involves identifying the DNA sequences that a TF binds *in vivo*, using chromatin immunoprecipitation (ChIP) assays. ChIP not only provides a tool to identify the *in vivo* location of DNA-binding proteins on the DNA but also complements many of the downfalls of EMSA and footprinting. The experimental steps necessary for implementing ChIP, or combinations of ChIP with the hybridization of microarrays (ChIP-chip) corresponding to entire genomes (tiling arrays) or just to the promoter space of a genome (promoter arrays), are the subject of this chapter.

In ChIP, intact tissues or cells are treated with a cross-linking agent that covalently links the protein with the DNA. The chromatin is then sheared (using enzymatic or mechanical methods) and the covalently linked protein–DNA complex is enriched by immunoprecipitation (IP) using the specific antibodies to the proteins (7). Multiple cross-linking agents that provide different spacer lengths are available for ChIP (*see* **Note 1**). Formaldehyde reacts with the amino groups of cytosines, guanines, and adenines and the imino groups of thymines and guanines on the DNA, although the reaction with imino groups is likely to be favored in single-stranded DNA regions. From the protein side, several amino acids are targeted by formaldehyde, including Lys, Arg, Trp, and His. The final product of the reaction is the joining of the two amino groups by a methylene bridge. The uniqueness of this reaction is that it is reversible in aqueous solutions, and the reversion of the cross-linking can be significantly increased by incubation at 65°C. The reversibility of the formaldehyde-mediated reaction furnishes an advantage that has made it the favorite reagent for cross-linking. It should be noted, however, that in some instances it is worth considering other cross-linking agents, if the desired results are not obtained with formaldehyde (8).

ChIP-chip (aka ChIP-on-chip or genome-wide location analysis) involves the hybridization of a microarray representing a fraction or the entire genome space with the DNA resulting from the ChIP experiment (9). In plants, several arrays have become available over the past couple of years for ChIP-chip experiments; for example, for *Arabidopsis*, promoter (10, 11) and complete tiling genome (10, 12, 13) arrays are available. In most instances, the ChIPed DNA needs to be amplified prior to microarray hybridization, because nanogram quantities of DNA are precipitated, for example when using antibodies against a specific TF. Several different amplification methods are available (14), and some are discussed later. A plethora of statistical approaches have been developed for the analysis of ChIP-chip results [e.g., (9, 15)]; their discussion and application are however beyond the scope of this chapter. ChIPed DNA, however, can also be analyzed by methods other than the hybridization to a microarray. For example, the ChIP-Paired End diTag (PET) method results in the generation of short sequence tags from the enriched target DNA after a ChIP experiment (16). Alternatively, the ChIPed DNA can be cloned and sequenced [e.g., (17)].

ChIP approaches are gaining significance in plants, particularly for those for which entire genome sequences are available (e.g., *Arabidopsis*). While ChIP-chip experiments have been performed on just a handful of *Arabidopsis* TFs including HY5 (11) and TGA2 (10), ChIP is becoming an increasingly popular method to validate *in vivo* TF–DNA interactions predicted by other methods. Moreover, ChIP-chip can be applied to identify the epigenetic control of the transcriptional regulation.

2. Materials

1. Salmon sperm/protein A-agarose (Upstate P/N 16-157)
2. PCI: phenol:chloroform:isoamylic alcohol (25:24:1)
3. Buffer A: 0.4 M sucrose, 10 mM Tris pH 8.0, 1 mM EDTA, 1 mM PMSF (*see* **Note 2**), 1% formaldehyde
4. Lysis buffer: 50 mM HEPES pH 7.5, 150 mM NaCl, 1 mM EDTA, 1% Triton X-100, 0.1% sodium deoxycholate, 0.1% SDS, 10 mM Na butyrate, 1 mM PMSF, 1X plant proteinase inhibitor cocktail (Sigma) (*see* **Note 2**)
5. LNDET: 0.25 M LiCl, 1% NP40, 1% sodium deoxycholate, 1 mM EDTA
6. Elution buffer: 1% SDS, 0.1 M $NaHCO_3$, 0.25 mg/ml proteinase K

7. PCR purification kit (QIAGEN), DNA Clean & Concentrator – 25 (Zymo Research)
8. GenomePlex Whole Genome Amplification kit (Sigma, P/N WGA-1): 10X library buffer, 10X library stabilization solution, library preparation enzyme
9. GeneChip® Arabidopsis Tiling 1.0R Array (Affymetrix)

3. Methods

3.1. Chromatin Immunoprecipitation (ChIP) in Plants

3.1.1. Cross-Linking Proteins to DNA

1. Immerse tissue into buffer A in a 50 ml falcon tube and keep it under vacuum for 20 min (*see* **Notes 3** and **4**).
2. Add 2 M glycine to a final concentration of 0.1 M and continue vacuum for 10 min.
3. Wash the tissue with excess amount of distilled water and remove as much water as possible.
4. Grind tissue in liquid nitrogen and resuspend in 400 μl of lysis buffer (*see* **Notes 5** and **6**).

3.1.2. Sonication of Chromatin (see Note 7)

1. Shear DNA by sonication to a fragment length that ranges between 100 bp and 1000 bp (~500 bp on average) in an eppendorf tube (*see* **Notes 8** and **9**).
2. Centrifuge at 10,000 × *g* for 10 min at 4°C.

3.1.3. Immunoprecipitation (see Note 10)

1. Pre-clear supernatant with 30 μl of salmon sperm/protein A-agarose for at least 60 min with rotation at 4°C.
2. Transfer 100 μl of supernatant into three new eppendorf tubes and add the antibodies (*see* **Notes 11** and **12**). Keep approximately 100 μl of extract as the input fraction.
3. Incubate overnight with rotation at 4°C.
4. Add 30 μl of salmon sperm/protein A-agarose slurry and continue incubation with rotation at 4°C for at least 2 h.
5. Centrifuge at 750 × *g* (3000 rpm for microcentrifuge) for 1 min at 4°C.

3.1.4. Washes

1. Add 0.5 ml of lysis buffer, invert six times, centrifuge at 750 × *g* for 1 min, and discard supernatant (*see* **Note 12**).
2. Add 0.5 ml of lysis buffer, rotate for 5 min, centrifuge at 750 × *g* for 1 min, and discard supernatant.
3. Add 0.5 ml of LNDET, invert six times, centrifuge at 750 × *g* for 1 min, and discard supernatant.
4. Add 0.5 ml of LNDET, rotate for 5 min, centrifuge at 750 × *g* for 1 min, and discard supernatant.

5. Add 0.5 ml of TE, invert six times, centrifuge at 750 × *g* for 1 min, and discard supernatant.
6. Add 0.5 ml of TE, rotate for 5 min, centrifuge at 750 × *g* for 1 min, and discard supernatant.

3.1.5. Reverse Cross-Linking

1. Add 40 μl of elution buffer and incubate at 65°C for 15 min.
2. Centrifuge at 750 × *g* for 1 min and transfer supernatant to new tube.
3. Repeat eluting steps. The final elution volume should be now 80 μl. In parallel, add 70 μl of elution buffer into 10 μl of input fraction for the input control, which represents 10% of the cross-linked DNA (*see* **Note 13**).
4. Incubate all samples overnight at 65°C.

3.1.6. DNA Isolation

Extract the DNA by using the PCR purification kit (QIAGEN). Elute in 30 μl of EB buffer (10 mM Tris–HCl, pH 8.5) (*see* **Note 14**).

3.1.7. Quantification of ChIPed DNA

We generally use 1 μl of eluted DNA samples for standard PCR (*see* **Note 15**) and normalization (*see* **Note 16**), although larger quantities can be used, if necessary.

3.2. ChIP-chip

3.2.1. DNA Amplification After ChIP (see ***Note 17****)*

1. Add 1 μl of 10X fragmentation buffer to 10 μl ChIPed DNA solution.
2. Place the tube in a thermal cycler at 95°C for exactly 4 min (*see* **Note 18**).
3. Immediately cool the sample on ice and then centrifuge briefly.
4. Add 2 μl 1X library buffer to 11 μl material (*see* **Note 19**).
5. Add 1 μl library stabilization solution. Mix by pipetting. Place at 95°C for 2 min in thermal cycler.
6. Add 1 μl library preparation enzyme. Mix by pipetting.
7. Incubate in thermal cycler as follows:
   16°C for 20 min24°C for 20 min

   37°C for 20 min

   75°C for 5 min

   4°C hold
8. Add the following reagents into the library-prepared sample:
   7.5 μl of 10X Amplification Master Mix

   47.5 μl nuclease-free H_2O

   5 μl WGA DNA polymerase
9. Incubate in thermal cycler using the following program:
   95°C for 3 min, then 14 cycles of

   94°C for 15 s

65°C for 5 min, then

4°C hold

10. Purify the sample using the DNA Clean & Concentrator – 25 system.
11. Quantify the amount of DNA by A_{260}. If the total DNA amount is less than 1 μg, reamplify the sample using GenomePlex WGA Reamplification kit starting from step 1.
12. Use 5–10 μg of amplified DNA for the hybridization of the array (*see* **Note 20**).

3.2.2. DNA Fragmentation, Labeling, Tiling Array Hybridization, Wash, and Detection

For DNA fragmentation, labeling, hybridization, wash, and detection, we follow the Affymetrix 100K protocol (http://www.affymetrix.com/support/technical/byproduct.affx?product=100k).

*3.2.3. Data Analysis (see **Notes 21** and **22**)*

The complete information on the Affymetrix tiling array is provided by the .CEL file. To analyze the data, several tools are currently available, with MAT (model-based analysis of tiling array) providing a convenient first step (15). To use MAT, a UNIX platform or equivalent is required. MAT requires the .CEL, .bpmap, and .lib files. The .CEL file contains the signals of all the probes on the array, the .bpmap files provide information on the probe locations and copy numbers, and the .lib file contains the repeat information.

4. Notes

1. These include formaldehyde, dimethyl adipimidate (DMA), dimethyl pimelimidate (DMP) dimethyl suberimidate (DMS), *N*-hydroxysuccinimide (NHS), tris-succinimidyl aminotriacetate (TSAT), disuccinimidyl suberate (DSS), disuccinimidyl glutarate (DSG), and ethylene glycol bis(succinimidylsuccinate) (EGS) (8, 18).
2. PMSF, which is unstable in aqueous solution, and the proteinase inhibitor are added just before use.
3. In general, we use approximately 240 mg tissue for three precipitations plus input, which consist of IgG for the negative control, histone 3 (H3) antibody for the positive control, the antibody against the specific TF or tag and input. Alternatively 1.2 g of tissue for three precipitations might be used in a large-scale experiment, if the small-scale experiment does not yield enough DNA.
4. We have used various tissues so far including *Arabidopsis* and maize seedlings, *Arabidopsis* and maize leaf tissues, *Arabidopsis* root tissue, *Arabidopsis* flower buds, maize Black Mexican

Sweet cells, and maize protoplasts transiently expressing epitope-tagged transcription factors. If not used immediately, cross-linked samples can be stored at –80°C.

5. The quality of grinding is very critical for the successful outcome of the experiments and the tissues must be ground very well.
6. We also have homogenized tissues in microcentrifuge tubes using a small plastic pestle in lysis buffer, in cases when only small quantities of tissue were available. The quality of the small-scale homogenization is as good as grinding larger quantities of tissue in liquid nitrogen, yet care must be used in not warming the extract more than necessary when holding the tube between the finger tips.
7. Alternatively, methods are available that use DNase I nuclease.
8. Sonication is the most critical step for the success of ChIP experiments. The ideal sonication conditions depend on a number of factors including the volume of extract, the sonicator tip size, and the sonicator itself. An optimal sonication condition should be identified prior to performing the ChIP experiment. For example, when using *Arabidopsis* seedling we have determined that in a Vibra Cell Sonicator (Sonics& Materials) five repeats of 15 s each at 10% amplitude provide the best results. To determine the optimal sonication conditions, various parameters should be tested such as the amplitude and duration of the sonication cycles (e.g., 5, 10, 40% of amplitude for 0, 10, 30, 60, 300 s) using cross-linked extract. Then, after reverse cross-linking and DNA purification, the size of the fragmented DNA is verified by electrophoresis.
9. Avoid making bubbles during sonication. Bubbles cause a significant reduction in sonication efficiency.
10. ChIP results strongly depend on the quality (affinity and specificity) of the antibody. We use antibodies against histones as experimental positive controls since commercially available antibodies that recognize a number of histone-tail modifications have been extensively used in ChIP experiment in plants, yeast, and animals.
11. We succeeded in obtaining reproducible signals by using antibodies that recognize acetylated H3 at position K9 (H3K9ac) (Upstate P/N 06-599) and anti-GFP (abcam P/N ab290). Antibody amounts are variable *(19–21)*. For example, we use 2 μg of IgG, 1 μg of H3K9ac, and 1 μl of anti-GFP antibodies for 100 μl of extract. However, monoclonal anti-myc epitope antibodies (line 9E10) have so far resulted in faint and irreproducible signals when using *Arabidopsis* extracts.

12. The washing steps should be ideally performed in the cold room.
13. Make sure that the final concentration of SDS, $NaHCO_3$, and proteinase K in the input sample is the same as in the other samples when adjusting the volume.
14. Elution volume depends on the amount of starting tissue. For example, if we start with 200 mg of plant tissue, we elute in 30 μl. For the isolation of DNA from the input extract, PCI extraction can be used when starting from large quantities of plant material. The purification using the QIAGEN columns is performed after the PCI extraction.
15. Since the ChIPed DNA is usually in very low amounts, the detection of the target DNA requires PCR. Therefore, there is a risk of PCR amplification bias; thus we strongly recommend quantitative PCR or semi-quantitative PCR to compare the enrichment of the target DNA with respect to the input control.
16. To accurately compare the quantity of ChIPed and input DNA, we recommend a double normalization using input DNA and a reference primer set. The ratio between the input and ChIPed DNA is a good index for the enrichment during the ChIP. However, it is always possible that there is a bias, for example because of non-specific binding of DNA to the beads. To rule out such artifacts, the reference primer set is used. The reference primer set should not to be a target of the TF in study, of course. For *Arabidopsis*, for example, we routinely use primers corresponding to ACT2/7 (15). The final normalization can then be done using the following formula:

$$\frac{(ChIPedDNA_{target}/InputDNA_{target})}{(ChIPedDNA_{reference}/InputDNA_{reference})}$$

17. Amplification is one of the most critical steps for ChIP-chip. Among several available amplification methods, we use the GenomePlex Whole Genome Amplification (WGA) kit (Sigma) with the modifications previously described (14). Using WGA, we have successfully obtained reproducible results with various tissues and mutants of *Arabidopsis.*
18. The incubation time is very critical.
19. The amount of ChIPed DNA is usually less than 1 ng/μl; thus accurately measuring the amount of DNA is challenging. We generally use 11 μl as start material for the amplification.
20. For the purpose of this manuscript, we are primarily referring to the Affymetrix GeneChip® Arabidopsis Tiling 1.0R Array. We have also used less than 5 μg of amplified DNA, yet the results have been variable.

21. The ideal way to perform a ChIP-chip experiment is by the side-by-side comparison of wild-type and mutant tissue, the latter corresponding to tissues lacking the specific TF. In such a case, the ChIP-chip analysis is performed by comparing the input and ChIP DNA from both the wild-type and mutant samples. Alternatively, if using plants expressing an epitope-tagged version of the TF, plants not expressing the transgene can be used as the mutant sample.
22. There are many methods to analyze ChIP-chip data. Generally, the analysis can be divided into two steps: normalization and detection of enriched signal region. The MAT algorithm takes care of both steps at the same time. It standardizes the probe value through the probe model, which consist of baseline probe behavior by considering probe sequence and copy number. Therefore, it eliminates the normalization step.

Acknowledgments

We thank Herbert Auer for helping us develop the ChIP-chip method for *Arabidopsis* and for comments on this manuscript, Manli Davis for helping us grow and collect plant materials, and Marko Djordjevic for assistance with the analysis of ChIP-chip data. This research was supported by National Science Foundation (NSF) grant MCB-0418891 and by NRI grant 2007-35318-17805 from the USDA-CSREES to E.G.

References

1. Grotewold, E. and Springer, N. (2009) Decoding the transcriptional hardwiring of the plant genome. In: Coruzzi, G. and Gutierrez, R. (eds). In Systems Biology. Blackwell Publishing. *In Press*.
2. Sablowski, R.W. and Meyerowitz, E.M. (1998) A homolog of NO APICAL MERISTEM is an immediate target of the floral homeotic genes APETALA3/PISTILLATA. *Cell*. **92**, 93–103.
3. Spelt, C., Quattrocchio, F., Mol, J., and Koes, R. (2002) ANTHOCYANIN1 of petunia controls pigment synthesis, vacuolar pH, and seed coat development by genetically distinct mechanisms. *Plant Cell*. **14**, 2121–2135.
4. Shin, B., Choi, G., Yi, H., et al. (2002) AtMYB21, a gene encoding a flower-specific transcription factor, is regulated by COP1. *Plant J*. **30**, 23–32.
5. Wang, D., Amornsiripanitch, N., and Dong, X. (2006) A genomic approach to identify regulatory nodes in the transcriptional network of systemic acquired resistance in plants. *PLoS Pathog*. **2**, e123.
6. Wellmer, F., Alves-Ferreira, M., Dubois, A., Riechmann, J.L., and Meyerowitz, E.M. (2006) Genome-wide analysis of gene expression during early Arabidopsis flower development. *PLoS Genet*. **2**, e117.
7. Wells, J. and Farnham, P.J. (2002) Characterizing transcription factor binding sites using formaldehyde crosslinking and immunoprecipitation. *Methods*. **26**, 48–56.
8. Nowack, D.E., Tian, B., and Brasier, A.R. (2005) Two-step cross-linking method for identification of NF-KB gene network by chromatin immunoprecipitation. *Biotechniques*. **39**, 715–725.

9. Buck, M.J., Nobel, A.B., and Lieb, J.D. (2005) ChIPOTle: a user-friendly tool for the analysis of ChIP-chip data. *Genome Biol.* **6**, R97.
10. Thibaud-Nissen, F., Wu, H., Richmond, T., et al. (2006) Development of Arabidopsis whole-genome microarrays and their application to the discovery of binding sites for the TGA2 transcription factor in salicylic acid-treated plants. *Plant J.* **47**, 152–162.
11. Lee, J., He, K., Stolc, V., et al. (2007) Analysis of transcription factor HY5 genomic binding sites revealed its hierarchical role in light regulation of development. *Plant Cell.* **19**, 731–749.
12. Zhang, X., Clarenz, O., Cokus, S., et al. (2007) Whole-genome analysis of histone H3 lysine 27 trimethylation in Arabidopsis. *PLoS Biol.* **5**, e129.
13. Zhang, X., Yazaki, J., Sundaresan, A., et al. (2006) Genome-wide high-resolution mapping and functional analysis of DNA methylation in arabidopsis. *Cell.* **126**, 1189–1201.
14. O'Geen, H., Nicolet, C.M., Blahnik, K., Green, R., and Farnham, P.J. (2006) Comparison of sample preparation methods for ChIP-chip assays. *Biotechniques.* **41**, 577–580.
15. Johnson, W.E., Li, W., Meyer, C.A., et al. (2006) Model-based analysis of tiling-arrays for ChIP-chip. *Proc. Natl. Acad. Sci. USA.* **103**, 12457–12462.
16. Wei, C.L., Wu, Q., Vega, V.B., et al. (2006) A global map of p53 transcription-factor binding sites in the human genome. *Cell.* **124**, 207–219.
17. Denissov, S., van Driel, M., Voit, R., et al. (2007) Identification of novel functional TBP-binding sites and general factor repertoires. *EMBO J.* **26**, 944–954.
18. Hermanson, G.T. (1996) *Bioconjugate Techniques.* San Diego: Academic Press.
19. Morohashi, K. and Grotewold, E. (2009) A systems approach reveals regulatory circuitry for Arabidopsis trichome initiation by the GL3 and GL1 selectors. *PLoS Genetics.* **5**, e1000396
20. Xie, Z. and Grotewold, E. (2008) Serial ChIP as a tool to investigate the co-localization or exclusion of proteins on plant genes. *Plant Methods.* **4**, 25.
21. Morohashi, K., Zhao, M., Yang, M., Read, B., Lloyd, A., Lamb, R., and Grotewold, E. (2007) Participation of the *Arabidopsis* bHLH factor GL3 in trichome initiation regulatory events. *Plant Physiol.* **145**, 736–746.

Chapter 2

Genome-Wide Analysis of RNA–Protein Interactions in Plants

Alice Barkan

Abstract

RNA–protein interactions profoundly impact organismal development and function through their contributions to the basal gene expression machineries and their regulation of post-transcriptional processes. The repertoire of predicted RNA binding proteins (RBPs) in plants is particularly large, suggesting that the RNA–protein interactome in plants may be more complex and dynamic even than that in metazoa. To dissect RNA–protein interaction networks, it is necessary to identify the RNAs with which each RBP interacts and to determine how those interactions influence RNA fate and downstream processes. Identification of the native RNA ligands of RBPs remains a challenge, but several high-throughput methods for the analysis of RNAs that copurify with specific RBPs from cell extract have been reported recently. This chapter reviews approaches for defining the native RNA ligands of RBPs on a genome-wide scale and provides a protocol for a method that has been used to this end for RBPs that localize to the chloroplast.

Key words: RNA–protein interaction, RIP-chip, RNA coimmunoprecipitation, microarray, RNA binding protein.

1. Introduction

Organismal development, homeostasis, and environmental adaptation require the regulated expression of large sets of genes. The rates of an array of post-transcriptional events are superimposed upon the transcription rate to determine the output of each gene, and in some cases, post-transcriptional steps play a dominant role. Thus, RNA binding proteins (RBPs) that influence the processing, nuclear export, stability, or translation of RNA subsets are likely to contribute to the large-scale coordination of gene expression (reviewed by (1, 2)). RBPs are also at the core of the machineries

Dmitry A. Belostotsky (ed.), *Plant Systems Biology, vol. 553*

DOI 10.1007/978-1-60327-563-7_2 Springerprotocols.com

that localize specific mRNAs within cells, thereby influencing the localization of protein synthesis to specific subcellular domains in plants, animals, and fungi (reviewed in (3, 4)).

It has long been appreciated that post-transcriptional mechanisms play a major role in determining gene expression levels in plant mitochondria and chloroplasts (reviewed in (5, 6–9)). Recently, the importance of post-transcriptional events in dictating other plant traits has been highlighted by the recovery of genes encoding nuclear/cytosolic RBPs and microRNAs in genetic screens for phenotypes affecting diverse processes such as flowering, circadian control, and hormone responses (reviewed in (10, 11–13)). Genome-wide assays to explore the impact of post-transcriptional regulatory mechanisms in plants have only recently begun, but the results thus far suggest that their impact is considerable. For example, the stabilities of mRNA subsets are under circadian control (14) and stress-induced changes in the translation of large sets of plant mRNAs have been reported (15, 16). Although large-scale analyses of regulated changes in splice isoform populations have not yet been reported in plants, there is evidence that both biotic and abiotic stresses influence the alternative splicing of plant pre-mRNAs (reviewed in (17)) (18, 19).

Interactions between RNAs and RBPs in ribonucleoprotein particles (RNPs) underlie the biogenesis, localization, translation, and turnover of mRNAs, the biogenesis of non-coding RNAs, and the regulation of all of these processes (20). However, even in the most intensively studied fungal and animal systems, the functions and RNA ligands for the vast majority of predicted RBPs remain unknown (1). The challenge is still greater in plants, whose repertoire of predicted RBPs is considerably larger than that in metazoa (10, 21–26). Gene families encoding homologs of proteins implicated in nuclear pre-mRNA splicing, polyadenylation, and mRNA decay are expanded in plants (21–23). Additional complexity is introduced by the maintenance of a third genetic compartment in plants, the chloroplast, where RNA editing, group I and group II intron splicing, endonucleolytic mRNA processing, and regulated translation and mRNA turnover are prevalent (reviewed in (6, 8, 9)). Furthermore, RNA metabolism in plant mitochondria is substantially more complex than that in metazoa, sharing many features with that in chloroplasts (reviewed in (5, 7, 8)). Indeed, the expansion of two RBP families specifically in the plant lineage (the CRM and PPR families) appears to be linked to the prominent roles of post-transcriptional aspects of gene expression in these organelles (24, 25, 27).

Identification of the RNAs with which each RBP is associated is at the core of understanding RNP interaction networks. The phenotypes conditioned by loss-of-function mutations in RBP genes can provide clues about their RNA ligands, especially when coupled with genome-wide assays for changes in mRNA profiles (see, for example, (28–30)); however, it can be difficult to distinguish the direct consequences of a mutation from secondary effects. SELEX assays (31) can be informative (see, e.g., (32, 33)), but require large quantities of folded recombinant protein, reveal only the highest affinity RNA ligands which do not necessarily reflect the physiological targets, and identify only short sequence motifs, whose in vivo correlates can be difficult to infer. Yeast-3-hybrid screens (34, 35) provide another tool for seeking proteins that bind a known RNA or vice versa, but false-negatives due to poor expression of either the protein or RNA component are common. Proteins that bind a specific target RNA can sometimes be purified from native extract by RNA affinity chromatography (e.g., (36)). However, such approaches are plagued by false-positives due to the typically non-specific binding of RBPs to RNA in vitro unless conditions are carefully optimized to reveal sequence-specific interactions.

Thus, although assays that take RNAs and RBPs out of their normal cellular context are important components of the toolkit for deciphering RNA/protein partners and recognition mechanisms, ideal starting points for such explorations are the RNA/protein complexes found in the native organism. Toward this end, affinity purification of specific RBPs from native extract coupled with unbiased methods for identifying copurifying RNAs would seem to have great potential; indeed, studies employing such approaches are being reported with increasing frequency. The strategy used most commonly thus far couples immunoprecipitation of native RNPs with microarray analysis to identify the coimmunoprecipitated RNAs; this approach is often referred to as "RIP-chip", after the related "ChIP-on-chip" method for DNA binding proteins. Although RIP-chip has so far been applied in plants only to chloroplast RNPs (37–40) and to identify ribosome-bound cytosolic mRNAs (41) (see also the chapter by Mustroph, Juntawong, and Bailey-Serres in this volume), the wealth of informative results obtained with nuclear-cytosolic RNPs in other systems (reviewed in (1, 2)) argue that analogous efforts will be similarly fruitful in plants.

Approaches for large-scale analysis of native RNPs are still developing and a variety of methods have been employed for affinity purification of the RBP and identification of the associated RNAs. Below is a discussion of key choices that must be made in designing experiments of this nature, followed by our protocol for "RIP-chip" in chloroplasts.

1.1. Custom Antibody Versus Expression of Tagged Protein?

Purification of an RBP from cell extracts can employ either custom antibodies or the affinity purification of a modified protein expressed in vivo with an affinity tag. Tagging approaches are ideal in organisms such as yeast where the endogenous gene can easily be modified to encode a tagged protein isoform; indeed, this approach has been the rule in RIP-chip studies in yeast (42–49). In less easily manipulated organisms, the benefits of using a tag should be carefully weighed against the effort involved in excluding potential artifacts. First, it is essential to demonstrate functionality of a tagged protein by complementation of a corresponding loss-of-function mutant. In addition, protein expression should be driven from the native promoter because aberrantly high or ectopic RBP expression can lead to artifactual interactions. For example, over-expression of the RNA binding protein FRG1 causes the aberrant splicing that underlies the disease facioscapulohumeral muscular dystrophy (50). Ectopic over-expression of SR proteins caused aberrant splicing in human cells (51) and resulted in aberrant development and auxin signaling in *Arabidopsis* (52). Likewise, over-expression of the poly(A) binding protein PABPN1 changed the accumulation of a large number of mRNAs in mammalian cells (53).

Thus, when working with complex organisms like plants and metazoa it may generally be more reliable, faster, and more cost-effective to use custom antibodies than to appropriately express and test the functionality of tagged proteins. The published studies of this nature are illustrative in this regard: 18 large-scale studies of RNAs that copurify with RBPs from metazoan extracts had been reported at the time of submission, and 16 of these used custom antibodies to immunoprecipitate RNPs (28, 54–68). All of the reports involving bacteria (69) and chloroplasts (37–40) have likewise used custom antibodies. The two exceptions (70, 71) used *Drosophila* cells and demonstrated the functionality of the tagged protein by complementation of the appropriate mutants. That being said, an interesting approach to gene expression profiling in complex tissues has been described that effectively exploits a tagging approach by using a tagged isoform of poly(A) binding protein to pull down mRNAs expressed in one cell type within a heterogeneous cell population (72).

1.2. To Crosslink or Not to Crosslink...

An ongoing discussion in this field concerns the relative merits of crosslinking macromolecules prior to cell lysis versus purifying native complexes without crosslinking. Incorporation of a crosslinking step would appear to be advantageous in that crosslinks can capture weak interactions and allow the use of stringent washing procedures to reduce contaminants and eliminate RBP–RNA interactions that may form after cell lysis (73, 74). Formaldehyde crosslinking, which is used routinely for chromatin immunoprecipitation experiments, has thus far been reported only for small-scale studies of RNA–protein interactions (75). In fact, formaldehyde crosslinking has been reported to give poor results in large-scale RIP-chip assays (76, 77), possibly due to deleterious effects on cell lysis and RNA recovery, and a high background of RNA–RNA crosslinks. On the other hand, ChIP-on-chip procedures involving formaldehyde crosslinking were used effectively to identify genomic DNA that is associated co-transcriptionally with two splicing factors (47, 61).

Ultraviolet (UV) light is used as a crosslinking agent in the "CLIP" (*c*ross*l*inking and *i*mmuno*p*recipitation) assay, which was used to identify RNAs associated with the Nova splicing factor (62, 74). CLIP assays begin with the exposure of living cells to UV light, which induces crosslinks between proteins and RNAs that are in close proximity. This is followed by a limited ribonuclease digestion to reduce the size of the RNAs associated with the RBP of interest, immunopurification of the target RBP, further purification of the RNA/RBP complexes by SDS-PAGE, and identification of the bound RNAs by linker ligation, RT-PCR, cloning, and sequencing. An advantage of CLIP is that only RNAs that are in direct contact with the "bait" protein will be identified; however, this can also be viewed as a disadvantage if the goal is to elucidate the higher-level organization of RNPs. Another advantage of CLIP is that the protein binding sites on the copurified RNAs are pinpointed to within a few hundred nucleotides. This resolution greatly simplifies the identification of sequence motifs recognized by the RBP of interest. It should be noted, however, that similar resolution has been obtained without crosslinking by using small, tiled hybridization probes to detect peaks of enrichment within a large RNA ligand (38) (Don Rio, personal communication). Disadvantages of CLIP include the low efficiency of UV crosslinking and the fact that UV crosslinking can capture only the subset of interactions in which specific bases are in a specific juxtaposition with specific amino acids.

Despite its apparent advantages, the number of large-scale studies that have used a crosslinking strategy (62, 66) is dwarfed by the number that have used uncrosslinked lysates (28, 37–40, 42–46, 48, 49, 54–60, 63–65, 67–72). The conclusions in most of these studies were validated in other ways (see below), highlighting the enormous potential of straightforward strategies that use native lysates without crosslinking. The relatively low impact of the CLIP approach thus far may be attributable to the technically challenging protocol and to the high false-negative rate that is a predicted consequence of the UV-crosslinking strategy. With the recent refinements of the CLIP protocol (74) and as new crosslinking strategies are developed, approaches that use crosslinking are likely to increase in prominence.

1.3. Distinguishing True from False-Positive

A challenge with all genome-wide methodologies is to identify true-positives from within the complex data set returned. One potential source of artifacts is the method used to detect the RNA ligands. Thus, if microarrays are used initially to identify RNAs that copurify with the RBP of interest, several of the putative positives are typically confirmed by RT-PCR or slot-blot hybridization assays. A more important source of background arises from non-specific interactions of RNPs with the antibody or affinity matrix. This type of background can be reduced by preclearing the lysate with a mock precipitation and by the use of stringent washing conditions. However, it is inevitable that abundant RNAs will contaminate the affinity preparation (except, perhaps, with the CLIP assay). Therefore, suitable negative controls are essential. Negative controls for immunoprecipitation experiments typically use a non-specific antibody and/or start with mutant cells lacking the bait RBP; for tagging approaches, genotype-matched cells that do not express the tagged isoform are used. A variety of statistical approaches have been used to identify sequences that are significantly more enriched in experimental samples than in control assays (e.g., (38, 43, 45, 58) and see the chapter by Gadbury, Garrett, and Allison in this volume). Rigorous statistical cutoffs enhance confidence that authentic interactions have been identified, but firm conclusions require additional genetic and/or biochemical validation, as summarized below.

A potential source of false-positives in studies that lack a cross-linking step is rearrangement of RNPs after cell lysis. One study showed that an RBP expressed in one cell population can associate after cell lysis with one of its native RNA ligands expressed in a different cell population (73). Two other studies, however, addressed this issue explicitly and did not see evidence of post-lysis exchange (49, 72). In our chloroplast RIP-chip assays, there has been excellent correspondence between the RNA ligands suggested by the RIP-chip data and the RNAs whose metabolism is disrupted in the corresponding mutant background ((37–40) and unpublished data). Likewise, positives to emerge in many large-scale studies of nuclear-cytosolic RNPs were validated in other ways (see below). Thus, although post-lysis exchange can occur with some proteins under some conditions, evidence thus far suggests that it does not generally contribute a substantial signal in large-scale assays.

Regardless of the approach employed, even the strongest "positives" should be considered tentative without additional validation. Aberrant metabolism of putative RNA ligands in the corresponding knock-down or mutant lines is particularly compelling validation (28, 37–40, 43–46, 48, 54, 56–58, 62, 65, 66, 69, 70) Biochemical and bioinformatic approaches can also be used to validate putative positives by identifying shared sequence motifs and/or functional relationships among the RNAs identified and by demonstrating that the bait protein binds with specificity in vitro to a sampling of these RNAs (43, 48, 54–56, 58, 60, 67, 71).

1.4. Identification of the Copurifying RNA Species

Large-scale identification of RNAs that copurify with the bait RBP can be accomplished in a variety of ways. SAGE approaches have been used successfully (62, 63) but array-based approaches have been the more common choice. The broadest coverage and best resolution is provided by genomic tiling arrays, if they are available. Generally, fluorescent dyes have been incorporated into cDNA generated from the recovered RNA, either with or without an amplification step. Certainly, amplification should be avoided if possible because it can change the relative representation of different sequences. During the development of our chloroplast RIP-chip assay, we compared the results obtained with labeled cDNAs to those obtained by direct labeling of RNA using commercially available platinum-based reagents. We found direct labeling of the RNA to be simpler than the generation of labeled cDNA and also to give superior results (see protocol below). Unfortunately, oligonucleotide arrays designed for gene expression profiling are often designed with the assumption that they will be probed with cDNA, so the strand specificity of the array must be considered before deciding on a labeling strategy.

It is likely that high-throughput sequencing technologies will supercede array-based approaches for detecting nucleic acids that copurify with RBPs, especially as the costs for these technologies come down. In fact, sequencing with the Illumina platform was recently used for genome-wide chromatin immunoprecipitation assays (ChIPSeq) (78–81).

1.5. Extract Preparation: A Special Challenge in Plants

Most large-scale analyses of RBP/RNA interactions have used whole-cell extracts from yeast and metazoa. Maintaining a highly concentrated extract is likely to be important for minimizing the dissociation of weak RBP/RNA interactions. In plants, it is straightforward to generate concentrated chloroplast and mitochondrial extracts by lysing the purified organelles in a minimal volume. However, the generation of extracts suitable for the analysis of nuclear and cytosolic RNPs from plants may be complicated by the presence of the cell wall and the large vacuole. For nuclear RNPs, preparation of an extract from a nuclear pellet may be advantageous: this approach can yield a concentrated extract harboring the RNPs of interest while also eliminating potential RNA contaminants from other compartments. In fact, use of a nucleoplasm fraction prepared as described by Pinol-Roma et al. (82), in conjunction with high-resolution genome tiling arrays, resulted in distinct and highly resolved data sets for four hnRNP proteins in *Drosophila* (Don Rio, personal communication). An alternative for nuclear RBPs that are recruited co-transcriptionally to nascent RNA is to use ChIP to identify the DNA sequences with which they are associated (47, 61). ChIP in plants is well established ((83), and see chapter by Morohashi, Xie, and Grotewold in this volume), so extrapolation of this method to plant RBPs may be relatively straightforward. Another promising avenue to explore with plants is the use of frozen tissue for the preparation of whole-cell extracts. Rapid freezing of yeast cells prior to RNP extraction was recently reported to be advantageous for the recovery of intact, uncontaminated RNPs (49). Indeed, RNPs extracted from pulverized frozen leaf tissue were used for the immunoprecipitation of maize chloroplast polysomes translating specific proteins (84) and for the affinity purification of cytosolic mRNAs in *Arabidopsis* associated with ribosomes (41).

2. Materials

2.1. Stock Solutions

To minimize ribonuclease contamination, solutions should be filtered through 0.2 μm nitrocellulose filters to remove trace proteins, filter tips should be used for micropipetting, and clean gloves should be worn at all times. All steps are performed at 4°C, unless otherwise indicated.

1. Hypotonic lysis buffer: 30 mM HEPES-KOH pH 8, 10 mM magnesium acetate, 60 mM potassium acetate, 2 mM dithiothreitol, 2 μg/ml aprotinin, 2 μg/ml leupeptin, 1 μg/ml pepstatin, 800 μM PMSF (added fresh).
2. CoIP buffer: 150 mM NaCl, 20 mM Tris–HCl, pH 7.5, 1 mM EDTA, 0.5% NP-40, 5 μg/ml aprotinin. The addition of 2 mM $MgCl_2$ is sometimes beneficial to stabilize Mg^{++}-dependent RNPs. If $MgCl_2$ is added, EDTA should be excluded.
3. Phenol–chloroform–isoamyl alcohol (25:24:1), equilibrated with aqueous buffer (pH ~8).
4. 10% SDS.
5. 0.2 M EDTA.
6. 20X SSC: 3 M NaCl, 0.3 M sodium citrate, pH 7.0.
7. 20X SSPE: 3 M NaCl, 0.2 M NaH_2PO_4, 0.02 M EDTA.
8. 95% ethanol.
9. 0.1 M Tris–HCl, pH 8.3.
10. 3 M NaAcetate, pH ~5.5.
11. 1 M Tris–HCl, pH 8.0.
12. 5 M NaCl.
13. "RNase-free" deionized H_20.

2.2. Reagents and Equipment

2.2.1. Extract Preparation

1. Disposable syringe (3 ml), 21 gauge needle
2. Hypotonic lysis buffer
3. Bradford protein assay (Bio-Rad)

2.2.2. Immunoprecipitation

1. Antibodies: Any antibody that can immunoprecipitate the bait RBP from native extract can be used. We have had excellent success using polyclonal antibodies generated to a recombinant

fragment of the target protein, whereas antibodies generated to synthetic peptides have been unreliable in our hands. For the antigen, select a hydrophilic segment of the RBP (> ~10 kDa) that lacks strong similarity (e.g., <40% identity, <60% similarity) to other proteins in the host species. Over-express this protein in *Escherichia coli* by using any standard expression vector. Expression as a 6x-histidine-tagged protein from a pET vector is simple and reliable. Many proteins expressed in this manner aggregate in *E. coli*, but this does not detract from their suitability as an antigen. Antibodies are affinity-purified on an antigen affinity matrix (85) prior to use in RIP-chip assays.

2. CoIP buffer.
3. RNAsin ribonuclease inhibitor (Promega).
4. Protein A matrix for collection of antigen–antibody complexes: Antigen–antibody complexes can be precipitated by Protein A coupled to any of a variety of materials. Many immunoprecipitation protocols use Protein A/G Sepharose beads, but we prefer formalin-fixed Staph A cells (IgG Sorb, The Enzyme Center) because they are inexpensive, effective, and form a more easily manipulated pellet. However, we have some data suggesting that residual nucleic acids from the Staph A cells may contribute a false "pellet" signal to array spots from mitochondrial rRNAs, so it may be best to use beads for assaying mitochondrial RBPs. Conjugating antibodies to magnetic beads provides another alternative (49).

2.2.3. RNA Purification and Labeling

1. 10% SDS
2. 0.2 M EDTA
3. Phenol–chloroform–isoamyl alcohol
4. 10 mM Tris–HCl pH 8, 1 mM EDTA, 100 mM NaCl, 0.25% SDS
5. 3 M sodium acetate
6. 95% ethanol
7. 5 mM Tris–HCl pH 8.3
8. GlycoBlue (Ambion)
9. Micromax ASAP RNA Labeling Kit (Perkin–Elmer)
10. Qiaquick Nucleotide Removal Kit (Qiagen)

2.2.4. Immunoblot Analysis to Test Immunoprecipitation

1. Standard materials, buffers, and equipment for SDS-PAGE and immunoblotting.
2. "One-Step Western Blot Kit" (GenScript Cat # L00204) is useful for antigens that comigrate with the IgG heavy chain.

2.2.5. Microarray Hybridization and Washing

1. Microarray: The choice of microarray platform will depend upon the species and genetic compartment under study. Our assays have used custom microarrays with tiled PCR products

of ~500 base pairs spotted onto glass slides. The Micromax ASAP labeling method is not compatible with oligonucleotide arrays that are designed to detect only the antisense strand (i.e., cDNA) rather than the mRNA itself. Our methods for printing and post-processing chloroplast microarrays are described in (38).

2. Heating blocks set to 85 and 60°C.
3. Hybridization oven or heated water bath for microarray hybridization, set to 58°C.
4. Slide warmer (LabLine Instruments) set to ~55°C.
5. Microarray hybridization chamber (Corning #2551).
6. LifterSlip Premium Printed Coverglass (Erie Scientific 24 × 601-2-4733).
7. Slide Staining Dishes and Slide Holders (Wheaton Glass # 900200).
8. 3X SSC.
9. 0.5X SSC, 0.01% SDS.
10. 0.06X SSC, 0.01% SDS.
11. 0.06X SSC.

2.2.6. Slot-Blot Validation

1. 1X and 2X SSPE.
2. Standard reagents and equipment for generating radiolabeled probes, and for RNA gel blot hybridization and washing.
3. MagnaNylon Membrane (0.45 μm pore) (GE Water and Process Technologies).
4. Slot-blot manifold.
5. UV-crosslinking device designed for RNA/DNA gel blots (e.g., Stratalinker).

3. Methods

3.1. Lysate Preparation

Chloroplasts are purified from seedling leaves according to any standard protocol (see, e.g., (86)). The chloroplast pellet is lysed by incubation for 15 min on ice in a minimal volume of hypotonic lysis buffer, punctuated with several rounds of vortexing. To complete the lysis, the material is drawn with a syringe through a 21 gauge needle and syringed up and down several times; bubbles should be avoided. Membranes are pelleted by centrifugation in a microfuge (4°C) at the highest setting for 30 min. The supernatant, which contains the stroma and some envelope membranes, is used for the RIP-chip assays (*see* **Note 1**). The protein concentration of the extract is determined with a Bradford assay (Bio-Rad) and should be between 5 and 20 mg/ml. The extract is flash-frozen and stored in aliquots of ~100 μl (~1 mg protein) at –80°C. Freeze–thaws prior to immunoprecipitation should be avoided. Each aliquot is used for between two and four RIP-chip assays, performed in parallel. A modification of this method that has succeeded for maize mitochondrial extracts is described in **Note 1**.

3.2. Experimental Design

Our procedure uses spotted microarrays probed simultaneously with differentially labeled RNAs from the immunoprecipitation pellet and supernatant. Although total input RNA or RNA from the pellet of a mock precipitation can be used as the reference sample, we have obtained better results by using the supernatant RNA as the reference (see (38)).

A negative control is essential to identify abundant RNAs or "sticky" RNPs that contaminate the immunoprecipitation pellet. An effective control for this purpose uses affinity-purified antibody to a different protein at a similar IgG concentration to that used for the experimental sample. An immunoprecipitation using extract prepared from material that lacks the target antigen but that is otherwise similar to the experimental extract (e.g., from a null mutant, or from genotype-matched cells not expressing the tagged isoform, if a tag is used) is an excellent control, if such material is available.

3.3. Immunoprecipitation

Formalin-fixed Staph A cells must be washed thoroughly to remove any dissociated Protein A because this will titrate out the antibody and reduce the yield of the precipitation:

3.3.1. Wash Staph A Cells

1. Shake the bottle of cells to give a homogeneous suspension. Remove a quantity of cell suspension that is sufficient for several days' use. Divide the suspension between several 1.5 ml microfuge tubes; place less than ~0.8 ml in each tube because large cell pellets can be difficult to resuspend.
2. Pellet the cells by centrifugation for ~1 min in a microfuge at ~10,000 rpm. Pipet off the supernatant and replace it with a similar volume of coIP buffer. Pipet up and down vigorously to resuspend the cells.
3. Repeat this washing procedure two more times, for a total of three washes. Resuspend the final cell pellet in coIP buffer to the initial volume. Combine the aliquots. Store at 4°C for up to 2 days.

3.3.2. Preclear Lysate

The quantity of stromal extract to use for a RIP-chip experiment will depend upon the abundance of the target RNPs. Aliquots of extract containing 0.5 mg of protein have yielded strong RIP-chip signals with a variety of low-abundance plastid RBPs (37–40). For more abundant RBPs, less lysate should be sufficient.

Typically, several immunoprecipitations are performed with each thawed aliquot of lysate. It is convenient to perform these pre-clearing steps before dividing the stroma for subsequent immunoprecipitations:

1. Place sufficient extract for control and experimental immunoprecipitations on ice and thaw slowly. Add RNAsin (Promega) to the thawing stroma (~20 units per 100 μl stroma) to reduce RNA degradation (*see* **Note 2**).
2. Centrifuge the extract for 10 min at ~12,000 rpm in a microfuge to remove insoluble particles. *This step is very important to reduce background.*
3. Further preclear the supernatant with washed IgSorb Staph A cells as follows. Centrifuge 100 μl of the washed Staph A cell suspension briefly to pellet the cells. Discard the supernatant. Resuspend the cell pellet in the cleared stromal lysate from step 2 above. Pipet up and down to resuspend the pellet, avoiding bubbles. After 10 min on ice, pellet the cells by centrifugation for ~5 min at ~10,000 rpm in a microfuge (4°C). Carefully pipet the supernatant into new tubes for use in the immunoprecipitation reactions. Reserve a small aliquot (e.g., 1/20th) of this supernatant for immunoblot analysis to assess the success of the immunoprecipitation.

3.3.3. Antibody Binding

1. Add affinity-purified antibody to the lysate. The optimal amount of antibody needs to be determined empirically (*see* **Note 3**) but will typically be between 2 and 10 μl.
2. Leave on ice for 1 h with occasional gentle mixing.

3.3.4. Precipitation of Antigen–Antibody Complexes

1. Shake a tube of washed Staph A cells to suspend the cells, and add 100 μl of the suspension to each immunoprecipitation. Mix gently. (If using crude serum, *see* **Note 3**.)
2. Store on ice for 30 min with occasional gentle mixing.
3. Pellet cells by microcentrifugation at ~10,000 rpm for 1 min.
4. Carefully pipet off the supernatant. Remove 1/10th of the supernatant to a separate tube to be used for immunoblot analysis; the remainder will be used for RNA extraction. Store both tubes of supernatant at –80°C until needed.
5. Resuspend cells thoroughly in ~0.5 ml of coIP buffer by pipetting up and down. (*See* **Note 4** for alternative wash buffers used to reduce non-specific binding.) Be sure to disrupt all visible cell clumps.
6. Pellet the suspended cells by microcentrifugation for 1 min (~10,000 rpm in a microfuge). Discard the supernatant.

7. Repeat this washing procedure two more times for a total of three washes.
8. Resuspend the final washed cell pellet in 250 μl of coIP buffer. Do not add Mg^{++} to this buffer even if it had been used in the immunoprecipitation.
9. Remove a 25 μl aliquot of the suspension to a separate tube for the immunoblot analysis described in **Section 3.3.6**. Pellet the cells in this aliquot by microcentrifugation for 1 min and resuspend the cells in 20 μl 1.5 X SDS sample buffer. Store at –20°C until ready for SDS-PAGE. For long-term storage (>2 days), store at –80°C.

3.3.5. RNA Purification

1. Increase the volume of the reserved immunoprecipitation supernatant to match that of the immunoprecipitation pellet sample (~225 μl) by adding co-IP buffer (lacking Mg^{++}).
2. To disrupt the RNPs, add 25 μl 10% SDS and 10 μl 200 mM EDTA to each pellet and supernatant sample.
3. Add 1 μl GlycoBlue (Ambion) to the pellet sample. The GlycoBlue enhances the recovery of the small amount of RNA in this sample by serving as a carrier and by making it easier to visualize the RNA during the subsequent purification steps.
4. Add ~250 μl of phenol–chloroform–isoamyl alcohol to each sample. Vortex thoroughly and separate the phases by microcentrifugation at 10,000 rpm for 5–10 min at room temperature.
5. Carefully remove the aqueous phases to new tubes, *being sure to avoid any interface material*. Aqueous phase left with the organic phase at this point will be recovered during the back-extraction that follows.
6. Back-extract the organic phase and interface by adding 150 μl of 10 mM Tris–HCl pH 8, 1 mM EDTA, 100 mM NaCl, and 0.25% SDS. Vortex thoroughly and centrifuge as above. Carefully remove the aqueous phase and combine with the corresponding aqueous phase from the first extraction.
7. Bring the sodium concentration to ~0.3 M by the addition of ~20 μl 3 M NaAcetate. Add 2.5 volumes of ethanol (~1 ml) to each tube. Vortex. Store at –20°C for at least 1 h. The RNA can be stored indefinitely at this step.
8. Pellet the RNA by microcentrifugation at 4°C at >12,000 rpm for 15 min.
9. Carefully pipet off the ethanol. (The RNA pellet from the immunoprecipitation pellet sample should be blue.) Rinse the RNA pellets by adding ~500 μl 70% ethanol, vortexing briefly, and microcentrifuging at ~12,000 rpm for 10 min.

Pipet off most of the ethanol. Air dry the pellets by inverting the tubes on to clean KimWipes for ~15 min. Alternatively, dry the pellets in a Speed Vac, but be careful not to over-dry the pellets as the RNA may become difficult to resuspend.

10. Resuspend the RNA from the immunoprecipitation pellet sample in 12 μl RNAse-free water. Resuspend the RNA from the immunoprecipitation supernatant sample in 36 μl RNAse-free water. Store at –80°C until ready for the labeling reaction.

3.3.6. Immunoblot Analysis to Evaluate the Immunoprecipitation

Before proceeding with RNA labeling, the recovery of the target RBP in the immunoprecipitation should be checked by SDS-PAGE and immunoblot analysis. An equal proportion of the pellet and supernatant material (e.g., 1/10th of each) and a corresponding amount of the starting extract should be analyzed. If the RBP comigrates with the IgG heavy chain, then its signal will be obscured with immunoblot detection methods that use an anti-IgG antibody as the secondary antibody. The "One-Step Western Blot Kit" (GenScript) gives excellent results in this situation.

3.4. RNA Labeling

RNAs purified from the pellet and supernatant fractions are differentially labeled with fluorescent dyes by using the Micromax *ASAP* RNA Labeling Kit (Perkin–Elmer). This procedure labels the guanosine bases in both RNA and DNA, so the RIP-chip protocol described here can be modified for use with DNA binding proteins (*see* **Note 5**). We routinely label the pellet RNA with Cy5 and the supernatant RNA with Cy3; an experiment in which chloroplast RNA labeled with each of the two dyes was competitively hybridized to a microarray demonstrated that dye-bias is minimal (38).

The oxidation of Cy5 by environmental ozone can severely decrease the fluorescence yield; even though this problem occurs only sporadically, it is safest to routinely take the following precautions. Minimize exposure to ozone by performing all steps possible in a nitrogen environment: Pour a 1″ layer of liquid nitrogen into a large styrofoam tub. Place a microfuge tube rack into the tub, positioned so that the tubes will be above and not in contact with the liquid nitrogen. Open all tubes containing Cy5 and do all manipulations of Cy5-containing solutions in this environment. *Do not leave tubes in the tub for prolonged periods as their contents will freeze.* We suspect that the most ozone-sensitive step is the first one, when the Cy5 stock solution is opened and the Cy5 reagent is added to the RNA.

One-half of the RNA recovered from the pellet and one-sixth of the RNA recovered from the supernatant is labeled and used for hybridization. A smaller fraction of the supernatant sample is used in order to reduce saturation of array fragments complementary to the highly abundant rRNAs and tRNAs:

1. Set heating blocks to 85 and 60°C. Place a plastic tub full of water into an oven or water bath set to 58°C for microarray hybridization. Set a Slide warmer (LabLine Instruments) to ~55°C.
2. Pipet 6 μl of each RNA into separate 0.5 ml microfuge tubes. To each tube add 3 μl of the labeling buffer supplied with the MicroMax ASAP kit (*see* **Note 6**).
3. Add 1 μl of the appropriate fluorescent labeling reagent supplied with the MicroMax kit to each sample. Cy dyes are light sensitive so return them to their dark, refrigerated storage area immediately after use.
4. Place the reactions in the 85°C heat block for 15 min (*see* **Note 7**).
5. Transfer tubes to ice and add 2.5 μl *ASAP* Stop Solution.
6. The RNA is separated from free dye by purification with a Qiaquick Nucleotide Removal Kit (Qiagen). Add 250 μl Buffer PN supplied with the Qiagen kit at room temperature. Transfer the mixture to a Qiaquick spin column. Let it stay for 1 min at room temperature. Centrifuge for 1 min at 10,000 rpm in a microfuge. Discard the flow-through (which should contain only free label).
7. Wash the column by adding 500 μl of Buffer PE supplied with the column. Centrifuge for 1 min at 10,000 rpm in a microfuge at room temperature. Remove the flow through and centrifuge the column for one additional minute to remove excess ethanol.
8. Elute the RNA from the column by transferring the column to a new tube, adding 40 μl 5 mM Tris–HCl pH 8.3 (*see* **Note 8**), and centrifuging for 1 min at 10,000 rpm in a microfuge. Repeat the elution step by adding an additional 40 μl of 5 mM Tris–HCl pH 8.3 to the same column in the same tube and centrifuging for 1 min at 10,000 rpm. The eluted supernatant RNA should be visibly pink due to the coupled Cy3, but some pink dye is typically retained in the column. The eluted pellet sample is not generally blue to the eye.
9. Concentrate the eluted RNA samples in a Speed Vac (without heat) until ~5 μl remains in each tube. *It is important not to dry the RNA to completion*. This step takes ~30 min. We use the labeled RNA immediately for hybridization, but instructions for long-term storage are provided by the manufacturer of the labeling reagents.

3.5. Microarray Hybridization

Hybridization is performed in a microarray hybridization chamber (Corning #2551). A constant temperature is maintained during hybridization by submerging the sealed microarray hybridization chamber in a water bath in a sealed plastic container. Several hours in advance, pre-warm the water bath in the plastic container by placing it in a hybridization oven or water bath set to 58°C:

1. Warm Hybridization Buffer III supplied with the MicroMax kit to 60°C in a heat block. Add 30 μl of the warm Hybridization Buffer III to each labeled RNA sample (~5 μl each).
2. Combine the pellet and supernatant RNA samples (total of ~70 μl). Heat for 3–4 min at 60°C.
   During this incubation
   (a) Place a LifterSlip Premium Printed Coverglass (Erie Scientific) on top of the microarray, being careful to place the tape side down so that there is a gap between the slide and coverslip.
   (b) Pre-warm the microarray slide/coverslip assembly and the microarray hybridization chamber by placing them onto the slide warmer (set to ~55°C).
3. Centrifuge the warmed RNA samples very briefly in a microfuge at room temperature to pellet condensation. Minimize cooling of the sample (*see* **Note 9**). Immediately pipet the warm RNA onto the pre-warmed array by placing the pipet tip next to the opening at either end of the coverslip. Pipet slowly, checking that the sample is being drawn under the coverslip by capillary action. Place the slide into the microarray hybridization chamber. Pipet ~15 μl of 3X SSC into each of the two wells in the microarray hybridization chamber; this is essential to keep the array from dehydrating.
4. Seal the chamber and place it in the 58°C water bath inside the plastic container. Seal the container and place it in the 58°C oven or water bath. Incubate overnight.

3.6. Microarray Washing

1. Fill a slide staining dish with 0.01% SDS, 0.5X SSC (room temperature).
2. Remove the microarray slide/coverslip assembly from the hybridization chamber. Place it in the slide holder that accompanies the slide-staining dish. Do NOT remove the coverslip manually. Instead, dunk the slide holder gently in the solution until the cover slip passively detaches from the slide. The cover slip should fall to the bottom of the dish.
3. Wash 1: Place the staining dish on a rotary shaker and shake gently (~50 rpm) for 15 min.
4. Wash 2: Transfer the slide in its holder to a new staining dish, containing 0.01% SDS, 0.06X SSC. Shake at ~50 rpm for 15 min.
5. Wash 3: Transfer the slide in its holder to a new staining dish, containing: 0.06X SSC. Shake at ~50 rpm for 15 min.

6. Tilt the slide to drain droplets, dabbing drops with a KimWipe. Dry the slide by centrifuging the slide in the slide holder in a table-top centrifuge for 3 min at 550 rpm. If ozone is a concern, seal the slide in a 50 ml conical centrifuge tube in a nitrogen atmosphere prior to this centrifugation.

3.7. Microarray Scanning and Data Analysis

Scan slides as soon as possible as the fluorescent signals diminish over time. It is useful to scan at several laser intensities: lower intensities reduce saturation for highly abundant RNAs (e.g., rRNAs and tRNAs) whereas higher intensities can be important to detect low-abundance RNAs. Scans at 532 nm (which elicits green fluorescence from Cy3) are typically performed at a PMT gain between 400 and 550. Scans at 635 nm (which elicits red fluorescence from Cy5) are typically performed at a PMT gain between 450 and 650. Store the slides in a 50 ml conical centrifuge tube covered in foil at 4°C. Slides can be rescanned several times within a few days with only a small loss of signal. The specifics of the scanning and data analysis procedures will vary with the array platform and facilities available. We import the data into GenePix Pro 6 (Molecular Devices) and filter out low-quality spots as described in Schmitz-Linneweber et al. (38).

3.8. Validation of Results by Slot-Blot Hybridization

Slot-blot hybridization can be used to validate positives to emerge from the RIP-chip data (*see* **Note 10**) and to pinpoint the RNA sequences associated with an RBP to greater resolution than is generally possible from the microarray data alone (*see* **Note 11**). RNAs purified from experimental and control immunoprecipitation pellets and supernatants are applied to slot blots and analyzed by hybridization to probes that correspond to array positives. If the number of putative positives is small, each of them can be validated in this way. In large-scale studies, a sampling of the positives should be validated:

1. Perform immunoprecipitations and extract RNA from the pellets and supernatants as for the RIP-chip assays.
2. Resuspend the RNA purified from each pellet and supernatant in 1200 μl 2X SSPE. 100 μl of the resuspended RNA samples will be applied to each slot. Heat the RNA to 70°C for ~10 min, while setting up the slot blotter.
3. Cut nylon hybridization membrane (MagnaNylon) to fit the slot-blot manifold and prewet it in 1X SSPE. Place the membrane into the slot blotter and place under vacuum for ~1 min to dry the membrane slightly.
4. Pipet 100 μl of each RNA sample into a separate slot while under vacuum. Allow the vacuum to pull the solution through the membrane. For each validation test, the following samples should be included: (i) pellet and supernatant RNAs for the experimental antibody; (ii) pellet and supernatant RNAs for a

negative control antibody; (iii) total RNA extracted from an amount of stroma equivalent to that used for each immunoprecipitation. Store unused RNA at –80°C.

5. Remove the nylon membrane from the slot-blot apparatus and place the side to which the RNA was applied face up on Whatman 3MM paper soaked with 1X SSPE. Crosslink the RNA to the membrane in a UV crosslinker (e.g., Stratalinker in "optimal crosslink" mode).
6. Air dry the membrane. Prehybridize and hybridize the membrane using probes corresponding to each validation test using standard conditions for RNA gel blots.

4. Notes

1. All of the chloroplast RBPs we have studied are in the soluble fraction. However, if the RNP of interest is membrane associated, it will need to be stripped from the membrane or the membrane will need to be solubilized with non-ionic detergent prior to immunoprecipitation. We have begun to modify this method for maize mitochondrial RBPs. Most mitochondrial RNAs and ribosomes pellet with the membrane fraction after organelle lysis. We have obtained interpretable RIP-chip data with a mitochondrial lysate generated by solubilization of the mitochondrial pellet with 1% NP-40. However, further optimization of this method is likely to be useful.
2. To increase the resolution of the assay so that the site of RBP binding within a large RNA ligand can be pinpointed, ribonuclease inhibitors should be omitted. This allows endogenous ribonucleases to reduce the size of the coimmunoprecipitated RNA molecules so that tethering of sequences distant from the binding site is reduced. We were able to pinpoint the binding sites of one RBP to within ~100 nucleotides by probing the RNAs coimmunoprecipitated in this manner with tiled 70-mer oligonucleotides (38). Hybridization of the coimmunoprecipitated RNA to an oligonucleotide tiling microarray and sequencing of the coimmunoprecipitated RNAs are alternative methods to identify enrichment peaks to high resolution.
3. The amount of antibody needs to be determined empirically. Antibodies that have been affinity purified against the antigen are ideal because the majority of the IgGs will be directed against the protein of interest, and the low quantity of IgGs ensures their quantitative precipitation by the Staph A cells. Titrations should be performed to determine how much

antibody is needed to recover most of the target RBP in the immunoprecipitation pellet; 5 μl of affinity-purified antibody is a good starting point. Crude serum can sometimes be used successfully. However, the quantity of Staph A cells may have to be increased to ensure quantitative binding of the IgGs, and the abundant IgGs may complicate the interpretation of immunoblot tests of immunoprecipitation efficiency.

4. The stringency of the immunoprecipitation can be increased by performing the first wash in a different buffer (e.g., coIP buffer supplemented with 500 mM NaCl or with 0.5% deoxycholate). However, these treatments may disrupt the RNPs of interest.

5. The Micromax *ASAP* RNA Labeling Kit modifies the N7 position of guanosine with the fluorescent dye in both DNA and RNA. DNA contributes only a small fraction of the total signal except for sequences that are not transcribed or that yield very low abundance RNAs. Thus, the contribution of DNA to the signal can generally be ignored. However, some proteins are bound to both RNA and DNA in chloroplast lysate (unpublished results), and in some cases, it may be desirable to use this method to identify the binding site of a DNA binding protein. To ensure that the signal arises from only RNA or only DNA, DNA or RNA can be removed from the immunoprecipitated material with DNAse or alkali hydrolysis, respectively. RNAse A is less effective than alkali hydrolysis to eliminate the RNA because rRNAs are highly abundant and structured, so they are resistant to ribonuclease digestion.

   Alkali hydrolysis is performed as follows. Resuspend the nucleic acids recovered from the immunoprecipitation pellet and supernatant each in 40 μl of dH_2O. Add 10 μl 1 N NaOH. Incubate at 70°C for 30–45 min. Neutralize the pH with the addition of 2.1 μl 4.8 N HCl and 2.5 μl 1 M Tris–HCl pH 7.5. Precipitate the DNA by adding 150 μl EtOH, placing the tubes at –20°C for at least 30 min, and microcentrifugation for 20 min. Resuspend each DNA pellet in 12 μl dH_20 and label this material using the MicroMax kit as described above for RNA. If the hydrolysis was successful, the fluorescence associated with rDNA fragments will be similar to that from other fragments.

   Elimination of signal arising from DNA is not necessary in most cases. However, if it is suspected that the bait protein interacts with both DNA and RNA, then this step can clarify the source of the fluorescent signal. After extracting nucleic acids from the immunoprecipitation pellet and supernatant, resuspend each sample in 48 μl dH_2O. Add 1 μl RNAsin

(Promega), 6 μl 10X RQ1 DNAse buffer, and 2 μl RQ1 DNAse, RNAse free (1 unit/μl) (Promega, # M6101). Incubate at 37°C for 30 min. Add 140 μl dH_2O, 20 μl 10% SDS, 5 μl 0.2 M EDTA, and 4 μl 5 M NaCl. Phenol–chloroform extract and back-extract as described in **Section 3.3.5**. Add 1 μl GlycoBlue to the pellet sample and ethanol precipitate as above. Resuspend the RNAs derived from the immunoprecipitation pellet and supernatant in 12 and 36 μl of water, respectively.

6. The volume of RNA in the labeling reactions can be varied. However, the 10 μl reaction must include at least 2 μl of labeling buffer.
7. This incubation time determines the proportion of the guanosines that will be labeled. Too short an incubation results in poor labeling, but too long an incubation yields RNA that is so heavily modified that it hybridizes poorly. A 15 min incubation is reported by the vendor to be optimal for an mRNA of "average" length. However, for short RNA fragments, it may be possible to further optimize this step.
8. A pH above 8 is critical for elution of RNA from the column. H_2O can be effective for elution, but it is prudent to add a low concentration of buffer to control the pH.
9. For DNA and highly structured RNAs, reheating the sample to 80°C for 30 s may better denature the nucleic acids. This additional heating also reduces viscosity, allowing the sample to slide more smoothly under the coverslip on the microarray. The dyes are heat stable so this additional heating is not detrimental.
10. Real-time PCR or any other quantitative assay could be used as an alternative.
11. To pinpoint sites of RNA interaction at selected loci, coimmunoprecipitated RNA can be applied to replicate slot blots and hybridized with tiled oligonucleotides to detect peaks of enrichment within a large RNA molecule (38). For such experiments, ribonuclease inhibitors should not be added to the immunoprecipitation reaction.

Acknowledgments

I would like to thank Rosalind Williams-Carrier and Christian Schmitz-Linneweber for their central contributions in developing our RIP-chip methodology, Eric Johnson and his laboratory for

---

Considerable progress was made in this field between the time of manuscript submission and the time of publication. Therefore, several important recent studies are not discussed.

guidance in microarray printing and hybridization, and Jana Prikryl, Yukari Asakura, Susan Belcher, and Kenneth Watkins for optimizing aspects of the protocols. I would also like to thank Rodger Voelker, Todd Mockler, and Don Rio for stimulating discussions and for their comments on the manuscript and Don Rio for communicating results prior to publication. This work was supported by grant DBI-0421799 from the National Science Foundation.

References

1. Hieronymus, H. and Silver, P.A. (2004) A systems view of mRNP biology. *Genes Dev.* **18**, 2845–2860.
2. Keene, J.D. (2007) RNA regulons: coordination of post-transcriptional events. *Nat. Rev. Genet.* **8**, 533–543.
3. Jambhekar, A. and Derisi, J.L. (2007) Cis-acting determinants of asymmetric, cytoplasmic RNA transport. *RNA.* **13**, 625–642.
4. Okita, T.W. and Choi, S.B. (2002) mRNA localization in plants: targeting to the cell's cortical region and beyond. *Curr. Opin. Plant Biol.* **5**, 553–559.
5. Knoop, V. and Brennicke, A. (2002) Molecular biology of the plant mitochondrion. *Crit. Rev. Plant Sci.* **21**, 111–126.
6. Bollenbach, T.J., Schuster, G., and Stern, D.B. (2004) Cooperation of endo- and exoribonucleases in chloroplast mRNA turnover. *Prog. Nucleic Acid Res. Mol. Biol.* **78**, 305–337.
7. Bonen, L. (2004) in *Molecular Biology and Biotechnology of Plant Organelles*, eds. Daniell, H. & Chase, C. (Springer, Dordrecht), pp. 323–345.
8. Barkan, A. (2004) in *Molecular Biology and Biotechnology of Plant Organelles*, eds. Daniell, H. & Chase, C. (Kluwer Academic Publishers, Dordrecht, The Netherlands), pp. 281–308.
9. Zerges, W. (2004) in *Molecular Biology and Biotechnology of Plant Organelles*, eds. Daniell, H. & Chase, C. (Springer, Dordrecht), pp. 347–383.
10. Belostotsky, D.A. and Rose, A.B. (2005) Plant gene expression in the age of systems biology: integrating transcriptional and post-transcriptional events. *Trends Plant Sci.* **10**, 347–353.
11. Fedoroff, N. (2002) RNA-binding proteins in plants: the tip of an iceberg? *Plant J.* **5**, 452–459.
12. Carrington, J. and Ambros, V. (2003) Role of MicroRNAs in plant and animal development. *Science.* **301**, 336–338.
13. Kuhn, J. and Schroeder, J. (2003) Impacts of altered RNA metabolism on abscisic acid signalling. *Curr. Opin. Plant Biol.* **6**, 463–469.
14. Lidder, P., Gutierrez, R.A., Salome, P.A., McClung, C.R., and Green, P.J. (2005) Circadian control of messenger RNA stability. Association with a sequence-specific messenger RNA decay pathway. *Plant Physiol.* **138**, 2374–2385.
15. Kawaguchi, R., Girke, T., Bray, E.A., and Bailey-Serres, J. (2004) Differential mRNA translation contributes to gene regulation under non-stress and dehydration stress conditions in Arabidopsis thaliana. *Plant J.* **38**, 823–839.
16. Branco-Price, C., Kawaguchi, R., Ferreira, R.B., and Bailey-Serres, J. (2005) Genome-wide analysis of transcript abundance and translation in Arabidopsis seedlings subjected to oxygen deprivation. *Ann. Bot. (Lond).* **96**, 647–660.
17. Reddy, A.S. (2007) Alternative splicing of pre-messenger RNAs in plants in the genomic era. *Annu. Rev. Plant Biol.* **58**, 267–294.
18. Iida, K., Seki, M., Sakurai, T., Satou, M., Akiyama, K., Toyoda, T., Konagaya, A., and Shinozaki, K. (2004) Genome-wide analysis of alternative pre-mRNA splicing in Arabidopsis thaliana based on full-length cDNA sequences. *Nucleic Acids Res.* **32**, 5096–5103.
19. Wang, B.B. and Brendel, V. (2006) Genomewide comparative analysis of alternative splicing in plants. *Proc. Natl. Acad. Sci. USA.* **103**, 7175–7180.
20. Dreyfuss, G., Kim, V.N., and Kataoka, N. (2002) Messenger-RNA-binding proteins and the messages they carry. *Nat. Rev. Mol. Cell. Biol.* **3**, 195–205.
21. Wang, B.B. and Brendel, V. (2004) The ASRG database: identification and survey of Arabidopsis thaliana genes involved in pre-mRNA splicing. *Genome Biol.* **5**, R102.

22. Belostotsky, D. (2003) Unexpected complexity of poly(A)-binding protein gene families in flowering plants: three conserved lineages that are at least 200 million years old and possible auto- and cross-regulation. *Genetics.* **163**, 311–319.
23. Lorkovic, Z. and Barta, A. (2002) Genome analysis: RNA recognition motif (RRM) and K homology (KH) domain RNA-binding proteins from the flowering plant *Arabidopsis thaliana. Nucleic Acids Res.* **30**, 623–635.
24. Barkan, A., Klipcan, L., Ostersetzer, O., Kawamura, T., Asakura, Y., and Watkins, K. (2007) The CRM domain: an RNA binding module derived from an ancient ribosome-associated protein. *RNA.* **13**, 55–64.
25. Lurin, C., Andres, C., Aubourg, S., Bellaoui, M., Bitton, F., Bruyere, C., Caboche, M., Debast, C., Gualberto, J., Hoffmann, B., Lecharny, A., Le Ret, M., Martin-Magniette, M. L., Mireau, H., Peeters, N., Renou, J.P., Szurek, B., Taconnat, L., and Small, I. (2004) Genome-wide analysis of Arabidopsis pentatricopeptide repeat proteins reveals their essential role in organelle biogenesis. *Plant Cell.* **16**, 2089–2103.
26. Walker, N.S., Stiffler, N., and Barkan, A. (2007) POGs/PlantRBP: a resource for comparative genomics in plants. *Nucleic Acids Res.* **35**, D852–D856.
27. Small, I. and Peeters, N. (2000) The PPR motif – a TPR-related motif prevalent in plant organellar proteins. *Trends Biochem. Sci.* **25**, 46–47.
28. Blanchette, M., Labourier, E., Green, R.E., Brenner, S.E., and Rio, D.C. (2004) Genome-wide analysis reveals an unexpected function for the Drosophila splicing factor U2AF50 in the nuclear export of intronless mRNAs. *Mol. Cell.* **14**, 775–786.
29. Blanchette, M., Green, R.E., Brenner, S.E., and Rio, D.C. (2005) Global analysis of positive and negative pre-mRNA splicing regulators in Drosophila. *Genes Dev.* **19**, 1306–1314.
30. Rehwinkel, J., Herold, A., Gari, K., Kocher, T., Rode, M., Ciccarelli, F.L., Wilm, M., and Izaurralde, E. (2004) Genome-wide analysis of mRNAs regulated by the THO complex in Drosophila melanogaster. *Nat. Struct. Mol. Biol.* **11**, 558–566.
31. Fitzwater, T. and Polisky, B. (1996) A SELEX primer. *Methods Enzymol.* **267**, 275–301.
32. Kim, S., Shi, H., Lee, D.K., and Lis, J.T. (2003) Specific SR protein-dependent splicing substrates identified through genomic SELEX. *Nucleic Acids Res.* **31**, 1955–1961.
33. Faustino, N.A. and Cooper, T.A. (2005) Identification of putative new splicing targets for ETR-3 using sequences identified by systematic evolution of ligands by exponential enrichment. *Mol. Cell. Biol.* **25**, 879–887.
34. Hook, B., Bernstein, D., Zhang, B., and Wickens, M. (2005) RNA–protein interactions in the yeast three-hybrid system: affinity, sensitivity, and enhanced library screening. *RNA.* **11**, 227–233.
35. Seay, D., Hook, B., Evans, K., and Wickens, M. (2006) A three-hybrid screen identifies mRNAs controlled by a regulatory protein. *RNA.* **12**, 1594–1600.
36. Miller, J.W., Urbinati, C.R., Teng-Umnuay, P., Stenberg, M.G., Byrne, B.J., Thornton, C.A., and Swanson, M.S. (2000) Recruitment of human muscleblind proteins to (CUG)(n) expansions associated with myotonic dystrophy. *EMBO J.* **19**, 4439–4448.
37. Asakura, Y. and Barkan, A. (2007) A CRM domain protein functions dually in group I and group II intron splicing in land plant chloroplasts. *Plant Cell.* **19**, 3864–3875.
38. Schmitz-Linneweber, C., Williams-Carrier, R., and Barkan, A. (2005) RNA immunoprecipitation and microarray analysis show a chloroplast pentatricopeptide repeat protein to be associated with the 5′-region of mRNAs whose translation it activates. *Plant Cell.* **17**, 2791–2804.
39. Schmitz-Linneweber, C., Williams-Carrier, R.E., Williams-Voelker, P.M., Kroeger, T.S., Vichas, A., and Barkan, A. (2006) A pentatricopeptide repeat protein facilitates the trans-splicing of the maize chloroplast rps12 pre-mRNA. *Plant Cell.* **18**, 2650–2663.
40. Watkins, K., Kroeger, T., Cooke, A., Williams-Carrier, R., Friso, G., Belcher, S., Wijk, K.V., and Barkan, A. (2007) A ribonuclease III domain protein functions in group II intron splicing in maize chloroplasts. *Plant Cell.* **19**, 2606–2623.
41. Zanetti, M.E., Chang, I.F., Gong, F., Galbraith, D.W., and Bailey-Serres, J. (2005) Immunopurification of polyribosomal complexes of Arabidopsis for global analysis of gene expression. *Plant Physiol.* **138**, 624–635.
42. Inada, M. and Guthrie, C. (2004) Identification of Lhp1p-associated RNAs by microarray analysis in Saccharomyces cerevisiae

reveals association with coding and noncoding RNAs. *Proc. Natl. Acad. Sci. USA.* **101**, 434–439.

43. Gerber, A.P., Herschlag, D., and Brown, P.O. (2004) Extensive association of functionally and cytotopically related mRNAs with Puf family RNA-binding proteins in yeast. *PLoS Biol.* **2**, E79.
44. Shepard, K.A., Gerber, A.P., Jambhekar, A., Takizawa, P.A., Brown, P.O., Herschlag, D., DeRisi, J.L., and Vale, R.D. (2003) Widespread cytoplasmic mRNA transport in yeast: identification of 22 bud-localized transcripts using DNA microarray analysis. *Proc. Natl. Acad. Sci. USA.* **100**, 11429–11434.
45. Hieronymus, H. and Silver, P.A. (2003) Genome-wide analysis of RNA–protein interactions illustrates specificity of the mRNA export machinery. *Nat. Genet.* **33**, 155–161.
46. Duttagupta, R., Tian, B., Wilusz, C.J., Khounh, D.T., Soteropoulos, P., Ouyang, M., Dougherty, J.P., and Peltz, S.W. (2005) Global analysis of Pub1p targets reveals a coordinate control of gene expression through modulation of binding and stability. *Mol. Cell. Biol.* **25**, 5499–5513.
47. Kotovic, K.M., Lockshon, D., Boric, L., and Neugebauer, K.M. (2003) Cotranscriptional recruitment of the U1 snRNP to intron-containing genes in yeast. *Mol. Cell. Biol.* **23**, 5768–5779.
48. Guisbert, K., Duncan, K., Li, H., and Guthrie, C. (2005) Functional specificity of shuttling hnRNPs revealed by genome-wide analysis of their RNA binding profiles. *RNA.* **11**, 383–393.
49. Oeffinger, M., Wei, K.E., Rogers, R., Degrasse, J.A., Chait, B.T., Aitchison, J.D., and Rout, M.P. (2007) Comprehensive analysis of diverse ribonucleoprotein complexes. *Nat. Methods.* **4**, 951–956.
50. Gabellini, D., D'Antona, G., Moggio, M., Prelle, A., Zecca, C., Adami, R., Angeletti, B., Ciscato, P., Pellegrino, M.A., Bottinelli, R., Green, M.R., and Tupler, R. (2006) Facioscapulohumeral muscular dystrophy in mice overexpressing FRG1. *Nature.* **439**, 973–977.
51. Gabut, M., Mine, M., Marsac, C., Brivet, M., Tazi, J., and Soret, J. (2005) The SR protein SC35 is responsible for aberrant splicing of the E1alpha pyruvate dehydrogenase mRNA in a case of mental retardation with lactic acidosis. *Mol. Cell. Biol.* **25**, 3286–3294.
52. Kalyna, M., Lopato, S., and Barta, A. (2003) Ectopic expression of atRSZ33 reveals its function in splicing and causes pleiotropic changes in development. *Mol. Biol. Cell.* **14**, 3565–3577.
53. Corbeil-Girard, L.P., Klein, A.F., Sasseville, A.M., Lavoie, H., Dicaire, M.J., Saint-Denis, A., Page, M., Duranceau, A., Codere, F., Bouchard, J.P., Karpati, G., Rouleau, G.A., Massie, B., Langelier, Y., and Brais, B. (2005) PABPN1 overexpression leads to upregulation of genes encoding nuclear proteins that are sequestered in oculopharyngeal muscular dystrophy nuclear inclusions. *Neurobiol. Dis.* **18**, 551–567.
54. Kiesler, E., Hase, M.E., Brodin, D., and Visa, N. (2005) Hrp59, an hnRNP M protein in Chironomus and Drosophila, binds to exonic splicing enhancers and is required for expression of a subset of mRNAs. *J. Cell. Biol.* **168**, 1013–1025.
55. Gama-Carvalho, M., Barbosa-Morais, N.L., Brodsky, A.S., Silver, P.A., and Carmo-Fonseca, M. (2006) Genome-wide identification of functionally distinct subsets of cellular mRNAs associated with two nucleocytoplasmic-shuttling mammalian splicing factors. *Genome Biol.* **7**, R113.
56. Lopez de Silanes, I., Galban, S., Martindale, J.L., Yang, X., Mazan-Mamczarz, K., Indig, F.E., Falco, G., Zhan, M., and Gorospe, M. (2005) Identification and functional outcome of mRNAs associated with RNA-binding protein TIA-1. *Mol. Cell. Biol.* **25**, 9520–9531.
57. Brown, V., Jin, P., Ceman, S., Darnell, J., O'Donnell, W., Tenenbaum, S., Jin, X., Feng, U., Wilkinson, K., Keene, J., Darnell, R., and Warren, S. (2001) Microarray identification of FMRP-associated brain mRNAs and altered mRNA translational profiles in Fragile X Syndrome. *Cell.* **107**, 477–487.
58. Reynolds, N., Collier, B., Maratou, K., Bingham, V., Speed, R.M., Taggart, M., Semple, C.A., Gray, N.K., and Cooke, H.J. (2005) Dazl binds in vivo to specific transcripts and can regulate the pre-meiotic translation of Mvh in germ cells. *Hum. Mol. Genet.* **14**, 3899–3909.
59. Tenenbaum, S., Carson, C., Lager, P., and Keene, J. (2000) Identifying mRNA subsets in messenger ribonucleoprotein complexes by using cDNA arrays. *Proc. Natl. Acad. Sci. USA.* **97**, 14085–14090.
60. Townley-Tilson, W.H., Pendergrass, S.A., Marzluff, W.F., and Whitfield, M.L. (2006) Genome-wide analysis of mRNAs

bound to the histone stem-loop binding protein. *RNA.* **12**, 1853–1867.

61. Swinburne, I.A., Meyer, C.A., Liu, X.S., Silver, P.A., and Brodsky, A.S. (2006) Genomic localization of RNA binding proteins reveals links between pre-mRNA processing and transcription. *Genome Res.* **16**, 912–921.
62. Ule, J., Jensen, K.B., Ruggiu, M., Mele, A., Ule, A., and Darnell, R.B. (2003) CLIP identifies Nova-regulated RNA networks in the brain. *Science.* **302**, 1212–1215.
63. Klimek-Tomczak, K., Wyrwicz, L.S., Jain, S., Bomsztyk, K., and Ostrowski, J. (2004) Characterization of hnRNP K protein–RNA interactions. *J. Mol. Biol.* **342**, 1131–1141.
64. Eystathioy, T., Chan, E.K., Tenenbaum, S.A., Keene, J.D., Griffith, K., and Fritzler, M.J. (2002) A phosphorylated cytoplasmic autoantigen, GW182, associates with a unique population of human mRNAs within novel cytoplasmic speckles. *Mol. Biol. Cell.* **13**, 1338–1351.
65. Mordes, D., Yuan, L., Xu, L., Kawada, M., Molday, R.S., and Wu, J.Y. (2007) Identification of photoreceptor genes affected by PRPF31 mutations associated with autosomal dominant retinitis pigmentosa. *Neurobiol. Dis.* **26**, 291–300.
66. Guil, S. and Caceres, J.F. (2007) The multifunctional RNA-binding protein hnRNP A1 is required for processing of miR-18a. *Nat. Struct. Mol. Biol.* **14**, 591–596.
67. Waggoner, S.A. and Liebhaber, S.A. (2003) Identification of mRNAs associated with alphaCP2-containing RNP complexes. *Mol. Cell. Biol.* **23**, 7055–7067.
68. Labourier, E., Blanchette, M., Feiger, J.W., Adams, M.D., and Rio, D.C. (2002) The KH-type RNA-binding protein PSI is required for Drosophila viability, male fertility, and cellular mRNA processing. *Genes Dev.* **16**, 72–84.
69. Zhang, A., Wassarman, K.M., Rosenow, C., Tjaden, B.C., Storz, G., and Gottesman, S. (2003) Global analysis of small RNA and mRNA targets of Hfq. *Mol. Microbiol.* **50**, 1111–1124.
70. Easow, G., Teleman, A.A., and Cohen, S.M. (2007) Isolation of microRNA targets by miRNP immunopurification. *RNA.* **13**, 1198–1204.
71. Gerber, A.P., Luschnig, S., Krasnow, M.A., Brown, P.O., and Herschlag, D. (2006) Genome-wide identification of mRNAs associated with the translational regulator PUMILIO in Drosophila melanogaster. *Proc. Natl. Acad. Sci. USA.* **103**, 4487–4492.
72. Penalva, L.O., Burdick, M.D., Lin, S.M., Sutterluety, H., and Keene, J.D. (2004) RNA-binding proteins to assess gene expression states of co-cultivated cells in response to tumor cells. *Mol. Cancer.* **3**, 24.
73. Mili, S. and Steitz, J.A. (2004) Evidence for reassociation of RNA-binding proteins after cell lysis: implications for the interpretation of immunoprecipitation analyses. *RNA.* **10**, 1692–1694.
74. Ule, J., Jensen, K., Mele, A., and Darnell, R.B. (2005) CLIP: a method for identifying protein–RNA interaction sites in living cells. *Methods.* **37**, 376–386.
75. Niranjanakumari, S., Lasda, E., Brazas, R., and Garcia-Blanco, M.A. (2002) Reversible cross-linking combined with immunoprecipitation to study RNA–protein interactions in vivo. *Methods.* **26**, 182–190.
76. Penalva, L.O., Tenenbaum, S.A., and Keene, J.D. (2004) Gene expression analysis of messenger RNP complexes. *Methods Mol. Biol.* **257**, 125–134.
77. Keene, J.D., Komisarow, J.M., and Friedersdorf, M.B. (2006) RIP-chip: the isolation and identification of mRNAs, microRNAs and protein components of ribonucleoprotein complexes from cell extracts. *Nat. Protoc.* **1**, 302–307.
78. Johnson, D.S., Mortazavi, A., Myers, R.M., and Wold, B. (2007) Genome-wide mapping of in vivo protein–DNA interactions. *Science.* **316**, 1497–1502.
79. Robertson, G., Hirst, M., Bainbridge, M., Bilenky, M., Zhao, Y., Zeng, T., Euskirchen, G., Bernier, B., Varhol, R., Delaney, A., Thiessen, N., Griffith, O.L., He, A., Marra, M., Snyder, M., and Jones, S. (2007) Genome-wide profiles of STAT1 DNA association using chromatin immunoprecipitation and massively parallel sequencing. *Nat. Methods.* **4**, 651–657.
80. Mikkelsen, T.S., Ku, M., Jaffe, D.B., Issac, B., Lieberman, E., Giannoukos, G., Alvarez, P., Brockman, W., Kim, T.K., Koche, R.P., Lee, W., Mendenhall, E., O'Donovan, A., Presser, A., Russ, C., Xie, X., Meissner, A., Wernig, M., Jaenisch, R., Nusbaum, C., Lander, E.S., and Bernstein, B.E. (2007) Genome-wide maps of chromatin state in pluripotent and lineage-committed cells. *Nature.* **448**, 553–560.
81. Barski, A., Cuddapah, S., Cui, K., Roh, T.Y., Schones, D.E., Wang, Z., Wei, G., Chepelev, I., and Zhao, K. (2007) High-resolution profiling of histone methylations in the human genome. *Cell.* **129**, 823–837.

82. Pinol-Roma, S., Swanson, M.S., Matunis, M.J., and Dreyfuss, G. (1990) Purification and characterization of proteins of heterogeneous nuclear ribonucleoprotein complexes by affinity chromatography. *Methods Enzymol.* **181**, 326–331.
83. Haring, M., Offermann, S., Danker, T., Horst, I., Peterhaensel, C., and Stam, M. (2007) Chromatin immunoprecipitation: optimization, quantitative analysis and data normalization. *Plant Methods.* **3**, 11.
84. Barkan, A. (1988) Proteins encoded by a complex chloroplast transcription unit are each translated from both monocistronic and polycistronic mRNAs. *EMBO J.* 7, 2637–2644.
85. Harlow, E. and Lane, D. (1988) *Antibodies: A Laboratory Manual* (Cold Spring Harbor Laboratory Press, Cold Spring Harbor).
86. Barkan, A. (1998) Approaches to investigating nuclear genes that function in chloroplast biogenesis in land plants. *Methods Enzymol.* **297**, 38–57.

Chapter 3

Whole-Genome Microarrays: Applications and Technical Issues

Brian D. Gregory and Dmitry A. Belostotsky

Abstract

DNA microarrays have become a mainstream tool in experimental plant biology. The constant improvements in the technological platforms have enabled the development of the tiling DNA microarrays that cover the whole genome, which in turn catalyzed the wide variety of creative applications of such microarrays in the areas as diverse as global studies of genetic variation, DNA-binding proteins, DNA methylation, and chromatin and transcriptome dynamics. This chapter attempts to summarize such applications as well as discusses some technical and strategic issues that are particular to the use of tiling microarrays.

Key words: Tiling arrays, transcriptome, tilemap, *Arabidopsis*, rice.

1. Overview of DNA Microarray Technology

The ever-increasing abundance of available genome sequences has enabled a wide variety of experimental and/or computational studies at the whole-genome level. In parallel with the advances in available sequence data, recent improvements in microarray technologies have made it feasible to interrogate a complete genome sequence with arrays through the use of high-density whole-genome tiling microarrays. These DNA microarrays serve as a powerful platform for numerous experimental approaches aiming to probe, in a single experiment, the depths of functional and structural information contained within an entire genome.

Dmitry A. Belostotsky (ed.), *Plant Systems Biology, vol. 553*

DOI 10.1007/978-1-60327-563-7_3 Springerprotocols.com

Two general types of high-density microarray platforms are used most widely. The first type of microarrays consist of short (up to ~100-mer) oligonucleotide probes, which are synthesized directly on the surface of arrays by photolithography using light-sensitive synthetic chemistry and photolithographic masks, an ink-jet device, or programmable optical mirrors. These types of arrays can be further distinguished based on the type of probes of which they consist. There are the so-called semi-whole-genome (non-tiling) expression arrays that represent only the predicted (annotated) features of a genome, such as exons or splice junctions. On the other hand, the truly whole-genome tiling arrays (hereafter referred to as WGAs) are designed to interrogate an entire genome in an unbiased fashion (1–3). This class of microarrays consists of non-overlapping or partially overlapping probes that are tiled or spaced at regular intervals to cover the entire genome from end to end. The WGAs are already being manufactured with over 6,000,000 discrete features per array, with every feature comprising millions of copies of the specific probe sequence. For instance, the Affymetrix® GeneChip® *Arabidopsis* tiling array is a single array comprised of over 3.2 million perfect match and mismatch probe pairs (~6.4 million probes total) tiled with 35 base pair spacing throughout the complete non-repetitive portion of the *Arabidopsis thaliana* genome.

The second array platform is made by mechanically printing/spotting probes, such as amplified PCR products, oligonucleotides, or cloned DNA fragments, onto the glass slides (referred to from this point as spotted arrays). Spotted arrays generally have a much lower feature density, usually on the order of approximately 10,000–40,000 spots per chip, than the in situ synthesized oligonucleotide arrays. Overall, the high reproducibility, the ability to synthesize probes representing virtually any sequence of a finished genome, and the increased feature density have made the WGAs the preferred platform for whole-genome analysis. Moreover, the ability to utilize relatively short probe lengths combined with the flexibility of using multiple overlapping probes representing every region of an organism's genome makes WGAs an ideal choice for detecting the broadest range of genomic features (including even small polymorphisms and splice variants), rivaled only by the ultradeep sequencing (discussed in this volume in the chapter by Fox et al.). Furthermore, the specificity gained from using short probes also allows repetitive regions or gene family members to be distinguished from one another. Here, we discuss several approaches using WGAs for transcriptome characterization, novel gene discovery, analysis of alternative splicing, mapping of regulatory DNA motifs using chromatin immunoprecipitation (ChIP-chip), methylome analysis, and sequence polymorphism discovery.

2. Applications of Whole-Genome Tiling Arrays (WGAs)

2.1. Using WGAs for Transcriptome Characterization and Gene Discovery

Although computational methods of gene prediction have steadily improved over the past decade, such methods alone still do not enable the accurate determination of the gene structure and/or identify all transcription units in an organism. Additionally, large-scale cloning and sequencing of complementary DNA (cDNA) molecules corresponding to expressed gene products, the traditional approach for identifying coding regions, often misses very low abundance transcripts. Furthermore, any given cDNA collection can be devoid of transcripts that are expressed only in a subset of tissues and/or in response to a specific physiological or environmental condition(s). Hence, the WGAs that cover the entire sequence of the genome of interest represent an attractive alternative that largely circumvents such issues. For example, to study the tissue-specific expression comprehensively, the targets for the WGA hybridization should be generated from a variety of tissues. In brief, total RNA samples from the selected set of tissues are used to make the first strand cDNAs using an oligo(dT) primer containing a linked promoter for T7 RNA polymerase (T7 RNAP), followed by the conversion into the double-stranded form and an in vitro transcription by T7 RNAP to generate as well as amplify the biotinylated complementary RNA (cRNA). This protocol, based on a method devised by Eberwine and colleagues (4), results in an unbiased representation of all expressed gene products contained in the total RNA samples, while allowing for an amplification of the targets in sufficient quantity for hybridization to WGAs.

Remarkably, the very first data sets addressing the transcriptional activity in the various tissues in *Arabidopsis* using WGAs identified a large number of novel sites of expression that were missed by computational gene prediction algorithms and cDNA collections (2, 5–7). To define such novel sites of transcriptional activity, the raw data were first pre-processed by dividing the intensity values for each probe by the median intensity value of all probes, including the perfect match (PM), mismatch (MM), and control probes present on the chip, thereby establishing the background noise level in a given experiment. Then, regions of transcriptional activity from the array data that did not correspond to annotated genic units within the most recent genome annotation were classified as novel "expressed" regions if the median intensity value of the probes in that region fell above a certain background cutoff threshold (operationally defined through a metric summarizing the signal emanating from the promoter regions). This approach gives an unbiased tally of novel genic units, which is based solely on the probe intensity values for the regions that fall outside of annotated gene structures.

Interestingly, many of these newly identified transcripts are expressed from the antisense strand relative to previously annotated transcripts (2), and many of them possess an intriguing regulatory potential. For instance, this study has revealed an antisense transcript overlapping the 3′ end of the mRNA for the key repressor that regulates flowering time, *FLOWERING LOCUS C* (*FLC*). This antisense transcript may act as a substrate for the biogenesis of the small interfering RNAs (siRNAs) responsible for the heterochromatization and subsequent silencing of this genomic region (8). Additionally, this study has uncovered evidence for the expression in centromeric regions, which were previously thought to be mostly devoid of active transcription (2). Thus, WGAs offer an extremely powerful platform for the discovery of novel transcription units.

2.2. Population Genomic Studies Using WGAs

The genomic content of individuals from the same species can vary in sequence as a result of diverse evolutionary processes. Comprehensive polymorphism data constitute a powerful resource for identifying the sequence variants that affect the phenotypic differences among the individuals (9). Although direct sequencing of individual populations is the most straightforward method for amassing the comprehensive polymorphism data, this methodology has not yet become cost-effective and widely accessible in most organisms (10). To circumvent these problems, Clark et al. (11) applied WGAs for comprehensive polymorphism detection in *Arabidopsis*, expanding upon the strategy used earlier to identify a large fraction of the SNP variation in human and mouse (12, 13). To this end, Clark et al. targeted 19 wild accessions of *A. thaliana*, selected so as to sample the maximal span of genetic diversity. Each DNA sample was whole genome amplified to generate sufficient DNA for hybridization, partially fragmented with DNase I, end-labeled with biotinylated dUTP and ddUTP, and used to probe WGAs spanning the entire *Arabidopsis* genome with single base resolution on both strands, hence interrogating nearly a billion features per experiment.

This WGA-based approach succeeded in capturing much of the common sequence polymorphism found in the worldwide *A. thaliana*population. Furthermore, this data enabled the systematic identification of the types of sequences that differ between accessions, as well as provided a high-resolution map of the genome-wide distribution of polymorphism in this reference plant. Altogether, more than 1 million non-redundant single nucleotide polymorphisms (SNPs) were identified, and ~4% of the genome was identified as being highly dissimilar (or even deleted) relative to the reference (Col-0) genome sequence. Curiously, the patterns of polymorphism between the 19 wild accessions and the reference genome sequence (Col-0) are highly non-random among the gene families, with genes mediating the interaction with the biotic environment exhibiting an exceptionally high polymorphism levels. Also, regional variation in polymorphism was readily apparent at the chromosome-level scale. This WGA-enabled polymorphism data set provides an unprecedented resource for further evolutionary, genetic, and functional genomic studies.

Two related studies used WGA hybridization of DNA samples from wild accessions of *A. thaliana* to measure the genetic diversity and intraspecific polymorphism between individuals (14, 15). These studies demonstrated that total and pairwise diversity was higher near the centromeres and the heterochromatic knob region, which are highly repetitive in nature and are less active in transcription. Furthermore, the overall diversity between the *Arabidopsis* accessions positively correlated with recombination rate. The combined data from the three studies (11, 14, 15) has enabled the production of an *Arabidopsis* genotyping array, which contains 250,000 SNPs and is commercially available from Affymetrix. This SNP array assures more than adequate coverage for the genome-wide association mapping studies in *Arabidopsis (15)*, thus providing the research community with the framework for the future in-depth studies of genetic variation in plants. Taken together, these studies demonstrate that, even in the absence of sequence data for a number of individuals from the same species, population genomic studies can still be carried out successfully using hybridization to WGAs.

2.3. ChIP-Chip Studies Using WGAs

Transcription represents the first major control point in gene expression pathways. Although the overall process of transcription can be regulated by a variety of mechanisms, the most prominent among them are those mediated by the DNA-binding transcription factors and by chromatin structure, which is largely modulated via covalent modifications of the histone N-terminal tails. Chromatin immunoprecipitation (ChIP) with an antibody specific to the protein or modification of interest, followed by the hybridization to WGAs of the DNA extracted from the captured chromatin fragments (i.e., ChIP-chip), has emerged as a powerful approach for gaining insight into the genome-wide distribution of the specific transcription factors or histone modifications ((16–20) and the chapter by Morohashi et al. in this volume). The quality of the antibody used in immunoprecipitation of the DNA-bound protein of interest is the major limiting factor of this technique, because it is critical for achieving an effective enrichment of the protein-bound DNA fragments for hybridization to WGAs.

In one instructive study of this kind, the antibodies against the sequence-specific transcription factor TGA2 were used to map its binding sites genome-wide after the treatment of *Arabidopsis* plants with the phytohormone salicylic acid (SA) (21).

The TGA2-crosslinked, immunoprecipitated DNA fragments were nonspecifically amplified to obtain enough material for the hybridization, fragmented with DNase I, end-labeled with biotinylated-ddATP using terminal transferase, and hybridized to two distinct types of WGAs. The first platform contained 190,000 probes representing 2 kb regions upstream of all annotated genes at a density of seven probes per promoter, while the second platform represented the entire *Arabidopsis* genome at a density of one probe per 90 bases. This study revealed 51 putative binding sites for TGA2, including the only previously identified (in the promoter of *At2g14610 PR-1* gene), as well as 15 putative binding sites that lie outside of presumed promoter regions. Additionally, when the effect of SA treatment on gene expression was measured using standard gene expression arrays, SA-induced transcripts were found to be significantly overrepresented among the genes neighboring the putative TGA2-binding sites. This example illustrates how the combined use of WGA platforms for ChIP-chip and gene expression studies can give important clues as to how sequence-specific transcription factors govern the key regulatory networks within plant cells.

Covalent modification of histones is another key mechanism controlling the eukaryotic genome dynamics. Motivated in part by the evidence that tri-methylation of lysine 27 of histone H3 (H3K27me3) plays critical roles in regulating development in animals (16, 22, 23) and plants (24–27), WGA-based profiling of the H3K27me3 in *Arabidopsis* was undertaken (28, 29). These analyses revealed, for the first time, that H3K27me3 is a major silencing mechanism in *Arabidopsis* that regulates an unexpectedly large number of genes located in mostly euchromatic regions. Furthermore, analysis of the H3K27me3 profiles in the relevant mutant backgrounds suggested that establishment and maintenance of this histone modification is largely independent of other epigenetic pathways, such as DNA methylation or RNA silencing. Interestingly, the genomic domains marked by H3K27me3 associate almost exclusively and co-extensively with binding sites for TERMINAL FLOWER 2/LIKE HETEROCHROMATIN PROTEIN 1, which is similar to the HETEROCHROMATIN PROTEIN 1 (HP1) of metazoans and *Schizosaccharomyces pombe* (28, 29). However, the genome-wide distribution of H3K27me3 was unaffected in *lhp1* mutant, suggesting that TFL2/LHP1 is not involved in the deposition of this chromatin modification but rather is a part of the epigenetic mechanism that represses the expression of genes that are marked with the H3K27me3. Therefore, ChIP-chip experiments with WGAs can be very powerful in revealing the key regulatory mechanisms controlling the complex dynamics of the genome activity in plants. As the number of WGA-based ChIP-chip experiments grows, a much more complete view of the transcriptional networks controlling plant growth and development will emerge.

2.4. Characterization of the Methylome Using WGAs

DNA methylation is a conserved epigenetic silencing mechanism involved in many important biological phenomena, including defense against transposon proliferation, genomic imprinting, and regulation of gene expression. DNA methylation is a heritable epigenetic modification that is perpetuated through DNA replication by DNA methyltransferases (30, 31). DNA methylation allows to regulate the expression of a number of coding regions without mutation to the DNA sequence, and it can occur in *cis* (i.e., the gene itself is methylated) or in *trans*(when the methylation event at another site in the genome regulates the target gene) (32–34).

In *Arabidopsis*, DNA methylation is established in all sequence contexts by DRM1/2, which are homologs of the mammalian DNMT3a/b de novo DNA methyltransferases (35, 36). DRM1/2 activity can be directed to a precise genomic location by RNA-directed DNA methylation (RdDM) system that involves 21–24 nucleotide small RNA (smRNA) generated in a DICER-LIKE3-dependent manner and acting in concert with ARGONAUTE4 (37–39). On the other hand, DNA methylation within the context of CpG dinucleotide is stably maintained maintained through genome replication by the DNA methyltransferase MET1, a homolog of mammalian DNA methyltransferase1 (40–42). Finally, the plant-specific DNA methyltransferase CMT3 primarily targets cytosines in the CHG sequence context (where H = A, C, T) (43).

WGAs allow to comprehensively map the methylome, i.e., the sum total of the sites of DNA methylation within the *Arabidopsis* genome (3, 44–46). In the pioneering study of this kind, an antibody against the 5-methyl cytosine was used to generate the target for interrogating the WGAs spanning the entire *Arabidopsis* genome (3). The resulting DNA methylation map reveals that approximately 19% of the genome is methylated, with the regions containing the highest density of methylation located in highly repetitive regions of the genome, such as centromeric heterochromatin. Predictably, the highest levels of methylation were seen in pseudogenes and unexpressed genes, but surprisingly, a considerable amount of methylation was distributed in euchromatin. However, only ~5% of expressed genes contained methylation upstream of their ORFs (promoters), while 33% of the transcribed regions of these genes were methylated (body methylation), consistent with an earlier smaller-scale study (47). Another surprising discovery from these WGA studies was that most of the genes that contain DNA methylation within their transcribed regions are highly expressed and constitutively active. Furthermore, the distribution of DNA methylation is clearly different between transposons and genes: while DNA methylation of transposons is distributed across their entire length, methylation density in genes was low in the promoter regions, but gradually increased within the transcribed region and dropped off again in the 3′

flanking sequences. This pattern may indicate negative selection against methylating the 5′ and 3′ ends of expressed genes, e.g., because of incompatibility with transcription initiation and termination.

The methyl groups in DNA are not static but can be removed by the DNA demethylases (48–51). *Arabidopsis* has four such DNA demethylases, REPRESSOR OF SILENCING1 (ROS1), DEMETER (DME), DEMETER-LIKE2 (DML2), and DEMETER-LIKE3 (DML3) (48, 49, 51). DME is required for genomic imprinting during *Arabidopsis* embryo development (52), while the closely related ROS1 is involved in transcriptional silencing of a transgene (51). WGAs were employed to globally map the sites of DNA demethylation within the *Arabidopsis* genome, via comparing the methylome in WT and mutant plants lacking three of the DNA demethylases (ROS1, DML2, DML3) (53). It appears that 179 loci are actively demethylated by one or all of these enzymes in *Arabidopsis*, and interestingly, demethylation in the coding regions primarily occurs at both the 5′ and 3′ ends, i.e., in a pattern opposite to the overall distribution of DNA methylation. This suggests that DNA methylation is highly dynamic and that the process of demethylation may act to protect the genes from potentially deleterious methylation events. Taken together, these first methylome studies provide important insights into the nature as well as the function of this important epigenetic mark.

3. Technical Considerations Regarding the WGA Analyses

3.1. Sequence-Specific Probe Effects

Although tiling microarrays are very powerful, as illustrated in the preceding sections, several limitations and important technical considerations must be taken into account. One major technical constraint that is inherent to the concept of WGAs lies in the severely limited freedom of choice in designing the probes. This limitation translates into the widespread sequence-based probe effects. For example, ~20% of probes located entirely within a known (i.e., experimentally proven) exon exhibit twofold or higher difference in the signal intensity relative to the average intensity of their two neighboring probes located within same exon (54). While in principle such probe behavior can result from alternative splicing or from cross-hybridization to other transcribed sequences that map to unrelated genomic locations, in reality the most significant source of such effects is the variability in thermodynamic properties of the probes themselves, as dictated by their respective sequences.

Several approaches have been used in an attempt to correct for such unevenness in the probe behavior. One alternative is to tackle the problem early, i.e., at the stage of the design of the array. For instance, maskless tiling arrays manufactured by NimbleGen are composed entirely of isothermal probes, whereby the length of each molecule is varied in order to attain a consistent melting temperature (usually set at 76°C). Although the isothermal arrays should produce uniform probe behavior, this advantage comes at a price of the decrease in the feature density as compared to the one afforded by the Affymetrix platform, as well as reduced resolution. Furthermore, while the theoretical design of isothermal probes is typically based on the nearest-neighbor behavior of the respective oligonucleotides in solution, in practice the behavior of the array probes is strongly influenced by additional factors, such as steric hindrance on the microarray surface, probe–probe interaction, and secondary structure formation (55, 56).

An alternative strategy to address the unevenness in the probe behavior relies on statistical methods. Such approaches extend the earlier efforts to model the sequence-specific probe behavior for gene expression microarrays (57, 58). For example, MAT (model-based analysis of tiling arrays) predicts the baseline probe behavior by considering the 25-mer sequence and copy number of all probes on a single Affymetrix tiling array (59). This approach standardizes the probe value through the probe model, eliminating the need for sample normalization. As opposed to estimating probe behavior from multiple samples (60–62), MAT can standardize the signals of each probe in each array individually. MAT approaches perform particularly well in ChIP-chip applications that measure the genome-wide transcription factor binding and can detect with high accuracy the enriched regions from a single or multiple ChIP samples. This is due to the fact that most probes in ChIP-chip analyses measure only unspecific binding, because transcription factors usually bind only to a small fraction of the genome. One variation on the MAT theme is to use an a priori sequence-dependent physical model of probe-specific intensity bias (occurring primarily due to unspecific binding), instead of estimating it from the data (63).

Finally, the third alternative, which may hold a particular appeal to experimentalists, is to empirically calibrate the behavior of the probes on the array against a suitable reference sample. For example, in the global mapping study of transcriptional activity in yeast, nonequivalencies in the probe behavior were corrected via experimental RNA/DNA hybridization-based model, by correcting for the background as well as adjusting the signal of each probe by sequence-specific parameters, which was estimated from a calibration set of genomic DNA hybridizations (64). The following normalization methods were evaluated: (1) dividing RNA signal by DNA signal and then taking base 2 logarithm; (2) background-subtracting the RNA signal, dividing it by DNA signal, then applying variance stabilizing normalization (vsn, log base 2); and (3) in addition to method 2, dropping the 5% weakest probes in the DNA hybridization. Method 3 yielded the highest gain in signal to noise ratio, which was estimated as follows. Noise was estimated from the median of absolute differences between each pair of probes on the Crick strand of chromosome IV whose start points were three intermediate probes apart. Signal was obtained from the difference between 99% and 1% quantiles of all these probes. The optimal method of normalization for the individual probe behavior increased the signal/noise ratio on average by 1.7-fold.

3.2. Distinguishing the Signal from Noise

Although the issue of distinguishing the signal from noise is not unique to the whole-genome microarrays in particular, it is particularly important in the case of WGAs, because as opposed to all other types of microarrays analyses, WGAs assume no underlying gene models or annotations. Hence, a radically different strategy is required to make the decisions on how to make the "present" calls (54). In the early studies, positive probes were called based on a probe signal cutoff (65), and the genomic regions containing a significant number of positive probes were designated as transfrags (transcribed fragments). At present, two major alternative strategies to identify the regions of significant signal on WGAs are based on either the sliding window approaches or on structured change point detection algorithm (66, 67). The latter approach aims to segment the genome using dynamic programming in an unbiased fashion into regions with different expression levels in such a way that the probe signals are similar within each region. Such methods are reported to give more accurate estimates of change point locations, as well as depend on fewer user-defined parameters (64). However, the sliding window-based approaches remain more common.

The authors of this chapter have experience with TileMap (60). This package was originally developed for ChIP-chip analysis, but it can be used to analyze other types of genome-wide data, such as that of the entire transcriptome or methylome. The distinctive feature of TileMap is that it treats every probe as a separate entity, rather than computing a metric for a particular gene. Therefore, rather than generating a gene-level measurement of intensity changes, TileMap enables an unbiased identification of those genomic regions that demonstrate significantly up- or down-regulated hybridization between two different sets of arrays. An additional advantage of this algorithm is that it goes beyond the ability of just making the "present" calls, but rather allows complex multiple-condition comparisons (e.g., mutant 1 > WT > mutant 2).

In the first step of the TileMap procedure, a t-like test statistic is computed separately for each probe on the array, using a hierarchical empirical Bayes model to pool information from all probes across the array. On the other hand, during the calculation of the conventional t-statistic, only the estimate of the probe's own standard deviation is accounted for. This significantly increases the sensitivity of the analysis in the very common circumstances when there are only a small number (2–3) of replicates available for each condition. In the second step, the test statistics of probes within a genomic region are used to infer whether the region is of interest or not (i.e., whether it shows transcriptional activities of interest). TileMap offers two different ways to do this: users can choose to combine neighboring probes by using either a hidden Markov model (HMM) or a moving average method (MA). Finally, TileMap uses unbalanced mixture subtraction (UMS) to provide approximate local false discovery rate (lfdr) estimates for MA and model parameters for HMM. Compared with the commonly employed permutation test, UMS performs better for complex multiple-sample comparisons, such as mutant 1 > WT > mutant 2. Importantly, while UMS estimates the lfdr for a null hypothesis H0: "not (mutant 1 > WT > mutant 2)", permutation test usually can only provide lfdr for a null H0: "mutant 1 = WT = mutant 2".

3.3. Primary vs. Secondary Effects in Tiling Microarray Experiments

One cautionary point is warranted concerning the widespread practice of using microarrays to reveal the expression changes in various mutant backgrounds compared to WT. In the case of constitutive mutants, the differentially expressed regions represent the sum of primary and secondary effects of inactivating the respective cellular factor. Distinguishing the direct from secondary consequences of a given mutation on the transcriptome can be challenging and requires special consideration during the design stage and/or extensive follow-up experimental and/or bioinformatic analyses. While this problem is by no means unique to WGA-based studies, it can become particularly acute in this case because of the sheer volume of data that such studies generate.

While there is no single universal solution to this problem, several considerations may be helpful. One of these concerns the analyses of transcription factors, which often tend to regulate other transcription factors, forming branched networks. For the sake of example, if several sets of genes (regulons) are coordinately affected upon inactivating the factor X, one can query the public microarray repositories (e.g., www.weigelworld.org/resources/microarray/AtGenExpress) and/or specialized transcription factor databases (e.g., AGRIS, arabidopsis.med.ohio-state.edu) for the transcription factors(s) that may directly control the expression of these gene sets. A reasonable hypothesis then would be that the effect of X on these otherwise disparate gene sets is indirect and mediated by its regulating the expression of these transcription factors(s).

A more radical approach to the problem of secondary effects would be to attempt to bypass this issue altogether. In one example, this was achieved by putting the transcription factor under study under exclusively posttranslational control, via fusing it to the rat glucocorticoid receptor (GR). Simultaneous treatment of the transgenic line expressing such a chimeric factor by the activating glucocorticoid dexamethasone and the translational inhibitor cycloheximide then led to the transcriptional induction of the direct target genes only, while the expression of any secondary effects was blocked (68). However, the most generally applicable tools allowing to filter out secondary effects are conditional mutants. In this case, one can apply the restrictive condition at will and monitor the real-time progression of the ensuing changes in the transcriptome by microarray analysis of the early timepoints of the response. In this case, the expectation is that the inactivation of the transcription factor should have a very rapid effect on its immediate target promoters, comparable with inactivation of general transcription. On the other hand, a considerably longer period of time would be required to develop secondary effects, because such effects must be preceded by significant alterations in the mRNA as well as in the protein levels of the immediate downstream targets of the transcription factor in question.

Unfortunately, temperature-sensitive mutations, which are widely used for this type of analysis in microorganisms, are rather rare in plants. However, the authors have been successful in implementing an inducible RNAi (iRNAi) for creating a conditional knockdown of the subunits of the exosome complex in *Arabidopsis.* The exosome is an essential and conserved RNA-degrading and RNA-processing complex that has multiple and diverse RNA targets that are yet to be comprehensively defined in any eukaryote. An iRNAi system was engineered by expressing the constructs containing the segments of the exosome complex subunits *RRP4* or *RRP41*as a pair of inverted repeats separated by an intron, under the control of an estradiol-regulated chimeric transactivator XVE (69). Growing such exosome iRNAi plants on estradiol-containing media induced the RNAi-mediated knockdown of *RRP4* ($rrp4^{iRNAi}$) or *RRP41* ($rrp41^{iRNAi}$) mRNA, resulting in the growth arrest and subsequent death of seedlings. Importantly, growth arrest was preceded by the highly specific molecular phenotype associated with the defective processing of the 5.8S ribosomal RNA, which is highly specific to exosome malfunction (70), and is never observed in WT plants exposed to estradiol (neither is growth inhibition). Thus validated conditional iRNAi knockdown system was subsequently used in conjunction with WGAs to comprehensively define the Arabidopsis exosome targets (71). In contrast, WGA analysis of a constitutive loss of function mutant of one of the subunits of this complex produced massive amounts of secondary effects, even though this particular mutant had little if any phenotype at the whole-plant level (71).

3.4. Inhibition of RNA Degrading/Processing Enzymes as a General Strategy to Uncover the Hidden Dynamics in the Transcriptome

Numerous studies during the past few years revealed the existence of the vast "dark matter" in eukaryotic transcriptomes, in the form of noncoding (nc) RNAs with unknown function (72). Although there is much debate as to what fraction of these ncRNAs have biological targets vs. merely represent spurious transcriptional activity (73), it is important to comprehensively understand such events. In this regard, it is instructive to consider the finding of the "deeply hidden" layer in the *Arabidopsis* transcriptome that is only detectable under the conditions of exosome knockdown (71). Such RNAs are largely composed of intergenic noncoding transcripts that emanate from the repetitive, heterochromatic regions of the genome. Apparently, these transcripts are tightly down-regulated by the constitutive exosome activity to the extent that they are virtually undetectable under normal conditions. On a more general note, the exosome is but one among many diverse RNA processing/degrading activities in the cell, and hence a logical extension of this strategy would be to undertake a systematic identification and categorization of the transcriptome-wide consequences of modulating the activities of a wide variety RNA decay and processing factors.

While in hindsight this approach may seem intuitive, in practice such analyses have been conducted only rarely, and mostly in the context of specialized studies focusing on specific class of transcripts and/or specific aspects of RNA metabolism. In one such example, He et al. used microarrays to investigate the RNAs in *Saccharomyces cerevisiae* that are stabilized upon mutating the key components of the nonsense-mediated mRNA decay (NMD) pathway – a specialized mechanism dedicated to the degradation of mRNAs containing the premature stop codons (74). In another approach to identify the transcripts directly regulated by NMD, the same group examined which RNAs become rapidly down-regulated upon restituting the NMD pathway in the NMD-defective cells, using conditional promoter (75). These combined studies succeeded in defining a near-comprehensive core set of cellular transcripts regulated by NMD, many of which have not been previously known, e.g., such as those RNAs that fail to splice and escape into the cytoplasm, mRNAs with abnormally long 3′ UTRs, mRNAs with upstream open reading frames, mRNAs that are subject to leaky scanning resulting in the use of out-of-frame initiator codons, mRNAs translated via +1 frameshifting, bicistronic mRNAs, transcripts encoded by pseudogenes, as well as those emanating from the transposable elements or from their LTR sequences. In another study, a conditional promoter was used to inhibit the expression of an essential subunit of the nuclear RNase

P in yeast, combined with the WGA-based monitoring of the ensuing changes in the transcriptome during the time course of depletion (76). This study led to the discovery of 73 novel ncRNAs, many of them antisense relative to the previously annotated ORFs – a surprisingly large number for the best-annotated eukaryotic genome. We therefore propose that manipulating the activities of the key factors of plant RNA metabolism may be a productive approach for mining the depths of the dynamic plant transcriptomes.

Acknowledgments

Research on plant gene expression in the Belostotsky lab is supported by USDA (award 2007-35301-18207), the National Science Foundation (awards MCB-0424651 and DBI-0724168), and BARD (award 3756-05). B.D.G. is a Damon Runyon Fellow, supported by the Damon Runyon Cancer Research Foundation.

References

1. Penterman, J., Zilberman, D., Huh, J.H., Ballinger, T., Henikoff, S., and Fischer, R.L. (2007) DNA demethylation in the Arabidopsis genome. *Proc. Natl. Acad. Sci. USA* **104**, 6752–6757.
2. Kakutani, T. (2002) Epi-alleles in plants: inheritance of epigenetic information over generations. *Plant Cell Physiol.* **43**, 1106–1111.
3. Kankel, M.W., Ramsey, D.E., Stokes, T.L., Flowers, S.K., Haag, J.R., Jeddeloh, J.A., Riddle, N.C., Verbsky, M.L., and Richards, E.J. (2003) Arabidopsis MET1 cytosine methyltransferase mutants. *Genetics* **163**, 1109–1122.
4. Clark, R.M., Schweikert, G., Toomajian, C., Ossowski, S., Zeller, G., Shinn, P., Warthmann, N., Hu, T.T., Fu, G., Hinds, D.A., Chen, H., Frazer, K.A., Huson, D.H., Scholkopf, B., Nordborg, M., Ratsch, G., Ecker, J.R., and Weigel, D. (2007) Common sequence polymorphisms shaping genetic diversity in Arabidopsis thaliana. *Science* **317**, 338–342.
5. Van Gelder, R.N., von Zastrow, M.E., Yool, A., Dement, W.C., Barchas, J.D., and Eberwine, J.H. (1990) Amplified RNA synthesized from limited quantities of heterogeneous cDNA. *Proc. Natl. Acad. Sci. USA* **87**, 1663–1667.
6. Zhang, X., Yazaki, J., Sundaresan, A., Cokus, S., Chan, S.W., Chen, H., Henderson, I.R., Shinn, P., Pellegrini, M., Jacobsen, S.E., and Ecker, J.R. (2006) Genome-wide high-resolution mapping and functional analysis of DNA methylation in arabidopsis. *Cell* **126**, 1189–1201.
7. Choi, Y., Gehring, M., Johnson, L., Hannon, M., Harada, J.J., Goldberg, R.B., Jacobsen, S.E., and Fischer, R.L. (2002) DEMETER, a DNA glycosylase domain protein, is required for endosperm gene imprinting and seed viability in arabidopsis. *Cell* **110**, 33–42.
8. Hanada, K., Zhang, X., Borevitz, J.O., Li, W.H., and Shiu, S.H. (2007) A large number of novel coding small open reading frames in the intergenic regions of the Arabidopsis thaliana genome are transcribed and/or under purifying selection. *Genome Res.* **17**, 632–640.
9. Daruwala, R.S., Rudra, A., Ostrer, H., Lucito, R., Wigler, M., and Mishra, B. (2004) A versatile statistical analysis algorithm to detect genome copy number variation. *Proc. Natl. Acad. Sci. USA* **101**, 16292–16297.
10. Keles, S., van der Laan, M.J., Dudoit, S., and Cawley, S.E. (2006) Multiple testing methods for ChIP-Chip high density

oligonucleotide array data. *J. Comput. Biol.* **13**, 579–613.

11. Kim, H., Snesrud, E.C., Haas, B., Cheung, F., Town, C.D., and Quackenbush, J. (2003) Gene expression analyses of Arabidopsis chromosome 2 using a genomic DNA amplicon microarray. *Genome Res.* **13**, 327–340.
12. Hekstra, D., Taussig, A.R., Magnasco, M., and Naef, F. (2003) Absolute mRNA concentrations from sequence-specific calibration of oligonucleotide arrays. *Nucleic Acids Res.* **31**, 1962–1968.
13. Zhang, X., Clarenz, O., Cokus, S., Bernatavichute, Y.V., Pellegrini, M., Goodrich, J., and Jacobsen, S.E. (2007) Whole-genome analysis of histone H3 lysine 27 trimethylation in Arabidopsis. *PLoS Biol.* **5**, e129.
14. Sablowski, R.W. and Meyerowitz, E.M. (1998) A homolog of NO APICAL MERISTEM is an immediate target of the floral homeotic genes APETALA3/PISTILLATA. *Cell* **92**, 93–103.
15. Tran, R.K., Henikoff, J.G., Zilberman, D., Ditt, R.F., Jacobsen, S.E., and Henikoff, S. (2005) DNA methylation profiling identifies CG methylation clusters in Arabidopsis genes. *Curr. Biol.* **15**, 154–159.
16. Cao, X. and Jacobsen, S.E. (2002) Role of the arabidopsis DRM methyltransferases in de novo DNA methylation and gene silencing. *Curr. Biol.* **12**, 1138–1144.
17. Wu, J., Smith, L.T., Plass, C., and Huang, T.H. (2006) ChIP-chip comes of age for genome-wide functional analysis. *Cancer Res.* **66**, 6899–6902.
18. Turck, F., Roudier, F., Farrona, S., Martin-Magniette, M.L., Guillaume, E., Buisine, N., Gagnot, S., Martienssen, R.A., Coupland, G., and Colot, V. (2007) Arabidopsis TFL2/LHP1 specifically associates with genes marked by trimethylation of histone H3 lysine 27. *PLoS Genet.* **3**, e86.
19. Lee, T.I., Jenner, R.G., Boyer, L.A., Guenther, M.G., Levine, S.S., Kumar, R.M., Chevalier, B., Johnstone, S.E., Cole, M.F., Isono, K., Koseki, H., Fuchikami, T., Abe, K., Murray, H.L., Zucker, J.P., Yuan, B., Bell, G.W., Herbolsheimer, E., Hannett, N.M., Sun, K., Odom, D.T., Otte, A.P., Volkert, T.L., Bartel, D.P., Melton, D.A., Gifford, D.K., Jaenisch, R., and Young, R.A. (2006) Control of developmental regulators by Polycomb in human embryonic stem cells. *Cell* **125**, 301–313.
20. Picard, F., Robin, S., Lavielle, M., Vaisse, C., and Daudin, J.J. (2005) A statistical approach for array CGH data analysis. *BMC Bioinformatics* **6**, 27.
21. Johnson, J.M., Edwards, S., Shoemaker, D., and Schadt, E.E. (2005) Dark matter in the genome: evidence of widespread transcription detected by microarray tiling experiments. *Trends Genet.* **21**, 93–102.
22. Bertone, P., Stolc, V., Royce, T.E., Rozowsky, J.S., Urban, A.E., Zhu, X., Rinn, J.L., Tongprasit, W., Samanta, M., Weissman, S., Gerstein, M., and Snyder, M. (2004) Global identification of human transcribed sequences with genome tiling arrays. *Science* **306**, 2242–2246.
23. Zilberman, D., Cao, X., Johansen, L.K., Xie, Z., Carrington, J.E., and Jacobsen, S.E. (2004) Role of Arabidopsis ARGONAUTE4 in RNA-directed DNA methylation triggered by inverted repeats. *Curr. Biol.* **14**, 1214–1220.
24. Shiu, S.H. and Borevitz, J.O. (2008) The next generation of microarray research: applications in evolutionary and ecological genomics. *Heredity* **100**, 141–149.
25. Chanvivattana, Y., Bishopp, A., Schubert, D., Stock, C., Moon, Y.H., Sung, Z.R., and Goodrich, J. (2004) Interaction of Polycomb-group proteins controlling flowering in Arabidopsis. *Development* **131**, 5263–5276.
26. David, L., Huber, W., Granovskaia, M., Toedling, J., Palm, C.J., Bofkin, L., Jones, T., Davis, R.W., and Steinmetz, L.M. (2006) A high-resolution map of transcription in the yeast genome. *Proc. Natl. Acad. Sci. USA* **103**, 5320–5325.
27. Li, W., Meyer, C.A., and Liu, X.S. (2005) A hidden Markov model for analyzing ChIP-chip experiments on genome tiling arrays and its application to p53 binding sequences. *Bioinformatics* 21 Suppl **1**, i274–i282.
28. Stolc, V., Samanta, M.P., Tongprasit, W., Sethi, H., Liang, S., Nelson, D.C., Hegeman, A., Nelson, C., Rancour, D., Bednarek, S., Ulrich, E.L., Zhao, Q., Wrobel, R.L., Newman, C.S., Fox, B.G., Phillips, Jr., G.N., Markley, J.L., and Sussman, M.R. (2005) Identification of transcribed sequences in Arabidopsis thaliana by using high-resolution genome tiling arrays. *Proc. Natl. Acad. Sci. USA* **102**, 4453–4458.
29. Zilberman, D., Gehring, M., Tran, R.K., Ballinger, T., and Henikoff, S. (2007)

Genome-wide analysis of Arabidopsis thaliana DNA methylation uncovers an interdependence between methylation and transcription. *Nat. Genet.* **39**, 61–69.

30. Zhu, J., Kapoor, A., Sridhar, V.V., Agius, F., and Zhu, J.K. (2007) The DNA glycosylase/lyase ROS1 functions in pruning DNA methylation patterns in Arabidopsis. *Curr. Biol.* **17**, 54–59.
31. Stam, M. and Mittelsten Scheid, O. (2005) Paramutation: an encounter leaving a lasting impression. *Trends Plant Sci.* **10**, 283–290.
32. Johnson, W.E., Li, W., Meyer, C.A., Gottardo, R., Carroll, J.S., Brown, M., and Liu, X.S. (2006) Model-based analysis of tiling-arrays for ChIP-chip. *Proc. Natl. Acad. Sci. USA* **103**, 12457–12462.
33. Li, C.F., Pontes, O., El-Shami, M., Henderson, I.R., Bernatavichute, Y.V., Chan, S.W., Lagrange, T., Pikaard, C.S., and Jacobsen, S.E. (2006) An ARGONAUTE4-containing nuclear processing center colocalized with Cajal bodies in Arabidopsis thaliana. *Cell* **126**, 93–106.
34. Agius, F., Kapoor, A., and Zhu, J.K. (2006) Role of the Arabidopsis DNA glycosylase/lyase ROS1 in active DNA demethylation. *Proc. Natl. Acad. Sci. USA* **103**, 11796–11801.
35. Swiezewski, S., Crevillen, P., Liu, F., Ecker, J.R., Jerzmanowski, A., and Dean, C. (2007) Small RNA-mediated chromatin silencing directed to the 3′ region of the Arabidopsis gene encoding the developmental regulator, FLC. *Proc. Natl. Acad. Sci. USA* **104**, 3633–3638.
36. Bickel, K.S. and Morris, D.R. (2006) Silencing the transcriptome's dark matter: mechanisms for suppressing translation of intergenic transcripts. *Mol. Cell* **22**, 309–316.
37. Chung, H.R., Kostka, D., and Vingron, M. (2007) A physical model for tiling array analysis. *Bioinformatics* **23**, i80–i86.
38. Hinds, D.A., Stuve, L.L., Nilsen, G.B., Halperin, E., Eskin, E., Ballinger, D.G., Frazer, K.A., and Cox, D.R. (2005) Whole-genome patterns of common DNA variation in three human populations. *Science* **307**, 1072–1079.
39. Cao, X., Aufsatz, W., Zilberman, D., Mette, M.F., Huang, M.S., Matzke, M., and Jacobsen, S.E. (2003) Role of the DRM and CMT3 methyltransferases in RNA-directed DNA methylation. *Curr. Biol.* **13**, 2212–2217.
40. He, F., Li, X., Spatrick, P., Casillo, R., Dong, S., and Jacobson, A. (2003) Genome-wide analysis of mRNAs regulated by the nonsense-mediated and 5′ to 3′ mRNA decay pathways in yeast. *Mol. Cell.* **12**, 1439–1452.
41. Kim, S., Plagnol, V., Hu, T.T., Toomajian, C., Clark, R.M., Ossowski, S., Ecker, J.R., Weigel, D., and Nordborg, M. (2007) Recombination and linkage disequilibrium in Arabidopsis thaliana. *Nat. Genet.* **39**, 1151–1155.
42. Finnegan, E.J. and Dennis, E.S. (1993) Isolation and identification by sequence homology of a putative cytosine methyltransferase from Arabidopsis thaliana. *Nucleic Acids Res.* **21**, 2383–2388.
43. Thibaud-Nissen, F., Wu, H., Richmond, T., Redman, J.C., Johnson, C., Green, R., Arias, J., and Town, C.D. (2006) Development of Arabidopsis whole-genome microarrays and their application to the discovery of binding sites for the TGA2 transcription factor in salicylic acid-treated plants. *Plant J.* **47**, 152–162.
44. Lippman, Z., Gendrel, A.V., Colot, V., and Martienssen, R. (2005) Profiling DNA methylation patterns using genomic tiling microarrays. *Nat. Methods* **2**, 219–224.
45. Martienssen, R.A., Doerge, R.W., and Colot, V. (2005) Epigenomic mapping in Arabidopsis using tiling microarrays. *Chromosome Res.* **13**, 299–308.
46. Yamada, K., Lim, J., Dale, J.M., Chen, H., Shinn, P., Palm, C.J., Southwick, A.M., Wu, H.C., Kim, C., Nguyen, M., Pham, P., Cheuk, R., Karlin-Newmann, G., Liu, S.X., Lam, B., Sakano, H., Wu, T., Yu, G., Miranda, M., Quach, H.L., Tripp, M., Chang, C.H., Lee, J.M., Toriumi, M., Chan, M.M., Tang, C.C., Onodera, C.S., Deng, J.M., Akiyama, K., Ansari, Y., Arakawa, T., Banh, J., Banno, F., Bowser, L., Brooks, S., Carninci, P., Chao, Q., Choy, N., Enju, A., Goldsmith, A.D., Gurjal, M., Hansen, N.F., Hayashizaki, Y., Johnson-Hopson, C., Hsuan, V.W., Iida, K., Karnes, M., Khan, S., Koesema, E., Ishida, J., Jiang, P.X., Jones, T., Kawai, J., Kamiya, A., Meyers, C., Nakajima, M., Narusaka, M., Seki, M., Sakurai, T., Satou, M., Tamse, R., Vaysberg, M., Wallender, E.K., Wong, C., Yamamura, Y., Yuan, S., Shinozaki, K., Davis, R.W., Theologis, A., and Ecker, J.R. (2003) Empirical analysis of transcriptional activity in the Arabidopsis genome. *Science* **302**, 842–846.

47. Chekanova, J.A., Gregory, B.D., Reverdatto, S.V., Chen, H., Kumar, R., Hooker, T., Yazaki, J., Li, P., Skiba, N., Peng, Q., Alonso, J., Brukhin, V., Grossniklaus, U., Ecker, J.R., and Belostotsky, D.A. (2007) Genome-wide high-resolution mapping of exosome substrates reveals hidden features in the Arabidopsis transcriptome. *Cell* **131**, 1340–1353.

48. Samanta, M.P., Tongprasit, W., Sethi, H., Chin, C.S., and Stolc, V. (2006) Global identification of noncoding RNAs in Saccharomyces cerevisiae by modulating an essential RNA processing pathway. *Proc. Natl. Acad. Sci. USA* **103**, 4192–4197.

49. Wu, Z. and Irizarry, R.A. (2005) Stochastic models inspired by hybridization theory for short oligonucleotide arrays. *J. Comput. Biol.* **12**, 882–893.

50. Kinoshita, T., Harada, J.J., Goldberg, R.B., and Fischer, R.L. (2001) Polycomb repression of flowering during early plant development. *Proc. Natl. Acad. Sci. USA* **98**, 14156–14161.

51. Saze, H., Mittelsten Scheid, O., and Paszkowski, J. (2003) Maintenance of CpG methylation is essential for epigenetic inheritance during plant gametogenesis. *Nat. Genet.* **34**, 65–69.

52. Xiao, W., Gehring, M., Choi, Y., Margossian, L., Pu, H., Harada, J.J., Goldberg, R.B., Pennell, R.I., and Fischer, R.L. (2003) Imprinting of the MEA Polycomb gene is controlled by antagonism between MET1 methyltransferase and DME glycosylase. *Dev. Cell* **5**, 891–901.

53. Zuo, J., Niu, Q.W., and Chua, N.H. (2000) Technical advance: an estrogen receptor-based transactivator XVE mediates highly inducible gene expression in transgenic plants. *Plant J.* **24**, 265–273.

54. Li, L., Wang, X., Sasidharan, R., Stolc, V., Deng, W., He, H., Korbel, J., Chen, X., Tongprasit, W., Ronald, P., Chen, R., Gerstein, M., and Wang Deng, X. (2007) Global identification and characterization of transcriptionally active regions in the rice genome. *PLoS ONE* **2**, e294.

55. Hudson, M.E. and Snyder, M. (2006) High-throughput methods of regulatory element discovery. *Biotechniques* **41**, 673, 675, 677 passim.

56. Schubert, D., Clarenz, O., and Goodrich, J. (2005) Epigenetic control of plant development by Polycomb-group proteins. *Curr. Opin. Plant Biol.* **8**, 553–561.

57. Bulyk, M.L. (2006) DNA microarray technologies for measuring protein–DNA interactions. *Curr. Opin. Biotechnol.* **17**, 422–430.

58. Shchepinov, M.S., Case-Green, S.C., and Southern, E.M. (1997) Steric factors influencing hybridisation of nucleic acids to oligonucleotide arrays. *Nucleic Acids Res.* **25**, 1155–1161.

59. Kapranov, P., Cawley, S.E., Drenkow, J., Bekiranov, S., Strausberg, R.L., Fodor, S.P., and Gingeras, T.R. (2002) Large-scale transcriptional activity in chromosomes 21 and 22. *Science* **296**, 916–919.

60. Borevitz, J.O., Hazen, S.P., Michael, T.P., Morris, G.P., Baxter, I.R., Hu, T.T., Chen, H., Werner, J.D., Nordborg, M., Salt, D.E., Kay, S.A., Chory, J., Weigel, D., Jones, J.D., and Ecker, J.R. (2007) Genome-wide patterns of single-feature polymorphism in Arabidopsis thaliana. *Proc. Natl. Acad. Sci. USA* **104**,12057–12062.

61. Boyer, L.A., Plath, K., Zeitlinger, J., Brambrink, T., Medeiros, L.A., Lee, T.I., Levine, S.S., Wernig, M., Tajonar, A., Ray, M.K., Bell, G.W., Otte, A.P., Vidal, M., Gifford, D.K., Young, R.A., and Jaenisch, R. (2006) Polycomb complexes repress developmental regulators in murine embryonic stem cells. *Nature* **441**, 349–353.

62. Grewal, S.I. and Klar, A.J. (1996) Chromosomal inheritance of epigenetic states in fission yeast during mitosis and meiosis. *Cell* **86**, 95–101.

63. Patil, N., Berno, A.J., Hinds, D.A., Barrett, W.A., Doshi, J.M., Hacker, C.R., Kautzer, C.R., Lee, D.H., Marjoribanks, C., McDonough, D.P., Nguyen, B.T., Norris, M.C., Sheehan, J.B., Shen, N., Stern, D., Stokowski, R.P., Thomas, D.J., Trulson, M.O., Vyas, K.R., Frazer, K.A., Fodor, S.P., and Cox, D.R. (2001) Blocks of limited haplotype diversity revealed by high-resolution scanning of human chromosome 21. *Science* **294**, 1719–1723.

64. Ji, H. and Wong, W.H. (2005) TileMap: create chromosomal map of tiling array hybridizations. *Bioinformatics* **21**, 3629–3636.

65. Bernstein, B.E., Mikkelsen, T.S., Xie, X., Kamal, M., Huebert, D.J., Cuff, J., Fry, B., Meissner, A., Wernig, M., Plath, K., Jaenisch, R., Wagschal, A., Feil, R., Schreiber, S.L., and Lander, E.S. (2006) A bivalent chromatin structure marks key developmental genes in embryonic stem cells. *Cell* **125**, 315–326.

66. Gehring, M., Huh, J.H., Hsieh, T.F., Penterman, J., Choi, Y., Harada, J.J., Goldberg, R.B., and Fischer, R.L. (2006) DEMETER DNA glycosylase establishes MEDEA polycomb gene self-imprinting

by allele-specific demethylation. *Cell* **124**, 495–506.

67. Southern, E., Mir, K., and Shchepinov, M. (1999) Molecular interactions on microarrays. *Nat. Genet.* **21**, 5–9.
68. Qi, Y., He, X., Wang, X.J., Kohany, O., Jurka, J., and Hannon, G.J. (2006) Distinct catalytic and non-catalytic roles of ARGONAUTE4 in RNA-directed DNA methylation. *Nature* **443**, 1008–1012.
69. Jackson, J.P., Lindroth, A.M., Cao, X., and Jacobsen, S.E. (2002) Control of CpNpG DNA methylation by the KRYPTONITE histone H3 methyltransferase. *Nature* **416**, 556–560.
70. Alleman, M. and Doctor, J. (2000) Genomic imprinting in plants: observations and evolutionary implications. *Plant Mol. Biol.* **43**, 147–161.
71. Johansson, M.J., He, F., Spatrick, P., Li, C., and Jacobson, A. (2007) Association of yeast Upf1p with direct substrates of the NMD pathway. *Proc. Natl. Acad. Sci. USA* **104**, 20872–20877.
72. Alleman, M., Sidorenko, L., McGinnis, K., Seshadri, V., Dorweiler, J.E., White, J., Sikkink, K., and Chandler, V.L. (2006) An RNA-dependent RNA polymerase is required for paramutation in maize. *Nature* **442**, 295–298.
73. Royce, T.E., Rozowsky, J.S., Bertone, P. Samanta, M., Stolc, V., Weissman, S., Snyder, M., and Gerstein, M. (2005) Issues in the analysis of oligonucleotide tiling microarrays for transcript mapping. *Trends Genet.* **21**, 466–475.
74. Kling, J. (2005) The search for a sequencing thoroughbred. *Nat. Biotechnol.* **23**, 1333–1335.
75. Mitchell, P., Petfalski, E., Shevchenko, A., Mann, M., and Tollervey, D. (1997) The exosome: a conserved eukaryotic RNA processing complex containing multiple 3′–>5′ exoribonucleases. *Cell* **91**, 457–466.
76. Yadegari, R., Kinoshita, T., Lotan, O., Cohen, G., Katz, A., Choi, Y., Katz, A., Nakashima, K., Harada, J.J., Goldberg, R.B., Fischer, R.L., and Ohad, N. (2000) Mutations in the FIE and MEA genes that encode interacting polycomb proteins cause parent-of-origin effects on seed development by distinct mechanisms. *Plant Cell* **12**, 2367–2382.

Chapter 4

Manipulating Large-Scale *Arabidopsis* Microarray Expression Data: Identifying Dominant Expression Patterns and Biological Process Enrichment

David A. Orlando, Siobhan M. Brady, Jeremy D. Koch, José R. Dinneny, and Philip N. Benfey

Abstract

A series of large-scale *Arabidopsis thaliana* microarray expression experiments profiling genome-wide expression across different developmental stages, cell types, and environmental conditions have resulted in tremendous amounts of gene expression data. This gene expression is the output of complex transcriptional regulatory networks and provides a starting point for identifying the dominant transcriptional regulatory modules acting within the plant. Highly co-expressed groups of genes are likely to be regulated by similar transcription factors. Therefore, finding these co-expressed groups can reduce the dimensionality of complex expression data into a set of dominant transcriptional regulatory modules. Determining the biological significance of these patterns is an informatics challenge and has required the development of new methods. Using these new methods we can begin to understand the biological information contained within large-scale expression data sets.

Key words: Clustering, microarray, gene expression, enrichment, gene ontology.

1. Introduction

1.1. Microarray Technology, Large-Scale Data Sets, High-Resolution Data

A microarray is a rectangular slide with thousands of short stretches of DNA chemically bonded to it. The Affymetrix^{TM} Arabidopsis ATH1 22 K microarray chip contains pieces of DNA, known as probes, corresponding to approximately 22,000 genes, and has become the quasi-standard in *Arabidopsis* expression profiling (1). Using this technology, two large data sets exist that have profiled expression at two different scales. The AtGenExpress data set contains three series of expression profiles, with each series examining a different set of cells or

Dmitry A. Belostotsky (ed.), *Plant Systems Biology, vol. 553*

DOI 10.1007/978-1-60327-563-7_4 Springerprotocols.com

conditions. The developmental timeline series contains profiles of expression in a different organ or developmental stage (2). The hormone series contains profiles of expression in seedlings in response to various hormones over time (3). Finally, the abiotic and biotic stress series contains profiles of expression from seedlings in response to biotic stimuli like pathogens and to environmental abiotic stimuli (4). In total, the AtGenExpress data set profiles the expression of approximately 22,000 genes in roots, shoots, seedlings or cell culture under many conditions. A second large data set contains fewer experiments, but contains expression profiles at much higher resolution than the AtGenExpress data set, with 19 different experiments profiling expression of nearly all cell types within the *Arabidopsis* root, in addition to 13 developmental time points (5–7). Together, these data sets contain massive amounts of expression information; however, extracting biological insights from this data is a real challenge. A useful approach is to look for groups of genes that are expressed in similar patterns across the different conditions. Expression similarity often implies co-regulation and can be used to identify transcriptional regulatory modules. Thus finding groups of genes with common patterns of expression is a good starting point for extracting biological insight from large expression data sets.

1.2. Identifying Dominant Expression Patterns and Associated Genes

The informatic task of grouping genes with similar expression patterns is commonly referred to as clustering. Clustering can be more rigorously defined as the task of separating a large set of elements (genes) into distinct subsets (groups/clusters of genes) such that all elements in a subset share a common feature (similar expression pattern). The similarity between two elements is defined by a distance metric, such as Euclidean distance or Pearson correlation. There are a wide variety of clustering algorithms which use different strategies to separate the full set of elements into subsets, with the most common being the hierarchical and K-means algorithms.

There are advantages and disadvantages to using one clustering algorithm over another. The K-means algorithm is useful for finding subsets with many members, but can force elements into subsets where they may not belong simply because all the other subsets are worse matches. Furthermore, the K-means algorithm begins with randomly chosen cluster centers, and different runs of the algorithm can result in subsets containing different members. Finally, K-means clustering will split the full set of elements into exactly K subsets, which is user-determined and may not be optimal. In contrast, hierarchical methods are good at identifying relationships between single elements or small groups but it is often difficult to determine the proper subsets as the set of elements grows large.

We present a method which utilizes both of these algorithms in an effort to identify a set of unique dominant expression patterns within a large gene expression data set. We define a "dominant expression pattern" to be a pattern that has strong support (i.e., many genes exhibit the pattern) within the data. Thus our method does not try to necessarily try to find all the different expression patterns present in a data set, only those which represent large groups of co-expressed genes. The method utilizes a variant of the K-means algorithm, fuzzy K-means (8), to create a preliminary set of groups, from which initial patterns can be defined. Using a K-means-based method ensures that the initial set of patterns will be "dominant", since the K-means algorithm will preferentially create subsets with many members. We then use hierarchical clustering, and its strengths in identifying relationships between a small number of elements, to cluster and collapse initial patterns that are similar to each other, resulting in a final set of unique dominant expression patterns. Once these unique dominant expression patterns have been identified, clusters of genes associated with each of the patterns can be assembled and analyzed for biological significance.

1.3. Analyzing Clusters of Genes for Biological Significance: Biological Information Resources

Once a cluster has been obtained, a second challenge is to determine the biological significance, if any, of this group of co-expressed genes. The "biological significance" of a gene or a group of genes can be subjective. It could mean the biochemical, developmental, or physiological process that the gene's product is involved in. Alternatively, if one is interested in the role of a gene in the context of transcriptional regulatory modules, which is particularly important when monitoring the output of the modules using microarray analysis, then biologically significant information could include *cis*-regulatory elements present within the upstream or downstream regulatory regions of the gene. If the gene is a transcription factor, it could be useful to know its transcription factor class and its binding site preference. Literature is often a good source for mining this type of information, and several consortia are responsible for mining literature for experiments that identify these biological features for individual genes and then storing these annotations in a database. The most popular of these consortiums is the Gene Ontology or GO consortia. Gene Ontology categories are controlled vocabulary terms that describe the biological process, molecular function, or cellular component that is associated with a gene in many model organisms (9). The Arabidopsis Information Resource (TAIR) curates the literature and maintains the *Arabidopsis* GO category database (10). Several databases exist that define and catalog *Arabidopsis* transcription factors as well as identified *cis*-regulatory elements. These databases include the database of Arabidopsis transcription factors (DATF), PLACE (plant *cis*-acting DNA regulatory elements), and the Arabidopsis gene regulatory information server (AGRIS) (11–13). Microarray expression analysis has become a useful tool for experimenters performing *Arabidopsis* research, and many publications exist that use microarrays to identify genes associated with particular pathways. For example, some of these studies have identified genes associated with the M-phase or S-phase of the cell cycle, genes expressed during root hair morphogenesis, or genes associated with primary or secondary cell wall biosynthesis (14–17). These studies often identify associated genes using rigorous statistical testing; however, TAIR and other resources do not currently compile genes associated with biological processes via microarray analysis in their databases. We have also mined the literature for these types of studies and have included them as a source of useful biological information to infer biological significance (7).

1.3.1. Testing for Enrichment: Multiple Hypothesis Testing

These consortia and statistical resources are rich in information. When given a set of genes exhibiting a similar expression pattern one can search this set for obvious relationships among the genes. For example, if one finds that several genes that are known to act in a particular biochemical pathway are coordinately expressed, it suggests that this biochemical pathway is of particular importance in the cell-type, developmental stage, or environmental condition experimentally tested. In essence, one can test for the enrichment of a biological feature within a data set. However, given the vast number of potential biological features, the statistical significance of enrichment when testing each individual hypothesis is of utmost importance. A common statistical method used to determine the significance of feature enrichment is the hypergeometric distribution (18). The hypergeometric distribution tests whether any feature or variable is found in a list at a frequency greater than would be expected by chance and calculates a *P*-value. This method has been used in particular to identify GO term enrichment and *cis*-regulatory enrichment in large-scale gene expression analysis (19, 20). *See* **Note 17** for alternative methods to determine the significance of these biological features.

1.3.2. Choice of Correct Background Distribution of GO Categories

Enrichment testing entails determining the significance of a number of categories present within a list of interest relative to the number of times these categories are found in a background list. Therefore the choice of correct background is imperative. Current web-based methods exist that test for enrichment of GO categories, including the Generic GO Term Finder and ATHENA (19, 20). These methods consider the background as all genes annotated with a GO category. In the context of microarray analysis, however, we must consider that our background is the set of genes being tested for enrichment, that is, all genes present on the microarray being used. Not all genes annotated in the *Arabidopsis* genome are present on the ATH1 22 K microarray. We have therefore developed a method that tests enrichment relative to only the genes whose expression we are testing (i.e., those present on the ATH1 microarray chip).

2. Materials

1. ***R Software***: R is a free software environment designed for statistical computing and is available for download from http://www.r-project.org/(21). The code for generating the set of unique dominant patterns is written for R.
2. ***R Scripts*** (**Table 4.1**)
   a. Code for generating initial set of clusters.
   b. Code for identifying unique dominant expression patterns and their associated genes.
   c. <http://www.arexdb.org/software.jsp>
3. ***Background Chip: ATH1 Chip***: Tab-delimited text file containing all Affymetrix™ probe sets in one column and all corresponding AGI chromosome locus identifiers in a second column.

   Singleton Chip: Tab-delimited text file containing all Affymetrix™ probe sets which map to only a single AGI chromosome locus identifier in one column and corresponding AGI locus identifiers in a second column.
4. ***Biological Information Files***
   a. ***GO Annotation File:*** Tab-delimited text file containing AGI locus identifiers in the first column, corresponding gene models in a second column, GO category descriptions in a third column, and GO IDs in a fourth column.
   b. ***Array Annotation File:*** Tab-delimited text file containing AGI locus identifiers in one column, and biological processes annotated to these genes as determined by mining the literature in a second column.
   c. ***Transcription Factor Family File:*** Tab-delimited text file containing transcription factor AGI locus identifiers in one column and the family that these transcription factors belong to in a second column. The transcription factor family file was created by querying three transcription factor databases: DATF, AGRIS, and Riken (7, 11, 13, 22). If a transcription factor was annotated as belonging to a particular family in two of three databases, it was included in this list.
   d. ***Query List:*** Tab-delimited text file containing the AGI locus identifiers in the first column. This is the list that will be tested for biological enrichment relative to the background.
5. Hypergeometric Distribution: http://jakarta.apache.org/commons/math

Table 4.1
Documentation of, and R source code for, software for clustering, pattern identification and gene assignment. Each portion of the table documents a particular section of the software and includes a short text description, with important features and considerations highlighted, and the relevant R source code (within the black boxes). The sections are presented in the order they should be run during an analysis. The source code can be downloaded from http://www.arexdb.org/software.

Code for Fuzzy K-Means clustering

User specified parameters

These seven parameters need to be set before each run of the method.

Parameter	Type	Description
inputFile	string	The name of the file with the tab-delimited expression data. See note 4.1.1 for required file format.
minExpFilter	decimal	This parameter is the minimum expression a gene must have in order to pass the low expression filter. Setting this to FALSE will skip the low expression filtering.
minVarFilter	integer	This parameter controls what percentage of genes, ranked by variance, are removed. A value of 75 will filter out the bottom 75% of the genes (thus retaining the top 25%). Setting this to FALSE will skip the low variance filter.
kChoice	integer	This integer sets the initial number of patterns found by the fuzzy K-means algorithm. See note 4.1.8.
fuzzyKmemb	decimal	This sets the membership exponent parameter in the fuzzy K-means algorithm. Read the `cluster` package documentation before adjusting this.
alreadyLog2	TRUE/FALSE	This flag tells the method if the data needs to be log2-fold transformed before computing the distance between each gene. Set this to TRUE if the input data is already in terms of fold change. See note 4.1.5.
methodResultFile	string	The name of the rData file where the results of the clustering will be stored.
diagnosticFile	string	The name of the output file where diagnostic and status information will be printed.

```
inputFile <- "inputData.txt"
minExpFilter <- FALSE
minVarFilter <- 50
kChoice <- 15
fuzzyKmemb <- 1.05
alreadyLog2 <- FALSE
methodResultFile <- "patternIdent_result.rDump"
diagnosticFile <- "clustering_diagnostic.txt"
```

Reading the input data

This code loads the required `cluster` package and reads the expression data. It also creates some book-keeping variables.

```
library(cluster)
removeLowE <- 0
removeLowV <- 0
dateRun<- date()
cat("Starting Fuzzy K-Means clustering on ",dateRun,"\n",file=diagnosticFile)
expressionData <-read.table(file=inputFile,sep="\t",header=TRUE)
rownames(expressionData)<-expressionData[,1]
expressionData <- expressionData[,2:ncol(expressionData)]
cat("Expression data read from ",inputFile,"\n",file=diagnosticFile,append=TRUE)
cat("\t",nrow(expressionData)," genes with ",ncol(expressionData)," observations.\n",file=diagnosticFile,append=TRUE)
expDataFiltered <- expressionData
filterSettings <- paste("Filter Settings:\n\tInput File = ",inputFile,sep="")
```

Running low expression filter

This is the code for the low expression filtering. If `minExpFilter` is not set to FALSE it will remove any genes which are not expressed above `minExpFilter` in any measurement (column).

```
cat("Removing low expressed genes - ",file=diagnosticFile,append=TRUE)
filterSettings <- paste(filterSettings,"\n\tLow Expression Filter = ",sep="")
if(is.numeric(minExpFilter)){
    keep <- c(1:nrow(expDataFiltered))[apply(expDataFiltered,1,max)>=minExpFilter]
    removeLowE <- nrow(expDataFiltered)-length(keep)
    expDataFiltered <- expDataFiltered[keep,]
    cat("Done\n",file=diagnosticFile,append=TRUE)
    filterSettings <- paste(filterSettings,minExpFilter," (",removeLowE," genes removed)\n",sep="")
}else{
    cat("Skipped\n",file=diagnosticFile,append=TRUE)
    filterSettings <- paste(filterSettings,"FALSE\n",sep="")
}
```

(continued)

Table 4.1 (continued)

Running low variance filter

This is the code for the low variance filtering. If **minVarFilter** is not set to FALSE it will remove the bottom **minVarFilter**% of genes ranked by variance.

```
cat("Removing low varying genes - ",file=diagnosticFile,append=TRUE)
filterSettings <- paste(filterSettings,"\tLow Variance Filter = ",sep="")
if(is.numeric(minVarFilter)){
    lowVarCut <- quantile(apply(expDataFiltered,1,var),probs=(minVarFilter/100))
    keep <- c(1:nrow(expDataFiltered))[apply(expDataFiltered,1,var)>=lowVarCut]
    removeLowV <- nrow(expDataFiltered)-length(keep)
    expDataFiltered <- expDataFiltered[keep,]
    cat("Done\n",file=diagnosticFile,append=TRUE)
    filterSettings <- paste(filterSettings,minVarFilter,"% (",removeLowV," genes removed)\n",sep="")
}else{
    cat("Skipped\n",file=diagnosticFile,append=TRUE)
    filterSettings <- paste(filterSettings,"FALSE\n",sep="")
}
cat(filterSettings,file=diagnosticFile,append=TRUE)
```

Normalizing the data

This will $\log_2$ transform the expression data if necessary. If **alreadyLog2** is set to TRUE the data will not be $\log_2$ transformed.

```
cat("Log2 Transforming expression data - ",file=diagnosticFile,append=TRUE)
if(!alreadyLog2){
    expDataFiltered[expDataFiltered==0] <- 1e-10
    expDataFiltered <- log2(expDataFiltered/apply(expDataFiltered,1,mean))
    expressionData[expressionData==0] <- 1e-10
    expressionData <- log2(expressionData/apply(expressionData,1,mean))
    cat("Done\n",file=diagnosticFile,append=TRUE)
}else{
    cat("Skipped\n",file=diagnosticFile,append=TRUE)
}
```

Building the distance matrix and clustering

This code will first build a R dist object, **distanceMat**, which holds the distances between each gene and every other gene. The distance between gene i and j is $\frac{1-\rho(i,j)}{2}$, where $\rho(i,j)$ is the Pearson correlation between i and j. This distance calculation ensures that genes which are perfectly correlated ($\rho(i,j) = 1$) have a distance of 0, and genes which are perfectly anti-correlated ($\rho(i,j) = -1$), have a distance of 1. See note 4.1.7 about using a different distance metric. The dist object is then used in the **fanny** implementation of the fuzzy K-means algorithm. The result of the clustering (**initClust**), the complete expression data (**expressionData**), and the filtered (and potentially $\log_2$ transformed) expression data (**expDataFiltered**), and other book-keeping variables are then saved into the **resultFile** file.

```
cat("Building distance matrix for clustering - ",file=diagnosticFile,append=TRUE)
distanceMat<-as.dist((1-cor(t(expDataFiltered),method="pearson"))/2)
cat(" Done\n",file=diagnosticFile,append=TRUE)

fuzzyKSettings <- paste("Fuzzy K-Means settings:\n\tK = ",kChoice,"\n\tmemb.exp = ",fuzzyKmemb,sep="")
fuzzyKSettings <- paste(fuzzyKSettings,"\n\tmaxit = ",(nrow(expDataFiltered)*4),"\n",sep="")

cat("Running fuzzy K-means to produce ",kChoice," clusters from ",nrow(expDataFiltered)," genes.",file=diagnosticFile,append=TRUE)
cat(fuzzyKSettings,file=diagnosticFile,append=TRUE)
cat("\n!!! This may take a long time to complete. !!!\n",file=diagnosticFile,append=TRUE)

initClust <- fanny(distanceMat, kChoice, diss = TRUE, memb.exp= fuzzyKmemb,maxit=nrow(expDataFiltered)*4)
cat("Done clustering!\nSaving results to ",methodResultFile,"\n",file=diagnosticFile,append=TRUE)
save(file=methodResultFile,expressionData,initClust,expDataFiltered,filterSettings,fuzzyKSettings, dateRun)
cat("Done. Initial clustering completed.\n\n",file=diagnosticFile,append=TRUE)
```

(continued)

Table 4.1 (continued)

Code for Pattern Identification & Gene Association

User specified parameters

These are the eight parameters which need to be set by the user for the method to identify, collapse and assign genes to patterns.

Parameter	Type	Description
autoSelectClusterCutoff	TRUE/FALSE	This flag tells the method whether or not to determine the **clusterCutoff** parameter automatically. If set to FALSE the user needs to specify the **clusterCutoff** parameter.
clusterCutoff	decimal	This is the minimum probability required of a gene belonging to a cluster in order for that gene to be used in building that clusters initial pattern. Ignored if **autoSelectClusterCutoff** is TRUE.
patternSimilarityCutoff	decimal	Maximum distance between patterns, under which the patterns are collapsed. When using Pearson correlation, 1-**patternSimilarityCutoff** corresponds to the Pearson correlation value above which patterns are collapsed.
pearsonCutoff	decimal	This sets the Pearson correlation above which a gene is assigned to a final dominant expression pattern.
methodResultFile	string	The filename of the rDump result file from the fuzzy K-means clustering.
userPatternInputFile	string	The filename of the user created patterns to be appended to the set of patterns (note 4.1.11). This file should be in the same format as the input expression data. If FALSE, no user patterns will be appended.
patternOutputFile	string	The filename of the tab-delimited output file, holding the $\log_2$ transformed dominant expression patterns.
groupOutputFile	string	The filename of the tab-delimited file containing the assignment of genes to each pattern.
diagnosticFile	string	The filename of the file where diagnostic and current status information will be printed.

```
autoSelectClusterCutoff <- TRUE
clusterCutoff <- 0.4
patternSimilarityCutoff <- 0.1
pearsonCutoff <- 0.85
methodResultFile <- "patternIdent_result.rDump"
userPatternInputFile <- FALSE
patternOutputFile <- "patternOutput.txt"
groupOutputFile <- "genesToPatterns.txt"
diagnosticFile <- "patternIdent_diagnostic.txt"
```

(continued)

Table 4.1 (continued)

Subroutines used in finding and collapsing similar patterns, assigning genes, and writing output

These four subroutines are used by the method to create the initial cluster patterns (**getClustProfiles**), collapse similar patterns (**makeCollapsedProfiles**), calculate the distance from each gene to each pattern (**correlateGenesToProfiles**), and write a table of genes assigned to each pattern (**writeMemberList**).

```
getClustProfiles <- function(expDataIn,clustIn,clusterCutoff){
    numClusters <- ncol(clustIn$membership)
    clustProfiles <- matrix(data=0,ncol=ncol(expDataIn),nrow=numClusters)
    for(i in 1:numClusters){
        genesInClust <- c(1:nrow(expDataIn))[clustIn$membership[,i]>=clusterCutoff]
        if(length(genesInClust)>0){
                    if(length(genesInClust)>1){
                            clustProfiles[i,]<-as.numeric(apply(expDataIn[genesInClust,],2,median))
                    }else{
                              clustProfiles[i,]<-as.numeric(expDataIn[genesInClust,])
                    }
        }
    }
    whichCluster <- c(1:numClusters)[apply(clustProfiles,1,var)>0]
    clustProfiles <- clustProfiles[apply(clustProfiles,1,var)>0,]
    return(list(profiles=clustProfiles,whichClust=whichCluster))
}


makeCollapsedProfiles <- function(expDataIn,clustProfsIn, clusterCutoff,grouping,origClust,clustIn){
    collapsedProfiles <- matrix(data=0,ncol=ncol(expDataIn),nrow=max(grouping))
    for(i in 1:max(grouping)){
        inClust <- c(1:length(grouping))[grouping==i]
        if(length(inClust)==1){
            collapsedProfiles[i,]<-clustProfsIn[inClust,]
            genesInClust <- c(1:nrow(expDataIn))[clustIn$membership[,origClust[inClust]]>=0.4]
        }else{
            cat("\tCollapsing profiles ",paste(inClust,collapse=","),"into new profile ",i,"\n")
            clustLookAt <- origClust[inClust]
            genesInClust <- c(1:nrow(expDataIn))[apply(clustIn$membership[,clustLookAt],1,max)>= clusterCutoff]
            collapsedProfiles[i,]<-apply(expDataIn[genesInClust,],2,median)
        }
    }
return(collapsedProfiles)
}

correlateGenesToProfiles <- function(expDataIn,profilesIn){
    profileCor <- matrix(data=0,ncol=nrow(profilesIn),nrow=nrow(expDataIn))
    for(i in 1:nrow(profilesIn)){
        profileCor[,i] <- apply(expDataIn,1,cor,y=profilesIn[i,])
        cat("\tCorrelating genes to profile ",i," out of ",nrow(profilesIn),"\n")
    }
    rownames(profileCor)<-rownames(expDataIn)
    colnames(profileCor)<-rownames(profilesIn)
    return(profileCor)
}


writeMemberList <- function(patternCor,minDist,fileOut){
    maxSize <- max(apply(patternCor>=minDist,2,sum))
    tempMat <- matrix(data="",ncol=ncol(patternCor),nrow=maxSize)
    colnames(tempMat)<-colnames(patternCor)
    for(i in 1:ncol(tempMat)){
        geneList <- rownames(patternCor)[patternCor[,i]>=minDist]
        if(length(geneList)>0){
                tempMat[1:length(geneList),i]<-geneList
        }
    }
    write.table(file=fileOut,tempMat,sep="\t",quote=FALSE,row.names=FALSE,col.names=colnames(tempMat))
}
```

(continued)

Table 4.1 (continued)

Creation of initial patterns

This code loads the results from the fuzzy K-means clustering and will build an initial set of patterns. If **autoSelectClusterCutoff** is TRUE, then the method will set **clusterCutoff** to a value at which the average gene is assigned to one cluster. The method then finds genes which do not belong to any cluster above a probability of **clusterCutoff** and flags those genes. The **getClustProfiles** subroutine then builds the initial set of patterns by finding the median (by column) expression pattern of all genes assigned to each cluster above **clusterCutoff**. The patterns are then stored in the **clustProfiles** variable.

```
cat("Starting Pattern Identification & Gene Assignment\n",file=diagnosticFile)
load(clusterResultFile)
if(autoSelectClusterCutoff){
    clusterCutoff <- initClust$membership[sort(initClust$membership,index.return=TRUE,decreasing=TRUE)$ix[nrow(initClust$membership)]]
}
cantAssignGenes<-(apply(initClust $membership,1,max)<clusterCutoff)
goodGenes<-(apply(initClust $membership,1,max)>=clusterCutoff)
initClust$cluster[cantAssignGenes]<-0
initClust$membership[cantAssignGenes,]<-0
cat("Number of genes assigned to a cluster above ",clusterCutoff," =",file=diagnosticFile,append=TRUE)
cat(sum(goodGenes),"/",length(initClust$cluster),"\n",file=diagnosticFile,append=TRUE)
cat("Number of genes not assigned to a cluster above ",clusterCutoff," =",file=diagnosticFile,append=TRUE)
cat(sum(cantAssignGenes),"/",length(initClust$cluster),"\n",file=diagnosticFile,append=TRUE)
cat("Generating inital set of patterns.\n",file=diagnosticFile,append=TRUE)
clustProfiles <- getClustProfiles(expDataFiltered,initClust,clusterCutoff)
whichClusters <- clustProfiles$whichClust
clustProfiles <- clustProfiles$profiles
```

Collapsing similar patterns

This code takes the initial set of patterns and collapses those that are similar to each other. The method calculates the distance (1-Pearson correlation) between all the patterns and then performs single-linkage hierarchical clustering using that information. The resulting tree is cut are a height of **patternSimilarityCutoff**, and patterns which are in the same cluster are collapsed in the **makeCollapsedProfiles** subroutine. The final set of unique dominant expression patterns is then written to **patternOutputFile** in the same format as the input expression data (see note 4.1.13 for output formats).

```
cat("Calculating distances between each pattern.\n",file=diagnosticFile,append=TRUE)
profileCor <- cor(t(clustProfiles))
diag(profileCor)<-NA
hClust <- hclust(as.dist(1-profileCor),method="single")
collapseGrp <- cutree(hClust,h=patternSimilarityCutoff)

cat("Collapsing similar patterns.\n",file=diagnosticFile,append=TRUE)
finalProfiles<-makeCollapsedProfiles(expDataFiltered,clustProfiles, clusterCutoff ,collapseGrp,whichClusters,initClust)
rownames(finalProfiles)<- paste("Pattern_",1:nrow(finalProfiles),sep="")
colnames(finalProfiles)<-colnames(expDataFiltered)
colnames(finalProfiles)[1] <- paste("\t",colnames(finalProfiles)[1],sep="")
if(is.character(userPatternInputFile)){
    cat("Adding user defined patterns from \"",userPatternInputFile,"\".\n",file=diagnosticFile,append=TRUE)
    userPatterns <- read.delim(file=userPatternInputFile,sep="\t",header=TRUE)
    tempFinal <- matrix(data=0,ncol=ncol(finalProfiles),nrow=nrow(finalProfiles)+nrow(userPatterns))
    tempFinal[1:nrow(finalProfiles),]<-finalProfiles
    tempFinal[(nrow(finalProfiles)+1):nrow(tempFinal),]<-as.matrix(userPatterns[,2:ncol(userPatterns)])
    rownames(tempFinal) <- c(rownames(finalProfiles),userPatterns[,1])
    colnames(tempFinal) <- colnames(finalProfiles)
    finalProfiles <- tempFinal
}
cat("Writing patterns to output file.\n",file=diagnosticFile,append=TRUE)
write.table(file=patternOutputFile,finalProfiles,quote=FALSE,sep="\t")
```

Assignment of genes to patterns

This code takes the final set of unique dominant expression patterns and calculates the pearson correlation between each gene used in the clustering (see note 4.1.12 for assigning genes not used in the clustering) to each pattern. This information is passed to the **writeMemberList** subroutine, and it creates a file (**groupOutputFile**), which is table where each column is a pattern, and each row contains the name of a gene which pearson correlates to that patterns above **pearsonCutoff**. Some genes may appear in multiple columns, and some may not appear at all. Finally the method deletes some bookkeeping and temporary variables and stores the rest of the results of the analysis in the rDump file specified by **methodResultFile**.

```
cat("Assigning genes to patterns.\n",file=diagnosticFile,append=TRUE)
geneToPatternCor <- correlateGenesToProfiles(expDataFiltered,finalProfiles)

cat("Writing gene to pattern output file.\n",file=diagnosticFile,append=TRUE)
writeMemberList(geneToPatternCor,pearsonCutoff,groupOutputFile)
rm(cantAssignGenes,goodGenes,profileCor,hClust,collapseGrp)
save(file=methodResultFile,list=ls())
cat("Pattern Identification and Gene Assignment completed.\n",file=diagnosticFile,append=TRUE)
```

6. ***Q-Value Test:*** John Storey's *Q*-value Method (23): http://faculty.washington.edu/jstorey/qvalue/
7. ***Biological Enrichment Software:*** http://www.arexdb.org/software

3. Methods

3.1. Identifying Unique Dominant Expression Patterns

3.1.1. Pre-clustering Data Filtering

Filtering input expression data is an important step in recovering dominant expression patterns of interest (*see* **Note 1** for expected input data format). Is it often the case that when an expression data set profiles only a subset of tissues or organs (i.e., the root is a subset of the whole plant) there will be many genes which are not expressed above a reasonable cutoff in any measurement in the subsystem being analyzed? When looking for unique dominant expression patterns, these genes should be removed. Additionally, there is often another subset of genes that are expressed uniformly across all the measurements. These genes are also not useful in identifying novel dominant expression patterns and should be removed (*see* **Note 2** for more information about flat expression patterns). Filtering the expression data is useful for three reasons: (1) it removes genes that are not informative in identifying dominant expression patterns, (2) genes with relatively flat expression profiles (i.e., all-off or non-varying) can behave poorly when compared using the Pearson correlation distance metric (*see* **Note 3**) and it reduces the size of the input to the clustering step which will decrease runtime of the fuzzy K-means algorithm. The rigor of this filtering can be adjusted in the code provided to suit the user's needs. Finally, if the input data set has already been filtered such that all the genes in the input data display some characteristic of interest, these filtering steps do not need to be applied.

3.1.2. Data Clustering

3.1.2.1. Normalization via $\log_2$ Transformation

When using non-fold normalized data (i.e., one-color Affymetrix™ arrays), it is important to normalize the data by a $\log_2$ transformation after filtering (*see* **Note 5** for definition of $\log_2$ normalization). This ensures that individual genes are compared on the basis of the shape of their expression patterns and not the absolute levels (*see* **Note 4** regarding comparing absolute levels). Two-color spotted arrays are usually measured in the form of a fold ratio between the observation and a control, and thus usually do not need to be $\log_2$ normalized.

3.1.2.2. Choice of Distance Metric

The code provided in **Table 4.1** calculates the similarity between expression profiles via Pearson correlation (*see* **Note 6** for the relationship between distance and similarity measures). Pearson correlation is used because it is sensitive to the shape of the expression profile while being relatively amplitude insensitive. Using other distance metrics is possible, but requires modification of the code (*see* **Note 7**).

3.1.2.3. Fuzzy K-means Clustering

The clustering method used in the provided code (**Table 4.1**) is an implementation of fuzzy K-means algorithm described in Chapter 4 of Kaufmann and Rousseeuw (8). The exact method is named "FANNY" and is implemented in the "cluster" package in R (24). The implementation, as it is used in the code, takes five parameters as input: *x*, *k*, *diss*, *memb.exp*, and *maxit*. The *x* variable contains the distance of each gene from every other gene. The *k* variable is the number of initial clusters desired (*see* **Note 8**). The *diss* variable should always be set to TRUE, indicating that the *x* variable contains distances. The *memb.exp* and *maxit* variables set the membership exponent and maximum iterations used by the implementation. Details on all the parameters can be found in the help files contained within the cluster R package, and can be accessed by typing "?fanny" in the R environment. The FANNY implementation returns a large object, with many components, all of which are detailed in the package help files. Our method uses only the output contained within the membership component of the returned object. The membership component is a matrix with one row for each gene and one column for each of the K clusters. The value in each cell is the probability that the gene belongs to a given cluster. We use this probability information to determine which genes belong to multiple clusters and which genes do not match well to any cluster.

3.1.3. Creating the Initial Patterns

The initial set of patterns is generated by taking the column-wise median of the expression profiles for all genes truly belonging to each cluster. The notion of truly belonging to a cluster is where the probability information in the membership component of the FANNY output is incorporated. Our method uses a probability cutoff variable, *clusterCutoff*, which is used to determine which genes belong to which cluster. A gene belongs to a cluster if the membership component of the fuzzy K-means clustering reports a probability of that gene being in that cluster at or above the value in the *clusterCutoff* variable. The code provided (**Table 4.1**) allows a user to set this cutoff themselves (autoSelectClusterCutoff = FALSE) or determine it automatically (autoSelectClusterCutoff = TRUE). If the *clusterCutoff* is determined automatically, the method finds a cutoff at which the average gene belongs to one cluster. Often, the cutoff will be below 0.5, allowing for some genes to truly belong to multiple clusters, and thus be included in the creation of multiple initial patterns. If another clustering method were used, the initial set of patterns could simply be the mean expression profile of all the genes assigned to each cluster by the algorithm (*see* **Note 9** for information on how to use a different clustering algorithm).

3.1.4. Collapsing Similar Patterns

The final step in creating the set of unique dominant expression patterns is to collapse similar patterns created in the step above. Collapsing similar patterns corrects for the tendency of the K-means clustering algorithm to split large, potentially similarly expressed groups of genes and ensures that one has obtained the full complement of distinct, dominant expression patterns. The distances between each of the patterns are computed using the same distance metric used in the initial clustering (Pearson correlation in the code provided). These distances are then used to hierarchically cluster the initial set of patterns. The resulting tree is cut at a user-defined cutoff (stored in the *patternSimilarityCutoff* variable), and the patterns within the resulting clusters are collapsed together. In this method, we have used single-linkage clustering as the default. Using single-linkage hierarchical clustering ensures that all patterns whose distance is less than the cutoff are collapsed together (*see* **Note 10**). The default value in the code is *patternSimilarityCutoff* = 0.1, which ensures that all patterns with a Pearson correlation at or above 0.9 (or a Pearson correlation "distance" of 0.1) are collapsed together. When multiple patterns are collapsed the new pattern is the column-wise median of all the collapsed patterns. After all the patterns are collapsed, the resulting rows are a set of unique dominant expression patterns.

3.2. Assignment of Genes to Clusters

Given this set of unique dominant expression patterns, is it useful to have lists of genes which exhibit each respective expression pattern? *See* **Note 11** for how to add additional user-defined patterns before gene assignment. Using the same distance measure employed to identify the dominant expression patterns, our method calculates the distance from each gene to each pattern. The code provided only examines those genes that were used in K-means clustering algorithm as genes of interest, but this can be modified (*see* **Note 12**). Given the calculated distances, the groups of genes assigned to a pattern are defined as the set of genes with a distance below a certain cutoff (or above a certain cutoff if measuring using a similarity measure like Pearson correlation). The default value in the code is *pearsonCutoff* = 0.85, which associates a gene with a pattern if the gene Pearson correlates to the pattern at or above 0.85. The advantage of doing assignment in this manner is that it allows genes to belong to more than one pattern if it is similar to multiple patterns, or, not be assigned to any pattern (25). The method produces a table with a set of genes assigned to each pattern. Having the complete set of genes matching an expression pattern is very desirable, especially when using this set to look for biological process enrichment. *See* **Note 13** for details of the method output formats.

3.3. Biological Process Enrichment

3.3.1. Process Background Chip

We utilize two types of background sets in our analysis. The first is the "ATH1 chip" and the second is the "Singleton Chip". (*See* **Section 2.3** and **Note 15** for additional details regarding the differences between these two backgrounds. *See* **Note 16** for information regarding nuclear, chloroplast- and mitochondrial-encoded genes.) This file contains a list of Affymetrix^{TM} probe set identifiers in the first column and a list of their corresponding AGI locus identifier in a second column (*see* **Note 17** for information regarding version releases of these conversions).

3.3.1.1. Singleton Chip Preprocessing

The singleton chip is first processed by removing Affymetrix^{TM} probe set identifiers that are mapped to multiple AGI locus identifiers. In the first column any rows with Affymetrix^{TM} probe set identifiers that contain <_x_at> or <_s_at> are removed (these identifiers indicate that a probe set matches to multiple loci). In the second column, any rows with multiple AGI locus identifiers <;> are also removed, in cases where <_x_at> or <_s_at> notation was not indicated (*see* **Note 15**).

3.3.1.2. Background Chip Reverse Mapping

Since we are testing for enrichment of features that are annotated to AGI locus identifiers, we must obtain a count of all AGI chromosomal loci found on the microarray chip. For ease of counting, the first and second columns are reversed. The number of AGI locus identifiers in the first column is then counted and stored for use in the hypergeometric distribution test.

3.3.2. Process Gene Descriptor Map (GO Annotations/Array Annotations/ TF family)

3.3.2.1. GO Annotation File Preprocessing

The GO annotation file requires additional processing relative to the array annotation file and transcription factor family file. The GO annotation file contains four columns: <AGI ID>, <model>, <description>, and <GO ID>. The second column, <model>, is disregarded in our analysis (*see* **Note 14** for additional details).

3.3.2.2. Gene Descriptor Map

As mentioned in Introduction, we only test for the enrichment of features relative to the appropriate background. We therefore filter the GO annotation list, array annotation, and transcription factor family lists so that they only contain the AGI locus identifiers present on the selected background chip. These filtered files are now named the <gene descriptor map> file. In the case of GO categories and array annotations, an AGI locus identifier can have multiple terms annotated to it; therefore, an AGI locus identifier can appear multiple times in the left column.

3.3.3. Process Enrichment Relative to Gene Descriptor Map

The query list is read in to the program. If multiple AGI locus identifiers are found in a single row, these locus identifiers are removed, as they were most likely obtained from an <_s_at> or an <_x_at> Affymetrix^{TM} probe set to AGI locus identifier conversion, and will bias the biological interpretation of results. The gene descriptor map is now also read into the program. A set of commands determines the measurements needed for input into the hypergeometric distribution test. For all further sections the term GO ID is interchangeable with <array annotation> or <TF family>. These methods can also be used for enrichment of any other biological features, including the enrichment of *cis*-elements.

Measurements needed for input into the hypergeometric distribution set:

A. The subset of AGI locus identifiers that are in common between these two files, and their associated GO IDs are filtered. For GO ID x, the number of AGI locus identifiers associated with it in the query list is counted. This is re-iterated for each GO ID present in the query list.

B. For GO ID x, the number of AGI locus identifiers associated with it in the background chip.

C. The number of AGI locus identifiers present on the query list.

D. The number of AGI locus identifiers present on the background chip.

3.3.4. Hypergeometric Distribution Test

These four numbers are provided to the hypergeometric distribution test script (*see* **Section 2.7**). The resulting *P*-values are then used to assess the false discovery rate (FDR) (*see* **Note 18**). We choose to accept a significant enrichment of $P \leq 10^{-3}$, although use of the FDR as a threshold is also recommended. Since we only consider enrichment in our query list, we only consider lists where the percentage of the query GO count/query set size is >background count/background set size.

3.3.5. Q-Value Testing

To assess the FDR we use John Storey's QVALUE program (23). This program runs in the R software environment. *See* **Section 2.1** for the direct link to this software. *P*-values must be stored in a tab-delimited text format to be read into QVALUE.

4. Notes

1. The input data should be a tab-delimited text file with each row being a gene and each column being a numeric measurement. The first column should contain a unique string identifying the gene in each row. The first row should contain unique strings identifying the measurements (tissues, developmental zones, etc.). Missing values should be reported as "NA".

2. The method has the ability to remove genes that are not varying across the measurements, either because they are constitutively on or off. As stated in **Section 3.1.1**, the reason for removing these genes is 3-fold: (1) there can be a very large number of these flat profiles that can obscure other patterns when performing the initial clustering, (2) certain distance metrics, such as Pearson correlation can behave erratically when comparing flat profiles (*see* **Note 4**), and (3) it reduces the size of the input to the clustering step which will decrease runtime. Additionally, if the user feels that the flat pattern is important, particularly in the cases where genes are ubiquitously expressed, it is easy to manually add it to the set of patterns and thus does not need to be computationally modified (*see* **Note 11**). Even if the user decides not to add a flat pattern into the final set, identifying genes which are expressed ubiquitously in all measurements can be biologically informative and could be another useful analysis. These genes can be identified by selecting genes with low variance and which are expressed above a given cutoff in every measurement.
3. Pearson correlation is a common measure of similarity between two genes. To a first approximation, two genes are well correlated if their expression patterns have the same overall shape across multiple experiments, even if the amplitude changes of the individual curves are different. As an example, two genes will be well correlated if they are both upregulated, even if one is upregulated 10-fold and the other is only 2-fold. However, this amplitude insensitivity is problematic when one of the genes has a profile that is nearly flat. A gene with a flat profile often has small unimportant up/down movements simply due to measurement noise. The problem arises from the fact that you can get large positive (or negative) Pearson correlations between noisy flat genes and other genes when the tiny random up/down movement in the noise happens to coincide with the biologically significant up/down movement of the other gene. In short, Pearson correlation is a very useful measure, but one should be careful when using it with genes whose expression changes are mostly due to measurement noise.
4. Some microarray platforms, such as the Affymetrix™ platform, will report gene expression in terms of absolute units as opposed to the fold change relative to a control which is reported for spotted two-color arrays. These absolute measurements are very useful as they can reveal the different biological concentrations of mRNA transcripts in the cell. However, the basal concentrations of two genes can be very different even though the genes are part of the same process.

This can make the co-expression of the two genes difficult to detect because of this difference in scale. Normalization using $\log_2$ transformation can help alleviate this problem.

5. In order to make the genes with expression patterns of very different amplitudes easily comparable, it is suggested that you $\log_2$ transform the expression data. The $\log_2$ normalization of a gene is done by dividing each of the gene's individual measurements by its mean value across all measurements, and then taking the log base 2 of that number. After the normalization, a value of 0 corresponds to a original measurement that was exactly the mean expression, a value of 1 corresponds to a original value that was twice or the mean expression, and a value of –2 corresponds to a original value that was 4-fold lower than the mean. This normalization is useful because it places all genes on the same measurement scale, even if they are expressed at very different levels in the cell.
6. Most clustering algorithms require a notion of distance, or dissimilarity, between two objects, where a larger number denotes larger distance or dissimilarity. The code provided uses Pearson correlation, which is a measure of similarity, and thus the method needs to convert from the similarity measure to distance. It does this by subtracting the Pearson correlation value from 1. This means that a high Pearson correlation value (1 = perfect correlation) will be converted to 0 (small distance) and a low Pearson correlation value (–1 = perfect anti-correlation) to 2 (high distance). The method then divides this number by 2, simply to scale the distances between 0 and 1. One needs to be careful when using similarity measures in conjunction with distance measures, to ensure that the interpretation of large and small values is consistent.
7. The code provided uses Pearson correlation as its distance metric. Other distance metrics can be substituted in its place, by changing the code at the appropriate places. Other distance metrics may be appropriate if one wants to define similarity not on the basis of correlation, but rather on the basis of absolute level (Euclidean distance would be appropriate), or other criteria for which a different measure is more suitable. When choosing a distance metric, be sure to consider whether or not the data should be normalized by $\log_2$ transformation. It is also important to consider how the patterns should be created and how similar patterns are collapsed (the column-wise median is usually appropriate, but may not be for some specialized applications).
8. The choice of K when using a K-means algorithm is very important. Although our method seeks to mitigate the impact of the choice by collapsing similar patterns, the

choice of K should still be considered carefully. Even though there are methods such as the gap statistic (26) which can be helpful in suggesting a good K, finding the correct choice of K can often be more of an art than a science. We suggest you start with a relatively high K initially. A good rule of thumb is to use a K 25% higher than the number of patterns you expect to find. It is also useful to run the method with higher and lower choices of K and compare the final sets of patterns.

9. The code provided uses a fuzzy K-means algorithm and its associated output to determine which genes should be used to build the initial patterns. The advantage of the fuzzy K-means algorithm is that the output can be used to detect genes that should be used in multiple initial patterns and genes that should not be used at all. You can use any clustering algorithm of your choice and rely on the hard cluster assignments to build the initial set of patterns. You will need to modify the code (in particular the getClustProfiles subroutine) to take the output of your clustering method and correctly build the initial set of patterns.

10. Linkage refers to the way distance between two clusters of elements is calculated. There are three common types of linkage used in hierarchical clustering: single, complete, and average. The distance between two clusters using single linkage is the minimum distance between any element in cluster 1 and any element in cluster 2. For complete and average linkage, the distance between two clusters is the maximum and mean distances between any element in cluster 1 to any element in cluster 2, respectively. Single linkage is used in the code provided.

11. If the user wishes, they can import an additional file of expression patterns that will be added to the final set of patterns. The user can define these patterns in a tab-delimited text file in the same format as the input data (*see* **Note 1**) except the first column will contain a unique pattern name. The user can then include them by setting the userPatternFile parameter to the name of that file. The method assumes the data in this file are in the same format and scale as the other patterns (pre-$\log_2$ transformed is necessary) and that the column order is the same. For adding in a flat pattern, the expression values could be set to all zeros. The user should be careful about adding in multiple additional patterns, because these patterns will not be collapsed. If the user chooses a distance metric which is sensitive to amplitude, then precise choice of values in the pattern file should be considered carefully.

12. The code currently calculates the correlation between each gene in the *expDataFiltered* variable and each pattern. The *expDataFiltered* variable contains all the genes and their expression which was used in the fuzzy K-means clustering. To associate a different set of genes, simply pass a different matrix of expression to the correlateGenesToProfiles subroutine. One common change would be to replace the line in the code: "*geneToPatternCor < - correlateGenesToProfiles(expDataFiltered, finalProfiles)*" with "*geneToPatternCor < - correlateGenesToProfiles(expressionData, finalProfiles)*". That would calculate the association between all genes in the input data to all the patterns. If one wishes to use a different subset, modify the code so that the subroutine is passed a matrix (in the same format as *expressionData*) containing only genes of interest.
13. The method will create three files. One file is an .rDump file (*methodResultFile*), which is a file that can be read into R using the load() command. This contains the information needed for and produced by the method and are not human readable. The method also produces a tab-delimited text file (whose filename is user specified with the *patternOutputFile* parameter), which has the same general format as the input expression data, but each row contains the expression for a dominant expression pattern, as opposed to a single gene. Finally, the method also produces a tab-delimited text file (whose filename is user specified with the *groupOutputFile* parameter) which defines the sets of genes associated with each pattern. Each column corresponds to a different pattern (with the pattern name defined in the first row), and each row contains the name of a gene which is associated with that pattern.
14. We download the GO annotation file from TAIR and modify it into the user format described. First, as described in **Note 16**, we only include nuclear-encoded genes.We then create, for each chromosome a list of AGI locus identifers, the corresponding gene models, GO description, and GO ID information as described in **Section 2.4.1.** When the GO IDs are associated with corresponding AGI locus identifiers (*see* **Section 3.3.3**), only unique AGI locus identifiers are considered in our counts for the hypergeometric distribution test. A number of gene models can exist for each AGI locus identifier (gene). These gene models can describe the protein-coding sequence, non-coding molecular species, or alternately spliced variants (10) associated with each gene. Although we ignore these models when we use this GO annotation file (*see* **Section 3.3.2.1**), after obtaining a statistically significant feature using the hypergeometric distribution test, one should cross-reference

the gene model sequence associated with each significant feature, with the corresponding probe sequences on the ATH1 22 K microarray chip to determine which gene model is most appropriate for further consideration.

15. As probe sets annotated to multiple genes can obscure the biological relevance of enrichment, in the majority of cases, we choose to use the "Singleton Chip".
16. In our analysis we have chosen to only consider nuclear-encoded genes. However, the ATH1 chip file does contain probe sets which map to mitochondrial- or chloroplast-encoded genes, and the GO annotation file which is available for download from TAIR also contains ontologies associated with mitochondrial- and chloroplast-encoded genes. We therefore remove these genes from the ATH1 chip file (and indirectly the singleton chip file), and from the GO annotation file. If one chooses to consider also enrichment among these mitochondrial- and chloroplast-encoded genes, one must include these in both the ATH1 chip file and the GO annotation file for proper comparison.
17. TAIR periodically releases new annotations of the *Arabidopsis* genome (10). With each new version of the *Arabidopsis* genome, a new file describing the Affymetrix^{TM} probe set to AGI locus identifier conversion is also released. We recommend using the most recent version of this release, although it is important to always use the same version for all rounds of data analysis when considering enrichment in multiple gene lists.
18. Alternate methods used to identify significance of enriched features are described in (27).
19. In essence, when we consider enrichment of biological features, we are testing the enrichment of each feature at a time, and thus, we are testing multiple hypotheses. False discovery rate (FDR) is a statistical method used in multiple hypotheses testing to correct for multiple comparisons (18). The FDR controls the expected proportion of incorrectly rejected null hypotheses (type I errors) (23). The *Q*-value of a test measures the proportion of false positives incurred (the FDR) when that particular test is called significant.

References

1. Busch, W. and Lohmann, J.U. (2007) Profiling a plant: expression analysis in *Arabidopsis*. *Current Opinion in Plant Biology* **10**(2), 136–141.
2. Schmid, M., Davison, T.S., Henz, S.R., et al. (2005) A gene expression map of *Arabidopsis thaliana* development. *Nature Genetics* **37**(5), 501–506.
3. Nemhauser, J.L., Hong, F., and Chory, J. (2006) Different plant hormones regulate similar processes through largely nonoverlapping transcriptional responses. *Cell* **126**(3), 467–475.
4. Kilian, J., Whitehead, D., Horak, J., et al. (2007) The AtGenExpress global stress expression data set: protocols, evaluation

and model data analysis of UV-B light, drought and cold stress responses. *Plant Journal* **50**(2), 347–363.

5. Birnbaum, K., Jung, J.W., Wang, J.Y., et al. (2005) Cell type-specific expression profiling in plants via cell sorting of protoplasts from fluorescent reporter lines. *Nature Methods* **2**(8), 615–619.
6. Birnbaum, K., Shasha, D.E., Wang, J.Y., et al. (2003) A gene expression map of the *Arabidopsis* root. *Science* **302**(5652), 1956–1960.
7. Brady, S.M., Orlando, D.A., Lee , J.-Y., et al. (2007) A high-resolution root spatiotemporal map reveals dominant expression patterns. *Science* **318**(5851), 801–806.
8. Kaufman, L. and Rousseeuw, P.J. (1990) *Finding Groups in Data: An Introduction to Cluster Analysis.* New York: Wiley.
9. Ashburner, M., Ball, C.A., Blake, J.A., et al. (2000) Gene ontology: tool for the unification of biology. The Gene Ontology Consortium. *Nature Genetics* **25**, 25–29.
10. Swarbreck, D., Wilks, C., Lamesch, P., et al. (2007) The Arabidopsis Information Resource (TAIR): gene structure and function annotation. *Nucleic Acids Research*, gkm965.
11. Guo, A., He, K., Liu, D., et al. (2005) DATF: a database of *Arabidopsis* transcription factors. *Bioinformatics* **21**(10), 2568–2569.
12. Higo, K., Ugawa, Y., Iwamoto, M., and Korenaga, T. (1999) Plant cis-acting regulatory DNA elements (PLACE) database. *Nucleic Acids Research* **27**(1), 297–300.
13. Palaniswamy, S.K., James, S., Sun, H., Lamb, R.S., Davuluri, R.V., and Grotewold, E. (2006) AGRIS and AtRegNet. A platform to link cis-regulatory elements and transcription factors into regulatory networks. *Plant Physiology* **140**(3), 818–829.
14. Brown, D.M., Zeef, L.A.H., Ellis, J., Goodacre, R., Turner, S.R. (2005) Identification of novel genes in *Arabidopsis* involved in secondary cell wall formation using expression profiling and reverse genetics. *Plant Cell* **17**(8), 2281–2295.
15. Jones, M.A., Raymond, M.J., and Smirnoff, N. (2006) Analysis of the root-hair morphogenesis transcriptome reveals the molecular identity of six genes with roles in root-hair development in *Arabidopsis. Plant Journal* **45**(1), 83–100.
16. Menges, M., de Jager, S.M., Gruissem, W., Murray, J.A.H. (2005) Global analysis of the core cell cycle regulators of *Arabidopsis* identifies novel genes, reveals multiple and highly specific profiles of expression and provides a coherent model for plant cell cycle control. *Plant Journal* **41**(4), 546–566.
17. Persson, S., Wei, H., Milne, J., Page, G.P., and Somerville, C.R. (2005) Identification of genes required for cellulose synthesis by regression analysis of public microarray data sets. *Proceedings of the National Academy of Sciences of the United States of America* **102**(24), 8633–8638.
18. Gadbury, G.L., Garrett, K.A., and Allison, D.B. Challenges and approaches to statistical design and inference in high dimensional investigations. **In this volume**.
19. Boyle, E.I., Weng, S., Gollub, J., et al. (2004) GO::TermFinder – open source software for accessing Gene Ontology information and finding significantly enriched Gene Ontology terms associated with a list of genes. *Bioinformatics* **18**, 3710–3715.
20. O'Connor, T.R., Dyreson, C., and Wyrick, J.J. (2005) Athena: a resource for rapid visualization and systematic analysis of *Arabidopsis* promoter sequences. *Bioinformatics* **24**, 4411–4413.
21. Team RDC. (2006) R: A Language and Environment for Statistical Computing. Vienna: R Foundation for Statistical Computing.
22. Iida, K., Seki, M., Sakurai, T., et al. (2005) RARTF: database and tools for complete sets of *Arabidopsis* transcription factors. *DNA Research* **12**, 247–256.
23. Storey, J.D. (2002) A direct approach to false discovery rates. *Journal of the Royal Statistical Society, Series B* **64**, 479–498.
24. Maechler, M., Rousseeuw, P.J., Hubert, M., and Hornik, K. (2007) *Cluster: Cluster Analysis Basics and Extensions.* In R package version 1.11. 9 ed.
25. Gasch, A. and Eisen, M. (2002) Exploring the conditional coregulation of yeast gene expression through fuzzy k-means clustering. *Genome Biology* **3**(11): research0059.1–research 22.
26. Tibshirani, R., Walther, G., and Hastie, T. (2000) Estimating the number of clusters in a dataset via the gap statistic. Technical Report 208. Department of Statistics, Stanford University.
27. Levine, D.M., Haynor, D.R., Castle, J.C., et al. (2006) Pathway and gene-set activation measurement from mRNA expression data: the tissue distribution of human pathways. *Genome Biology* **7**(10), R93.

Chapter 5

Applications of Ultra-high-Throughput Sequencing

Samuel Fox, Sergei Filichkin, and Todd C. Mockler

Abstract

The genomics era has enabled scientists to more readily pose truly global questions regarding mutation, evolution, gene and genome structure, function, and regulation. Just as Sanger sequencing ushered in a paradigm shift that enabled the molecular basis of biological questions to be directly addressed, to an even greater degree, ultra-high-throughput DNA sequencing is poised to dramatically change the nature of biological research. New sequencing technologies have opened the door for novel questions to be addressed at the level of the entire genome in the areas of comparative genomics, systems biology, metagenomics, and genome biology. These new sequencing technologies provide a tremendous amount of DNA sequence data to be collected at an astounding pace, with reduced costs, effort, and time as compared to Sanger sequencing. Applications of ultra-high-throughput sequencing (UHTS) are essentially limited only by the imaginations of researchers, and include genome sequencing/resequencing, small RNA discovery, deep SNP discovery, chromatin immunoprecipitation (ChIP) and RNA immunoprecipitation (RIP) coupled with sequence identification, transcriptome analysis including empirical annotation, discovery and characterization of alternative splicing, and gene expression profiling. This technology will have a profound impact on plant breeding, biotechnology, and our fundamental understanding of plant evolution, development, and environmental responses. In this chapter, we provide an overview of UHTS approaches and their applications. We also describe a protocol we have developed for deep sequencing of plant transcriptomes using the Illumina/Solexa sequencing platform.

Key words: Ultra-high-throughput DNA sequencing, HTS, UHTS, microread, sequencing, transcriptome, 454, Illumina, Solexa, SOLiD.

1. Introduction

Over the past few decades, Sanger DNA sequencing has dramatically changed the nature of biological research and ushered in the era of functional genomics. To an even greater extent, ultra-high-throughput sequencing (UHTS) is redefining the genome and the

Dmitry A. Belostotsky (ed.), *Plant Systems Biology, vol. 553*

DOI 10.1007/978-1-60327-563-7_5 Springerprotocols.com

ways in which genomes are studied. Recent technological developments in UHTS platforms have reduced the time, cost, complexity, and effort involved in sequencing projects, while providing an unprecedented amount of sequence information. UHTS has numerous applications such as in genome and targeted resequencing, metagenomics, deep SNP discovery, whole-genome de novo sequencing, gene expression profiling, ChIP/RIP studies, and transcriptome analysis (1, 2). Furthermore, the massive amount of sequence data generated by UHTS is already having a major impact on the field of bioinformatics through the development of new assembly algorithms and new concepts in data storage. With the newfound ability to conduct genomic studies on a truly global scale, our knowledge of genomes, gene structure, function, and regulation will advance markedly.

In the post-genomic era we are now able to more thoroughly analyze and characterize gene regulation, structure, and function in a global and high-throughput manner. UHTS provides the experimental tool that will make it possible to study all aspects of the genome as interconnected parts of the whole. For example, UHTS approaches will fundamentally alter the ways in which biologists analyze the gene expression networks guiding plant development and environmental responses from many directions including transcriptional regulation and post-transcriptional RNA processing. Ultra-high-throughput sequencing of the direct DNA and RNA targets of transcription factors and RNA-binding proteins, respectively, will be crucial to the elucidation of complex gene regulatory networks. Transcriptome analysis is of particular importance in the elucidation of the role of gene regulation in plant development. With UHTS, we are now able to empirically annotate transcription units in a plant genome and then interrogate their spatial and temporal expression patterns with unprecedented resolution and dynamic range.

1.1. Sanger/Dideoxy Sequencing

Sanger/dideoxy sequencing technology revolutionized biology and launched the genomics era. Sanger sequencing has many disadvantages, most of which revolve around the requirement to target, isolate, and amplify a single target gene or region, via PCR or bacterial cloning. It is a time-consuming and work-intensive process, requiring the cloning and bacterial propagation of clones to be sequenced rendering the method particularly inefficient for generation of genome-size data sets. In comparison to the new UHTS approaches, Sanger sequencing is relatively expensive and less efficient. However, Sanger sequencing reads are longer (700–1000 bp) and of higher quality (fewer errors) than those generated by the UHTS technologies and are therefore better for resolving repeat sequence structures. To date, most genome and transcriptome sequencing projects have utilized Sanger sequencing.

1.2. UHTS Platforms

The key to transforming genome sequencing was to maximize the throughput while minimizing costs and maintaining the high accuracy of sequence reads. Several platforms have recently made great strides in this area, with many other technologies proposed or underway. A large number of recent studies have utilized UHTS for many aspects of genome analysis. UHTS approaches can be broken down into two classes: short-read and microread UHTS. These technologies include massively parallel sequencing-by-synthesis short-read approaches such as Roche/454 pyrosequencing (http://www.454.com/) and the microread approaches of Illumina's (formerly Solexa; http://illumina.com/) "Clonal Single Molecule Array" technology and ABI's sequencing by ligation (ABI SOLiD: http://www.appliedbiosystems.com). A key benefit of all three UHTS approaches over Sanger sequencing is that there is no need to clone and propagate the DNA in bacteria. The clone-free approach has multiple benefits such as little or no bias in sequence representation and decreased time and cost for library construction. The 454 pyrosequencing system generates reads of up to 200–300 bp, but is presently more expensive per base than Illumina or SOLiD microread approaches. Due to the relatively longer read lengths of the 454 system, it is currently the best of the three platforms for de novo genome sequencing.

The Illumina and SOLiD systems currently generate approximately 35 bp reads and are best suited for resequencing or applications in which a reference genome is known or for applications such as gene profiling where the short length of the microread is not a concern. Microread technologies are presently used for transcriptome analysis, ChIP-seq, chromatin methylation studies, microRNA, and expression profiling experiments. In general, the 454 short-read system generates an order of magnitude-less sequence, around 100 megabases (MB) per run, compared to the microread platforms which are capable of delivering one to several gigabases (GB) of sequence per run. The 454 platform requires as little as 8 h for a single run, whereas 3 or more days are required for microread sequencers, and the material cost per run is similar across all three platforms at about \$5 K–\$10 K per run. Other ultra-high-throughput sequencing systems are being developed including platforms by Helicos Biosciences, VisiGen Biotechnologies, Pacific Biosciences, and Genovoxx, although these new systems have not yet been commercialized.

1.2.1. Roche/454 Genome Sequencer 20 and FLX Systems

The 454 sequencing platform performs sequencing by synthesis through pyrosequencing (3) on a PicoTiterPlate™ within which hundreds of thousands of emulsion-based PCR reactions amplify DNA strands attached to beads (4). Following amplification, these beads are separated and microscopic wells on the PicoTiterPlate are loaded with a single bead. To accomplish pyrosequencing, a single dNTP (e.g., dTTP) is added to the plate per polymerase cycle.

If this nucleotide is incorporated, the pyrophosphate released is recorded as a luminescent signal due to a luciferase-based reaction, and if multiple repeated bases are incorporated (e.g., three dTTP in sequence), a proportionally higher amount of light is released and recorded. The light generated during each nucleotide addition cycle is recorded and displayed in a pyrogram. Current generation 454 GSFLX sequencers, which are rapidly replacing the first generation 454 GS20 platforms, produce approximately 100 MB of sequence per 8 h run at a cost of approximately \$5 K–\$6 K per run. This cost is >10-fold less expensive than Sanger sequencing and potentially generates 5- to 10-fold more sequence per day. Presently, the GSFLX can provide 200–300 bp reads with a projected read length of 500 bp in the near future. However, the 454 sequencing system has limitations including difficulties in sequencing through homopolymers (5). Ultimately, the 454 *FLX* is best suited for applications that require longer sequence reads, such as de novo genome sequencing.

1.2.2. Illumina 1G Genome Analyzer

Illumina (formerly Solexa) sequencing is based on solid phase amplification followed by sequencing by synthesis of randomly fragmented DNA. The technology involves attachment of a short DNA fragment to a solid surface called a flow cell. The attached DNA fragments are PCR amplified to create clusters at a very high density (>10 million DNA clusters per lane) on the surface of the transparent sequencing flow cell. Amplified fragments representing a cluster are then sequenced and imaged with each reaction step. The system uses dNTPs containing fluorescently labeled 3′-reversible terminators, each emitting a different fluorescence signal. As the sequencing reaction occurs, all four dNTPs with their corresponding fluorescently labeled reversible terminator are added to the reaction, imaged, and the 3′-terminator is removed to allow for the next sequencing step. This sequencing process is repeated for multiple cycles. The Illumina platform generates a much greater amount of sequence (~1500 MB/run) at a similar cost (~\$4 K–\$6 K) to 454. This represents an approximately 10-fold increase in sequence information over 454 per run. However, the key disadvantages of the Illumina platform are shorter read lengths (25–36 bp) and a longer run time (3–4 days) than with the 454. Illumina's sequencing-by-synthesis approach provides high accuracy (<1.5% error per base). While the 454 platform generates longer reads (~200–300 bp), the key advantage of the Illumina platform is that it is capable of generating 10-fold more sequence data, and nearly 500-fold more independent reads for approximately the same cost per run. Although the Illumina platform generates very short reads, the total amount of sequence generated per run (>1 GB) makes the technology a great choice for applications that benefit from deep sampling and in studies where the shorter read length is not detrimental. Therefore, the Illumina genome analyzer is ideally suited for resequencing applications, such as RIP-seq, ChIP-seq, transcriptome sequencing, small RNA discovery, and expression profiling.

1.2.3. Applied Biosystems SOLiD

Applied Biosystems' sequencing by oligonucleotide ligation and detection (SOLiD) technology is based on a sequencing-by-ligation chemistry. Like the 454 platform, emulsion PCR is used to amplify DNA fragments on beads. However, the beads are then attached to a slide and the DNA fragments are interrogated through ligation by interactions of labeled oligonucleotide probes. Each oligonucleotide probe interrogates two bases at a time, and the base-pair combination is recorded as a color. The resulting DNA sequence is encoded in 2 bp color space depending upon the oligonucleotide primer probe. After the color has been recorded, the probe is cleaved, releasing the label. The process is repeated for multiple cycles with oligonucleotides that anneal offset by one base in each cycle. This means that each base is interrogated twice, allowing increased accuracy (a claimed 0.2% error rate), and provides added power in deep SNP detection for such applications as rare allele testing. The SOLiD platform has a sequence output comparable to that observed in Illumina sequencing, and supports sequencing from both DNA fragment libraries (35 bp) and mate-paired libraries (variable region between the paired ends). A single DNA fragment library sample can be sequenced on two slides to generate ~3 GB of 35 bp reads in 7 days at a cost of ~\$6 K or a mate-paired library can be sequenced to generate ~4.5 GB of paired 25 bp (50 bp) reads in about 10 days at a cost of ~\$8 K. Currently, 1–16 samples can be run at a time, with multiplexing enabling the differentiation of up to 256 mixed samples per run, thus providing the ability to test many samples simultaneously, which makes this technology competitive in terms of cost with microarrays for detecting transcript abundance.

1.3. UHTS Technologies in Genome Studies

Although there are significant obstacles in making UHTS technologies useful for de novo sequencing of complex plant genomes because of the difficulties in assembling the short reads, assembly of small microbial genomes has proven manageable (4, 6, 7). All three commercial UHTS platforms are making use of the power of paired-end reads (mate-pairs), where the ends of a larger DNA fragment are sequenced. Mate-pair sequencing was a key innovation that allowed shotgun sequencing of large complex genomes such as human and drosophila (8, 9). Mate-pair technology holds the promise of allowing UHTS to be used for de novo sequencing in the near future. Additionally, the utility of a 454-short-read/Sanger hybrid approach was investigated in several studies including the sequencing of a grape genome (10) and in the sequencing of marine microbes (11). The combination of the two technologies is highly amenable to the sequencing of small genomes with a low repeat content and as the lengths of UHTS reads increases, the utility of these approaches for de novo assembly will likely improve.

The feasibility of applying 454 pyrosequencing technology to sequencing complex plant genomes was investigated in many recent studies. In one study, 454 pyrosequencing was used for the determination of repeat sequences in the soybean genome (12). A second study used 454 reads to characterize families of repeat sequences in pea, demonstrating that 454-based UHTS proves useful for certain aspects of complex genome analysis (13). Another study used the technology to sequence four barley BAC clones (14). This proved to be an efficient approach for the gene-rich regions, but encountered difficulties in sequencing over repeats. It was concluded that 454 sequencing could be used with a BAC-by-BAC approach to sequencing of complex plant genomes, especially if the technology was to be used in combination with Sanger sequencing. A third study used the 454 GS20 to sequence the plastid genomes of two angiosperms *Nandina domestica* and *Platanus occidentalis* at greater than 99% coverage (15). Thus, it is clear that UHTS has utility for de novo sequencing of small genomes and can also be useful for certain applications in eukaryotic genome sequencing. However, with the present read length limitations, even in the case of the 454, the technology remains best suited for resequencing applications in plants or de novo genome sequencing of species in which a completed reference genome is available.

1.3.1. Identification of Polymorphisms by UHTS

Single nucleotide polymorphisms are useful as genetic markers in population genetics studies and for mapping of mutations in the laboratory. UHTS has been used for the detection of SNPs and mutations (16–18). In one study, the transcriptomes of two inbred maize lines were sequenced using 454 and the resulting short reads were used to identify >4900 putative SNPs (16). Another group has developed a novel SNP discovery approach termed complexity reduction of polymorphic sequences (CRoPS; (18)). In the CRoPS approach, tagged complexity-reduced libraries of two genetically distinct maize samples were prepared by amplified fragment length polymorphism and sequenced on a 454 FLX genome sequencer. This enabled the identification of SNPs in maize with applications among other plant species. Additionally, mutation detection can also be accomplished using UHTS. For example, 454 short-read sequencing was used to detect sequence variations in lung adenocarcinoma samples (17) in which previously known mutations were verified and additional mutations previously defined as wildtype by Sanger sequencing were identified. Thus, even the error-prone pyrosequencing technology is useful for detecting rare SNPs due to the massive amount of sequence that can be generated.

1.3.2. UHTS for Discovery of Small RNAs

UHTS is particularly well suited for the detection, quantification, and characterization of small RNAs (sRNA). Small RNAs play an important role in plant defense and development and regulate the expression of a diverse array of genes. Identification of small RNAs is important for the elucidation of gene regulatory networks guiding plant development. Previously, tag-based methods such as serial analysis of gene expression (SAGE; (19)) and massively parallel signature sequencing (MPSS; (20)) have been used to identify small RNAs (21, 22). One sRNA study used both 454-based UHTS and MPSS to discover and characterize small RNAs in an *Arabidopsis* RNA-dependent RNA polymerase 2 mutant (23). Other studies have further utilized the 454 sequencing platform to catalog and characterize small RNAs in *Arabidopsis* (24–27), *Populus trichocarpa* (28), California poppy (29), and wheat (30). Due to sufficient read lengths, the large number of reads, and the ability to barcode/multiplex samples, UHTS is ideally suited for the discovery and characterization of small RNAs in plants.

1.3.3. ChIP Sequence for Identification of DNA–Protein Interactions

Gene expression is regulated directly by transcription factor binding and indirectly influenced by chromatin packaging. Presently, microarrays remain the dominant method for analyzing DNA sequences interacting with proteins such as transcription factors in vivo. In such microarray studies, typically called ChIP-chip ((31); also *see* **Chapter 1**), a transcription factor is isolated by immunoprecipitation along with the DNA fragment to which it is bound. The co-immunoprecipitated DNA is labeled, hybridized to a DNA microarray, and the resulting hybridization signal data are analyzed. UHTS technologies offer an alternative to microarray hybridization, namely, direct sequencing of the DNA bound to the immunoprecipitated transcription factor protein (ChIP-seq). ChIP-seq offers several important advantages over microarrays including increased sequence information, sensitivity, and the need for less starting material (32). A handful of studies have recently implemented the use of ChIP-seq to identify DNA sequences bound by immunoprecipitated proteins. The procedure involves the immunoprecipitation of proteins followed by the isolation and sequencing of the physically interacting or bound DNA fragments. Using the Illumina 1G microread sequencing platform, two groups have recently created genome-wide profiles of binding sites for the transcription factors NRSF and STAT1 in Jurkat and HeLa cell culture systems, respectively (33, 34). The ChIP-seq method has also been used in chromatin mapping (35), a nucleosome positioning study (36), and methylation studies (37, 38). UHTS should broaden our ability to study sites of transcription factor binding, enabling the deciphering of networks of interactions and transcriptional regulatory cascades guiding plant development, and the genome-wide responses to environmental conditions. With the current cost, sensitivity, and sequencing capabilities per day of the Illumina and SOLiD platforms, ChIP-seq will likely replace array-based ChIP assays.

1.3.4. RIP Sequence for Identification of RNA–Protein Interactions

On another level of gene regulation, RNA-binding proteins (RBPs) participate in all facets of RNA metabolism. RBPs guide processes including the synthesis, splicing, transport, localization, translation, and degradation of RNA molecules. In an approach analogous to ChIP-seq, UHTS can be used to identify the RNA molecules bound by specific RBPs in vivo. Antibodies against specific RBPs or tagged versions of RBPs are used to immunoprecipitate the proteins and the interacting RNA is isolated, converted to cDNA, and sequenced (*see* **Chapter 2**). The RIP-seq technique has not yet been described in published studies, but is anticipated to elucidate the networks of RNA–protein interactions underlying post-transcriptional regulation of gene expression and transcript processing in plants.

1.3.5. UHTS Has a Wide Range of Applications

The diversity of applications of UHTS cannot be overstated. For example, 454 sequencing has been used in a several metagenomics/biodiversity studies (39–41). Additionally, the immense depth of sequencing is particularly well suited to sequencing of damaged ancient DNA (aDNA). UHTS technologies were recently implemented in studies involving the sequencing of wooly mammoth aDNA (42, 43) and DNA extracted from Neanderthals (44, 45). Finally, DNA microarray technologies were adapted and used in UHTS to enrich for and selectively sequence a specific subset of an entire genome (46, 47).

1.3.6. UHTS Analysis of Transcriptomes

UHTS can be used both for gene expression profiling and transcriptome sequencing. Previously, expression profiling was conducted with microarrays or tagging approaches such as SAGE or MPSS. SAGE was developed to quantitatively assess transcript expression and uses restriction enzymes to create short cDNA tags that are quantified by sequencing (19). However, a disadvantage of SAGE is that many sequences cannot be unambiguously mapped onto a genome due to their short lengths (48). Also, SAGE is a labor- and time-intensive protocol and some transcripts may not contain the restriction enzyme site required for tagging and sequencing the transcript. MPSS similarly utilizes a restriction endonuclease cleavage step, but sequence determination is accomplished with repeated steps in which the ligation of an adapter is followed by the hybridization of a labeled decoder probe (20). Perhaps the greatest limitation of tagging approaches is the entire sequence of the target is not determined. Microarray-based approaches have the disadvantages of associated high cost, being labor intensive, and requiring a larger amount of starting DNA material (32). Therefore, several recent studies have utilized UHTS for gene expression and EST sequencing applications.

A laser capture microscopy – 454 sequencing technique was used to analyze the transcriptome of maize stem apical meristem (49). A novel strategy to profile gene expression was conducted in maize ovaries which harnessed the specificity of the 3′-UTR enabling the resolution of transcripts possessing similar sequences (50). In another maize study, a technique called "robust analysis of 5′-transcript ends" (5′-RATE) in which 5′-oligocapping followed by restriction enzyme tagging and sequencing was used to profile gene expression (51). Finally, another group conducted gene expression profiling using "polony multiplex analysis of gene expression" (PMAGE) in which they sequenced millions of cDNA molecules through polony sequencing by ligation (52, 53).

Several recent studies have used UHTS to analyze transcriptomes from a variety of organisms and conditions. For example, transcriptome sequencing of a prostate cancer cell line LNCaP (54) and gene expression profiling in *Drosophila* were both accomplished using the 454 platform (55). Analysis of the wasp transcriptome, an organism for which the genome has yet to be sequenced, was also conducted using 454, generating nearly 400,000 brain cDNA sequence reads (56). UHTS-based plant transcriptome analysis has also been conducted in *Arabidopsis* (57) and *Medicago trunculata* (58), and deep 454-based sequencing of ESTs from two inbred maize lines was used to identify SNPs (16).

Additionally, ultra-high-throughput sequencing of transcriptomes is a powerful approach for the empirical annotation of exon structures and splice junctions in plant genomes. For example, we have used Illumina sequencing to validate and/or improve computationally predicted gene models in the model grass species *Brachypodium distachyon* (**Fig. 5.1**) and to empirically define transcription units otherwise overlooked by gene prediction algorithms. In addition to the deciphering of exon structures, we are able to determine alternative splice variants with unprecedented power.

In conclusion, UHTS is more efficient than classical Sanger sequencing for transcriptome discovery and validation, and provides a much higher signal-to-noise ratio than other approaches such as whole-genome tiling microarrays.

1.4. Considerations for Sample Preparation and Design of UHTS Experiments

There are many points to consider when designing UHTS experiments that will vary depending upon the biological questions being asked. Different studies such as the analysis of transcription-binding sites, transcriptome analysis, or gene profiling require different sample collection and preparation procedures. Plant growth conditions and tissue collection are sampling aspects

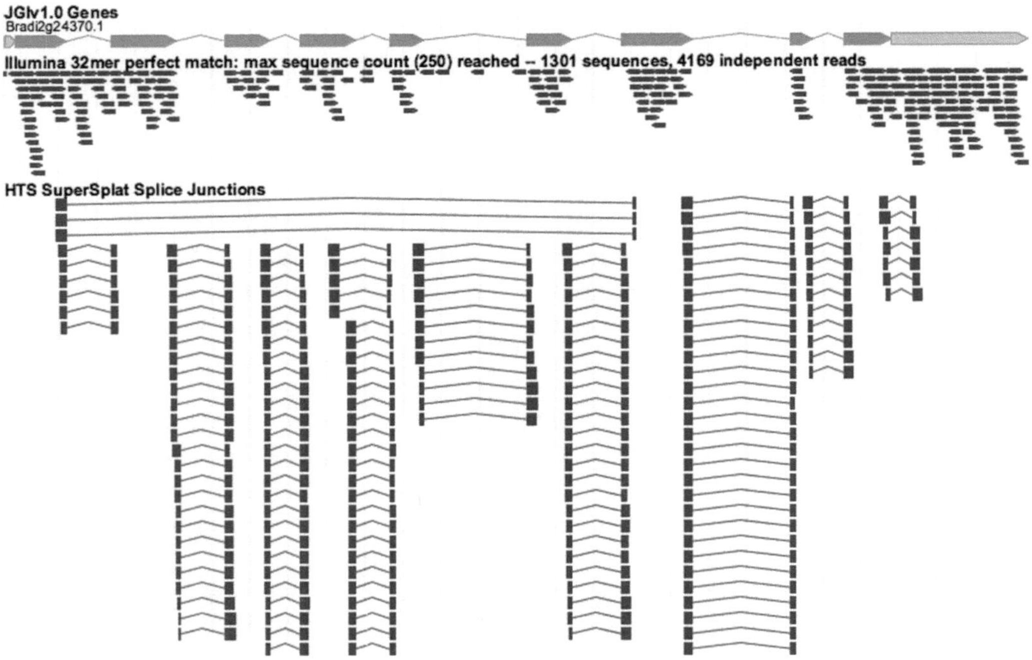

Fig. 5.1. Illumina sequencing of the *Brachypodium distachyon* transcriptome. A Gbrowse screenshot depicting empirical microread validation of a predicted gene. 32mer Illumina reads (arrowheads) aligned to the *Brachypodium* genome define the exon structure and splice junctions, including inferred alternative splicing events (http://www.brachybase.org).

that need to be addressed in the design of the experiments. For example, since the abundance of most, if not all transcripts fluctuates over the day in plants (59), periodic sampling over the entire day and night may be necessary to maximize cDNA library diversity in a transcriptome sequencing experiment. For transcriptome analysis, the type of cDNA library prepared depends upon the amount of RNA available and the results desired. In general, cells contain both extremely abundant mRNAs and rare transcripts that occur at levels of only a few copies per cell. Such a range in expression levels can make it extremely difficult to discover and sequence rare transcripts. Thus, to obtain a full representation of all transcribed genes, it may be necessary to normalize cDNA libraries. However, cDNA library normalization would not be applicable for gene expression profiling, and it may not be feasible to normalize a cDNA library when RNA quantities are particularly limited. We discuss a cDNA normalization technique in detail below.

Other considerations to address are whether or not to multiplex (i.e., "barcode") the samples so that several distinct samples can be analyzed in a single run or whether paired-end reads may be necessary if sequencing larger DNA fragments. In designing experiments, proper controls may need to be sequenced in parallel. Additionally, methods for the validation of discoveries, especially regarding SNPs, ChIP peaks, RIP peaks, and alternative splicing variants need to be considered. Finally, the importance of a well-planned computational infrastructure and experimental design cannot be understated for UHTS experiments (some of these are discussed in **Section 1.6.**). Ideally, a computational pipeline should be implemented prior to beginning the sequencing experiments, and simulated sequence data sets can be used to test and develop computational tools prior to sequencing efforts.

1.5. Which UHTS Platform Should Be Used for Sequencing?

Different UHTS approaches possess unique strengths and shortcomings for particular applications. The key performance differences are throughput, read lengths, the number of independent reads, and total nucleotides sequenced. Higher throughput may be more desirable than read length for some applications such as genome resequencing or transcriptome analysis, but longer read lengths and/or paired-end reads are beneficial for de novo genome assembly. Ultimately, microread approaches may be inefficient or inappropriate for most de novo sequencing applications, particularly when sequence repeats are an issue (e.g., eukaryotic genomes). In contrast, microread approaches may be superior for resequencing or when the number of reads is the key determinant of experimental success (e.g., transcript analysis or profiling).

Given a high-quality reference genome, it is possible to use bioinformatics approaches to empirically determine, beforehand, the unique single-copy K-mers in a genome; thus, the probability of being able to correctly map a sequence of a particular length of either micro- or short-read UHTS output can be predetermined. For example, it may be acceptable in a resequencing application for 85% of all 32mers in a genome to occur as unique single copy sequences because simply aligning microreads to the reference genome will unambiguously cover most of the genome. It is also notable that some computational analyses can be substantially more intensive with microread approaches due to the larger number of shorter sequences that need to be aligned to a reference sequence.

While individual platforms offer a tremendous amount of information, an attractive alternative may be a hybrid approach applying a combination of sequencing methods. For example, a Sanger/pyrosequencing hybrid approach was utilized for the genome sequencing of marine microbes (11) and the complex grape genome (10). The hybrid approach was found to be a very good and highly cost-effective method for draft genome assembly. Furthermore, a combination of 454 and Illumina platforms should allow researchers to benefit from the synergy of the large collection of Illumina microreads and the increased ability to assemble sequences with the longer reads of the 454 platform. We are currently pursuing this UHTS hybrid approach for de novo sequencing of cDNA libraries from non-model species of vertebrates.

1.6. Computational Considerations

UHTS can generate hundreds of millions of short-sequence reads per run. This enormous amount of information must be stored and manipulated in an efficient and cost-effective manner. Therefore, a well-designed computational pipeline is necessary to analyze large UHTS data sets. Thus, ultra-high-throughput sequence data have created a need for new sequence analysis algorithms and greater computational power and storage capacity. Computational analysis begins with preprocessing which may include error detection/correction. If a quality (Q) value is available, it can be used to detect and discard low-quality sequence reads. One significant problem is the issue of sequence accuracy. Sequencing errors are problematic in that they may be either incorporated into the contigs during de novo assembly or possibly create alignment errors in resequencing applications. Many assembly algorithms are available for the assembly of larger (~500 bp) Sanger sequences, including PHRAP (60), the Celera and TIGR assemblers (61, 62), and ARACHNE (63) among others. These long-read assembly algorithms are not suited to handling millions of short-read sequences and typically do not run at all on shorter read data. There have been many recent algorithmic and software tool developments for use in analyzing the short-read sequencing data, including several short-read assemblers with error-handling capabilities. Below we briefly outline several of the computational tools recently developed for analyzing UHTS data.

1.6.1. Brief Summary of Computational Tools for Analyzing UHTS Data

In addition to the software tools provided by the sequencing platform vendors, several tools useful for analyzing UHTS data sets have been developed by the community.

BLAT: The BLAST-like alignment tool (BLAT) scans an index of all non-overlapping K-mers in sequence database for short matches and extends these into high-scoring pairs. BLAT is different from BLAST, in that it builds an index of K-mers in the database in memory and scans through the query sequence for the matches. BLAT then combines matches into longer alignments (64).

Velvet: Velvet is a de novo assembler for short-read sequences that uses a de Bruijn graphs approach. Developed by Daniel Zerbino and Ewan Birney at the European Bioinformatics Institute (EMBL-EBI), Velvet is specifically designed for UHTS technologies (http://www.ebi.ac.uk/~zerbino/velvet/).

SSAKE: SSAKE cycles through reads stored in a hash table and searches a prefix tree for the longest possible match between any two sequences, extending matches to build a contig (65).

VCAKE: The Verified Consensus Assembly by K-mer Extension (VCAKE) uses a K-mer extension approach very similar to that applied in the SSAKE assembler. However, compared to

SSAKE, VCAKE has an improved ability to handle the sequencing errors observed in UHTS reads. VCAKE extends the seed sequence one base at a time relying upon the most commonly observed base from all matching reads (66).

SHARCGS: The short-read assembler based on robust contig extension for genome sequencing (SHARCGS) is another short-read assembler able to handle millions of reads and the erroneous base-calls in UHTS data sets. SHARCGS was demonstrated by assembling Illumina 36mer reads from the genome of *Helicobacter acinonychis*, yielding 937 contigs covering 98% of the genome (7).

MUMmerGPU: MUMmerGPU is an open-source high-throughput parallel pairwise local sequence alignment program that runs on graphics processing units (GPUs) in common work-stations. MUMmerGPU uses the nVidia Compute Unified Device Architecture (CUDA) to align multiple query sequences against a single reference sequence stored as a suffix tree. MUMmerGPU dramatically outperforms (10-fold faster) a serial CPU version of the MUMmer sequence alignment kernel (67).

EULER-SR: The Eulerian assembler was used for the analysis of 454 data from two bacterial genomes and Illumina short-read data from a human BAC. Using the proprietary 454 Newbler software for comparison, the Eulerian assembler was shown to assemble nearly optimal short-read assemblies (6).

RGA: Reference guided assembler (RGA) aligns microreads to their best match in a reference sequence, and then creates a guided consensus sequence from the aligned overlapping reads. RGA outputs the resulting contigs, singletons, the real coverage of each base in the assembly, and identifies SNPs and INDELs in the assembled sequence compared to the reference (Shen and Mockler, manuscript in preparation).

HashMatch: HashMatch rapidly aligns perfect matching micro-reads against a reference sequence. HashMatch is optimized for fixed length microreads (e.g., 25mers and 32mers) and exact matching and rapidly mines Illumina data to identify reads that hit a genome or any annotated feature within a genome. These features can include splice junctions, exons, introns, UTRs, and intergenic regions (Shen and Mockler, manuscript in preparation).

SPLAT: SPLAT (*spl*iced *a*lignment *t*ool) exhaustively aligns microreads against a reference sequence assuming a gapped alignment, which allows a read to span an intron. SPLAT predicts unannotated or novel splice junction reads, taking into consideration intron characteristics including intron length and sequence context and filters out microreads with low complexity sequences, or reads that match the genome over their entire length (Shen and Mockler, manuscript in preparation).

QCGA: Q-value and consensus guided assembler (QCGA) assembles short reads through a progressive K-mer search of sequence data organized in a prefix tree and stored in a hash, similar to SSAKE and VCAKE. Contigs are created and grown from the short reads taking into consideration the quality values associated with each base in the read and read multiplicity to resolve ambiguities table (Bryant, Wong and Mockler, manuscript in preparation).

1.7. Conclusions and Perspectives

After 20 years and the sequencing of the first complex genomes, Sanger sequencing has changed the way we study biology. UHTS holds the promise of ushering us into the next era of functional genomics where DNA sequence is not just a catalog but a guide to the extraordinary biology encoded in an organism's genome. The speed, cost, and depth of sequencing provided by UHTS changes the types of questions that biologists can ask, and potentially changes how we define a genome sequence. Soon, the DNA sequence alone may not be sufficient to describe the nuclear genome. A description of nucleosome positions, chromatin modifications, methylation events, coding and non-coding RNAs, alternative splicing, and natural antisense transcripts will be commonplace and essential for describing the functional genome of an organism.

2. Materials

During plant development, genes may be differentially expressed or alternatively spliced depending upon the tissue type and temporal and environmental cues. Transcriptome analyses are necessary to fully characterize gene structure including the identification of alternative splicing. These analyses are also important for the elucidation of gene expression in different cell types under different developmental conditions to better understand gene regulatory networks guiding plant development. The method below describes the preparation of a plant cDNA library for ultra-high-throughput sequencing on the Illumina platform.

2.1. Precautions and Stock Solutions

Special precautions should be taken to minimize RNA degradation by ribonucleases and to obtain libraries with high proportion of full-length cDNAs. To minimize RNase contamination, the workspace, centrifuge rotor, pipettors, and other equipments should be treated with RNase decontamination agents such as RNaseZap (Ambion). Plastic ware such as pipette tips and microcentrifuge tubes should be RNase-free grade. All RNA manipulations at room temperature should be performed in the shortest possible time. Frozen tissue powder should be placed directly into the ice-cold Concert reagent (Invitrogen), and the RNA solution should be treated with RNAsecure reagent (Ambion) before the cleanup step. All stock solutions should be prepared using RNase-free deionized water. Handling of the Concert reagent, phenol, chloroform, diethylpyrocarbonate (DEPC) and β-mercaptoethanol should be done in a fume hood. All RNA manipulations should be performed at 4°C, except when indicated otherwise.

2.2. Stock Solutions

1. 80% Ethanol
2. 2-Propanol
3. 3 M Sodium acetate, pH 5.5
4. 1 M Tris–HCl, pH 8.0
5. 5 M NaCl
6. RNase-free deionized water treated with diethylpyrocarbonate
7. 10% SDS

2.3. RNA Purification

1. RNaseZap (Ambion, cat. # AM9780)
2. RNase-free DNase I (Ambion, cat. # AM2238)
3. Concert Plant RNA Reagent (Invitrogen, cat. # 12322012)
4. RNase-free DNase I (Ambion, cat. # AM2238)
5. RNAsecure reagent (Ambion, cat. # AM7005)
6. RNeasy plant mini RNA kit (Qiagen, cat. # 74904)
7. Poly(A) Purist kit (Ambion, cat. # 1919)
8. Microcentrifuge
9. Vortex mixer, rotating platform
10. Heating block or PCR thermal cycler
11. Spectrophotometer (NanoDrop Technologies)
12. Bioanalyzer (Agilent Technologies)

2.4. cDNA Synthesis Using SMART Protocol and DSN Library Normalization

1. BD SMART cDNA Library Construction kit (BD Biosciences Clontech, cat. # 634901)
2. TRIMMERDIRECT cDNA Normalization kit (Evrogen, cat. # NK002)
3. Phenol:chloroform:isoamyl alcohol (25:24:1) mixture
4. TE buffer (10 mM Tris–Cl, pH 7.5, 1 mM EDTA)
5. Qiagen PCR Purification kit (Qiagen, cat. # 28106)
6. Microcentrifuge

7. PCR thermal cycler
8. Spectrophotometer (NanoDrop Technologies)
9. Horizontal agarose gel electrophoresis

2.5. SMART/DSN cDNA Preparation for Solexa/Illumina Sequencing

1. DNA Polymerase I, Large (Klenow) Fragment (NEB)
2. T4 DNA Polymerase (Invitrogen)
3. T4 Polynucleotide Kinase (NEB)
4. Klenow Fragment (3′–5′ *exo*–) (NEB)
5. Adenosine 5′-triphosphate (ATP) (NEB)
6. 100 mM dNTPs (Invitrogen)
7. Phusion Hot Start High-Fidelity DNA Polymerase (NEB)
8. T4 DNA Ligase (Invitrogen)
9. NuSieve GTG Agarose (Lonza, cat. # 50081)
10. Qiagen kits: PCR Purification (cat. # 28106); MinElute PCR Purification (cat. #28004); MinElute Reaction Cleanup (cat. # 28204); and MinElute Gel Extraction (cat. #28604)
11. Genomic DNA Sample Prep Oligo Only kit (Solexa/Illumina cat. # FC-102-1003/1002579).
12. Microcentrifuge
13. PCR thermal cycler
14. Spectrophotometer (NanoDrop Technologies)
15. Nebulizers (Invitrogen)
16. Tank with compressed nitrogen
17. Horizontal agarose gel electrophoresis system

2.6. cDNA Synthesis Using Random Priming Protocol

1. Superscript III First-Strand Synthesis kit (Invitrogen, cat. # 11904-018)
2. 100 mM dNTPs (Invitrogen)
3. DNA Polymerase I, Large (Klenow) Fragment (NEB)

2.7. Randomly Primed cDNA Preparation for Solexa/Illumina Sequencing

1. As used in **Section 2.5**.

2.8. Sequencing Using Solexa/Illumina 1G Genome Analyzer

1. Illumina Standard Cluster Generation kit (cat. # FC-103-1001/0801-0304)
2. 36 Cycle Solexa/Illumina Sequencing kit (cat. # FC-104-1003 /1001461)

3. Illumina Cluster Station
4. Illumina Genome Analyzer

3. Methods

3.1. Total RNA Isolation

This protocol has been used successfully for *Arabidopsis*, rice, poplar, and *Brachypodium* and yields high-quality intact RNA suitable for a synthesis of cDNA libraries enriched with full-length cDNAs. Approximately 200 mg of ground tissue yields up to 60–100 μg of total RNA. To prevent contamination with genomic DNA, RNA should be digested with DNase I followed by a cleanup on Qiagen mini-column. The procedure can be scaled up without changing tissue/reagents ratio if higher amounts of the total RNA are desired.

3.1.1. Extraction of the Total RNA and Genomic DNA Digestion

1. Grind flash-frozen tissues in liquid nitrogen using mortar and pestle or in stainless steel jars using Mixer Mill MM 301 (Retsch).
2. Transfer approximately 200 mg of frozen tissue powder directly into 1 mL of ice-cold Concert Plant RNA Reagent, immediately vortex for ~20 s and shaken for 5 min at room temperature (RT).
3. Centrifuge at ~21,000 × *g* for 2 min and transfer the supernatant to a new tube on ice.
4. Add 200 μL of cold 5 M NaCl and centrifuge at ~21,000 × *g* for 2 min.
5. Transfer the supernatant to new tube, add 500 μL of chloroform, and mix by inverting. Centrifuge at ~21,000 × *g* for 2 min and transfer the aqueous/top layer to a pre-chilled 2 mL tube. Repeat the chloroform extractions two to three times until the aqueous phase is clear.
6. After the final chloroform extraction step, transfer the aqueous layer to a pre-chilled tube and add 0.8 volumes of 2-propanol. Precipitate the RNA for 10 min at RT.
7. Centrifuge at ~21,000 × *g* for 10 min, remove supernatant, and wash RNA pellet with cold 80% ethanol.
8. Air dry the pellet for approximately 5 min and re-suspend RNA in 178 μL of 1 × RNAsecure reagent. To inactivate RNases, incubate for 10 min at 65°C.
9. Add 20 μL of 10 × Turbo-DNase buffer, 2 μL of Turbo-DNase, and digest the DNA at 37°C for 10 min.

3.1.2. RNA Cleanup

1. Add 700 μL of RLT buffer from RNeasy plant mini RNA kit to the digestion reaction.
2. Mix with 500 μL of 95% ethanol and proceed with RNA cleanup according to the manufacturer's protocol.
3. Retain 2 μL for quantification by NanoDrop spectrophotometer and 100–500 ng in 2 μL of water to check RNA integrity using Agilent 2100 Bioanalyzer (**Fig. 5.2**; *see* **Note 1**).

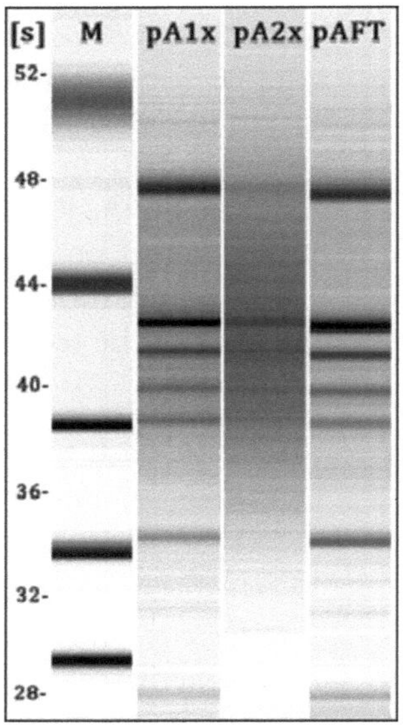

Fig. 5.2. Bioanalyzer analysis of polyadenylated mRNA fractions. The second lane, pA1x, is poly(A) RNA purified one time using oligo(dT) cellulose (Ambion RNA Purist kit). The third lane, pA2x, is poly(A) RNA purified two times using oligo(dT) column. Note the nearly complete lack of bands (ribosomal RNA) in the pA2x poly(A) sample. Thus, the pA2x poly(A) sample is suitable for random-primed cDNA synthesis protocol for the Illumina sequencing. The fourth lane, pAFT, is the flow through fraction of the pA1x sample through oligo(dT) column. RNA markers (M) are shown along with their relative RNA sizes [s]. RNA was analyzed using Agilent 2100 Bioanalyzer.

3.2. Purification of poly(A) RNA

In order to obtain high-quality mRNA essentially free of other cellular RNAs, two cycles of oligo(dT) purification using Ambion's Poly(A) Purification kit are recommended.

1. Bring the sample volume to 250 μL with nuclease-free water, add 250 μL of 2 × binding solution and mix thoroughly.
2. Add each sample to the oligo(dT) cellulose tube, mix well, and incubate at 72°C for 5 min. Then incubate the sample on a rocker for 60 min at RT with periodic "flick-mixing".
3. Centrifuge the sample at 4000 × *g* at room temperature for 3 min and remove the supernatant. Add 500 μL of Wash Solution I to the RNA-oligo(dT) cellulose, mix by vortexing, and transfer to the column in provided tube.

4. Centrifuge the sample/column at 4000 × *g* at RT for 3 min, discard the supernatant, and repeat the process with Wash Solution I.
5. Add 500 μL of Wash Solution II to the column, vortex briefly, and centrifuge at 4000 × *g* at RT for 3 min, discard the supernatant, and repeat the process with another 500 μL of Wash Solution II.
6. Place the spin column into new collection tube, add 100 μL of preheated RNA Storage Solution (to 72°C), vortex briefly, and centrifuge at 5000 × *g* at RT for 2 min.
7. Add a second 100 μL volume of RNA Storage Solution to column and repeat the RNA elution.
8. Transfer the sample to 1.5 mL microcentrifuge tube. Add 20 μL of 5 M ammonium acetate, 1 μL of 5 mg/mL glycogen, and 550 μL of 100% ethanol to the eluted mRNA. Mix by inverting and precipitate at –80°C for at least 1 h.
9. Centrifuge at maximum speed for 30 min at 4°C. Carefully remove supernatant, add 1 mL of 80% cold ethanol, vortex briefly, and centrifuge for 10 min at 4°C. Discard the supernatant and centrifuge for 2 min to remove all traces of ethanol.
10. Allow the pellet to air dry for no longer than 5 min. Dissolve pellet in ~15–50 μL of preheated (60° C) RNA Storage Solution.
11. Check the RNA quantity and integrity using NanoDrop spectrophotometer and Agilent 2100 Bioanalyzer.
12. Pool approximately 4 μg of 1X purified poly(A) RNA, bring sample volume to 250 μL with water, and repeat the above-described cycle of purification using single oligo(dT) column.
13. Retain the flow through fraction for the Bioanalyzer analysis (**Fig. 5.2**). Typically, the second cycle of oligo(dT) purification starting from 4 μg of 1X purified poly(A) RNA yields about 1 μg of mRNA essentially free of other cellular RNAs when starting from ~4 μg of 1X purified poly(A) RNA.

3.3. Construction of cDNA Libraries

To obtain cDNA libraries suitable for Illumina sequencing, we have used two different approaches. The first method is based on amplification of the full-length-enriched cDNA libraries using the SMART technology (BD Biosciences Clontech, (68)). The second approach is to generate cDNA libraries from highly enriched poly(A) mRNA using random hexamer priming. The advantages of the SMART protocol include, a requirement for only a small amount of starting RNA, which is essential when the amount of tissue or RNA available is a limiting factor, an ability to generate full-length cDNAs both from total or poly(A) RNA, and the ability to couple the procedure with library normalization using duplex-specific nuclease (DSN) treatment (69). The DSN normalization corrects for the bias in rare transcript coverage observed in non-normalized cDNA libraries. Potential pitfalls of the SMART cDNA preparation for Illumina sequencing may be over-amplification of the most abundant transcripts or preferential amplification of the shorter molecules and/or under-representation of the 5′-UTRs. In addition, microreads obtained from SMART libraries should be filtered for the sequences of SMART primers that flank both the 5′- and the 3′-cDNA ends. Some target mRNAs containing strong transcriptional pauses may also be under-represented or lost during the synthesis of the first strand of the full-length cDNA.

Randomly primed cDNA libraries have the advantage of unbiased representation of the 5′-cDNA ends including 5′-untranslated regions (UTRs). The average first cDNA strand fragment length can also be controlled by amount and/or by length (i.e., hexa-, hepta-, octamers, or their mixtures) of random primers. Therefore, the nebulization step can be omitted from the Illumina cDNA preparation. Random priming is essential for RIP-sequencing applications. The disadvantages of random priming include a requirement for the high purity of poly(A) RNA (to avoid contamination with non-polyadenylated cellular RNAs) and a requirement for larger starting amounts of tissues to obtain highly purified mRNA in microgram quantities.

3.3.1. Construction of the SMART Prepared Full-Length-Enriched cDNA Libraries

This protocol is a modification of BD Clontech SMART cDNA Library Construction method and utilizes SMART adapter primers (68).

1. In a PCR tube, add 1 μL of each primer (CDS III/3′-PCR primer to capture the poly(A) tail and 5′-SMART IV Oligonucleotide). Add 250–500 ng of poly(A) RNA sample and bring volume to total of 5 μL with nuclease-free water.
2. Incubate at 72°C and placed on ice for 2 min.
3. Add 2 μL of 5X First-Strand Synthesis Buffer (Clontech kit) and 1 μL of 20 mM dithiothreitol (DTT) and 1 μL of 10 mM dNTPs and 1 μL of moloney murine leukemia virus reverse transcriptase (M-MLV RT).
4. Incubate at 42°C for 1 h in the thermal cycler and proceed to the amplification step or store at –20°C.
5. Prepare a PCR reaction with the following reagents: 80 μL of sterile water, 10 μL of 10X Advantage 2 PCR Buffer, 2 μL of 50X dNTPs (10 mM each), 4 μL of 5′-PCR primer II A, 2 μL of 50X Advantage 2 PCR Polymerase Mix, and 2 μL of the control first-strand cDNA (provided with BD Biosciences Clontech kit).
6. PCR amplify in a thermocycler [95°C for 5 min (95°C for 20 s, 65°C for 30 s, 68°C for 6 min) × 15 cycles, 68°C for 7 min].

7. Remove 5 μL aliquots at cycles 7, 9, 11, and 13 for gel electrophoresis (*see* **Note 2**).
8. Separate PCR products on a 1% agarose gel to determine optimal cycle number for library amplification (**Fig. 5.3**).

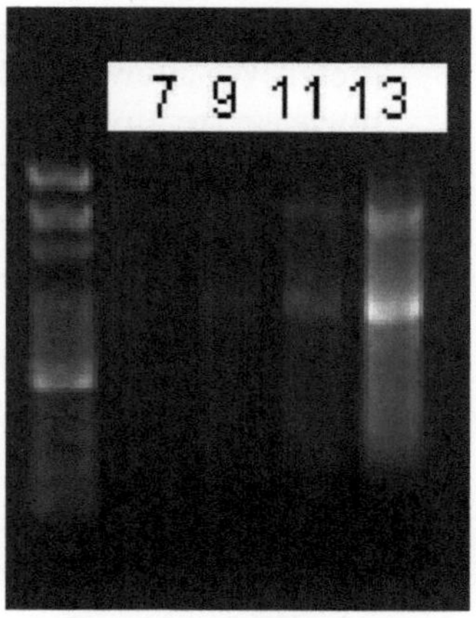

Fig. 5.3. Optimization of PCR cycling for a SMART cDNA library. The cDNA library was amplified for 7, 9, 11, and 13 cycles (see text for thermal cycler conditions). After each set of cycling, 5 μL aliquots were set aside for a gel and the remaining PCR reaction was repeatedly run for two additional cycles. The 5 μL of PCR product from each cycling point was run on a 1% agarose gel along with a 100-bp DNA marker.

9. Amplify the experimental cDNA library using the optimized cycling conditions (determined in step 7) and verify the PCR amplified products quality on a 1% agarose gel.
10. Purify the PCR products using QIAquick PCR Purification kit.

3.3.2. Library Normalization by DSN Treatment

This normalization method uses a modified protocol for double-strand cDNA removal by treatment with duplex-specific nuclease (DSN) isolated from the Kamchatka crab (Evrogen).

DSN exhibits a strong preference for cleaving dsDNA in either DNA–DNA or DNA–RNA hybrids, thereby making it useful for the removal of highly abundant transcripts (69). In this normalization procedure, the dsDNA is first denatured, then re-annealed for a brief period of time allowing the high-copy molecules to re-associate. Then DSN is added to digest the high-copy dsDNAs that have re-annealed. Through this process, the relative abundances of high- and low-copy transcripts are normalized, making cDNAs representing rare transcripts more likely to be sequenced. The key to the DSN treatment is optimization. We have worked out precise methods with several modifications for the optimization of this normalization procedure (discussed below). *Note*: The two most important factors to consider when performing this technique are the time allowed for DNA renaturation and the concentration of the DSN enzyme. The different lots provided by the company may have varying enzyme activity, thereby requiring an optimization of the DSN dilutions for each lot purchased. We have found that several key factors play a role in the optimization of the DSN procedure. The annealing time is a critical parameter. We allow 1.5, 2, 3, and 4 h for cDNA re-association before nuclease treatment. Additional optimization can be achieved by increasing dilutions of the DSN enzyme. We suggest using 1/4, 1/8, 1/16, and 1/64 dilutions of the enzyme to optimize DSN treatment (*see* **Note 3**).

1. Combine 1100–1200 ng of cDNA in a total volume of 12 μL (bring to volume with water if necessary) and add 4 μL of 4 × hybridization buffer. Divide each sample into two, 8 μL aliquots in PCR tubes (treatment and control).
2. Incubate the tubes at 98°C for 2 min, then at 68°C for 1.5, 2, 3, and 4 h to optimize re-association conditions.
3. While the cDNA is incubating, prepare the DSN dilutions using 50 mM Tris–HCl, pH 8.0, and preheat the 2 × DSN master buffer at 68°C.
4. Following the incubation (re-annealing time), add 10 μL preheated master buffer and incubate at 68°C for 10 min.
5. Quickly add 2 μL of the diluted DSN enzyme to the sample tube and incubate at 68°C for 25 min.
6. Stop the reaction by adding 20 μL of 5 mM EDTA and bring the final volume to 100 μL with 60 μL of sterile water.
7. Extract the normalized cDNA with an equal volume of phenol:chloroform and precipitate the DNA by adding 1/10th volume of 3 M sodium acetate, and 2.5 volumes of 100% ethanol.
8. Re-suspend DNA pellet in 12 μL of sterile water. Purify the DSN products using QIAquick PCR Purification kit and use 2 μL to determine DNA quantity.
9. Amplify the normalized cDNA with the Advantage 2 Polymerase mix. Add the following reagents to a PCR tube: 39 μL of sterile water, 5 μL of 10X Advantage 2 PCR Buffer, 1 μL of 50X dNTPs (10 mM each), 2 μL of 5′-PCR primer II A (provided in Clontech SMART cDNA Synthesis Kit), 1 μL of Advantage 2 Polymerase Mix, and 2 μL of template.
10. Run seven cycles with the SMART cDNA synthesis thermal cycler program described above (Point 6 in **Section 3.3.1**) and repeat the PCR optimization procedure (Point 8 in **Section 3.3.1**).

11. Cycle the non-normalized (no DSN treatment) samples by increasing two additional cycles for a total of 7, 9, 11, and 13 cycles and determine the optimum number of PCR cycles on a gel (**Fig. 5.3**).
12. Cycle the experimental (DSN-treated) samples to the optimized number of cycles. Normalized samples usually require two additional cycles as compared to the non-normalized sample (**Fig. 5.4**).

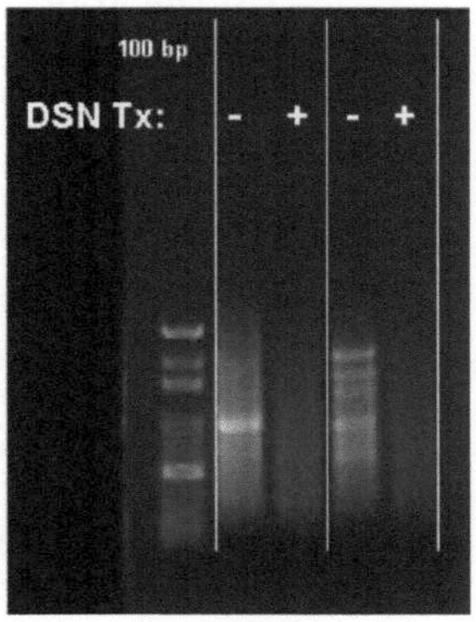

Fig. 5.4. cDNA libraries normalized by DSN treatment. This gel depicts two normalized libraries alongside their non-normalized counterparts. Five microliters of non-normalized control (–) and 5 μL normalized (+) cDNA libraries were separated on a 1% agarose gel. Note the lack of abundant transcripts (intense bands in the non-normalized (–)) in the normalized libraries (+). A 100-bp DNA ladder is used for size comparison.

3.3.3. Generation of Random Hexamer-Primed Libraries

1. Combine 1 μg of mRNA (*see* **Note 4**) in 4 μL of water and add 6 μL of random hexamers (50 mg/mL).
2. Heat the mixture at 75°C for 5 min and place on ice for 5 min.
3. Add the following components: 4 μL of 5X Superscript III Buffer, 0.5 μL of RNase inhibitor (40 U/mL), 2 μL of 10 mM dNTPs, and 1 μL of Superscript III RT.
4. Incubate at 25°C for 10 min and then 42°C for 1 h. Inactivate the reverse transcriptase by incubating at 70°C for 10 min.
5. Combine the following: 20 μL of the first-strand reaction, 8 μL of 10X Klenow Buffer, 1 unit of RNase H, 68.8 μL of water, and 3 μL of DNA Polymerase I (Klenow fragment).
6. Incubate at 15°C for 90 min and stop the reaction by adding 5 μL of 0.5 M EDTA, pH 8.0.
7. Purify cDNA using Qiaquick PCR purification kit. Elute the sample into 30 μL of EB buffer.

3.4. Preparation of cDNA for Solexa/Illumina Sequencing

We adapted a general procedure developed by Solexa/Illumina for genomic DNA preparation (Illumina Sample Preparation Protocol Version 2.3) with modifications described below.

1. Transfer the cDNA sample (~5 μg recommended) in a 50 μL volume to a nebulizer and add 750 μL of Illumina nebulization buffer (*see* **Note 5**).
2. Fragment the DNA using compressed nitrogen at 32–35 psi for 7 min and centrifuge the nebulizers at 450 × *g* for 2 min to collect the sample from the walls.
3. Purify the sheared DNA using a QIAquick PCR Purification Kit and elute into 32 μL of EB.
4. Mix the following in a PCR tube (*see* **Note 6**): 30 μL of hexamer/SMART cDNA, 10 μL of 5X T4 DNA ligase buffer with 10 mM ATP (Invitrogen), 4 μL of 10 mM dNTP mix, 2.5 μL of T4 DNA polymerase (3 U/μL), 1 μL of Klenow DNA polymerase (5 U/μL), and 2.5 μL of T4 polynucleotide kinase (10 U/μL).
5. Incubate for 30 min at 20°C. Purify the sample using the QIAquick PCR Purification kit and elute in 32 μL of EB.
6. To the 32 μL DNA from above, add 5 μL of 10X Klenow buffer, 10 μL of 1 mM dATP, and 3 μL of Klenow exo– (3′ to 5′ exo minus) polymerase (5 U/μL). Incubate for 30 min at 37°C (*see* **Note 7**).
7. Purify the DNA using a QIAquick MinElute Reaction Cleanup kit and elute into 12 μL of EB.
8. Prepare the following reaction mix (*see* **Note 8**): 10 μL of cDNA from above, 5 μL of 5X T4 DNA ligase buffer, 6 μL of adapter oligo mix (provided by Illumina), and 4 μL of T4 DNA ligase.
9. Incubate for 15 min at room temperature.
10. Purify with a QIAquick MinElute PCR Purification Kit eluting in 10 μL of EB.
11. Prepare 3.5% (w/v) NuSieve agarose in 1X TBE buffer (*see* **Note 9**).
12. Run the gel electrophoresis at 5 V/cm and stain the gel in 1 μg/mL of ethidium bromide in water in the dark for 10 min.
13. Excise the area in the range of 120–200 bp quickly to limit the exposure to UV light to 30 or less seconds to minimize DNA damage.
14. Purify the DNA from the gel slice using QIAquick Gel Purification Kit and elute in 32 μL of EB buffer.
15. Prepare the following PCR reaction mix (*see* **Note 10**): 2 μL of DNA from above, 1 μL of PCR primer 1.1 (Illumina), 1 μL of PCR primer 2.1 (provided by Illumina), 1 μL of 10 mM dNTPs, 44 μL of water, and 1 μL of Phusion DNA polymerase.

16. Amplify using the following PCR protocol: 30 s at 98°C, then (10 s at 98°C, 30 s at 65°C, 30 s at 72°C) for 18 cycles, followed by 10 min at 72°C.
17. Purify using the QIAquick PCR Purification Kit, elute in 30 μL of EB and run 5 μL of product on a 2% agarose gel (**Fig. 5.5**).

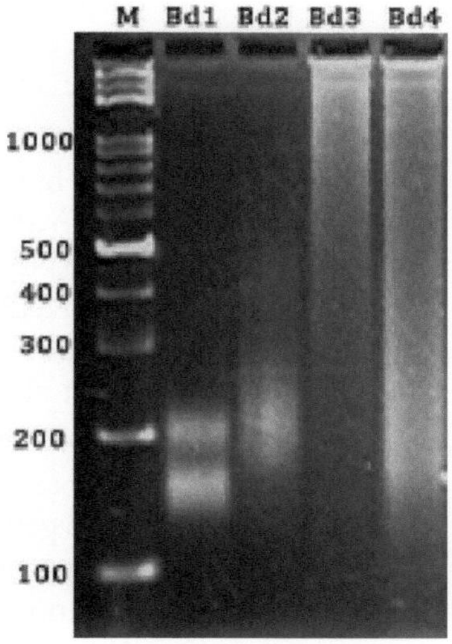

Fig. 5.5. Gel fractionation of hexamer-primed cDNA libraries. Lanes Bd1 and Bd2 are PCR-enriched *Brachypodium* cDNA libraries of average sizes 160 and 220 bp, respectively. Lanes Bd3 and Bd4 are cDNA libraries before gel fractionation and PCR enrichment. M = 100 bp DNA ladder markers (sizes in bp are indicated on *left*).

18. Measure the concentration of cDNA using a Nanodrop spectrophotometer.
19. Dilute the cDNA to 10 nM final concentration by approximating the average MW of the fragments to ~160 bp (an average size of cDNA extracted from gel).
20. At this point, the cDNA may be used directly for Illumina cluster generation or stored at –20°C.

4. Notes

1. A maximum of 100 μg of RNA can be bound to the Qiagen mini-column. Therefore, multiple columns may be needed if the amount of RNA exceeds this limit. A 260:280 nm wavelength ratio for the RNA obtained by this method should be 2.0 or higher. Store RNA at –80°C.

2. During the cDNA PCR amplification, overcycling of the cDNA results in highly undesirable nonspecific PCR amplification. Therefore, it is necessary to optimize the number of cycles necessary to amplify a quality cDNA library.
3. Other important points to consider for optimal DSN normalization: start with a consistent 1100–1200 ng of cDNA; add all reagents/enzyme simultaneously via multi-channel pipette; treat the cDNA with the DSN enzyme for precisely 25 min; perform a phenol/chloroform extraction followed by ethanol precipitation and Qiagen column purification of DSN-treated cDNA library to entirely eliminate the enzyme and salts.
4. The random hexamer approach requires the isolation of poly(A) mRNA that is of high purity and essentially free of other cellular RNAs. This is achieved with an additional round of poly(A) mRNA purification on oligo(dT) cellulose (*see* **Sections 3.2** and **3.3**). To decrease an average size of fragments, the first cDNA strand is synthesized using a high ratio of hexamer primers (300 ng per each μg of poly(A) mRNA).
5. The SMART prepared cDNA must be sheared using a nebulizer in order to generate fragments less than 800 bp. The cDNA prepared by hexamer priming contains a significant population of double-stranded fragments in the range 120–220 bp and therefore does not require an additional fragmentation via nebulization. For the random-primed libraries proceed directly to Point 4 of **Section 3.4**.
6. The nebulization/fragmentation process creates 5′- and 3′-overhangs. This step is implemented to convert the overhangs to blunt ends with phosphorylated 5′-termini.
7. A single "dA" nucleotide must be added to the 3′-blunt end of the templates to accommodate the ligation of the adapters which have a single "T" base overhang at their 3′-ends. A single "dA" is added to the ends of double-stranded cDNA molecules by employing the polymerase activity of exo minus (3′–5′) Klenow fragment.
8. The ligation reaction requires adapters supplied by Solexa/Illumina. The molar ratio of adapter to double-strand cDNA fragments should be maintained approximately 10:1.
9. The gel purification step ensures proper size selection of cDNA fragments and removal of the excess of free adapters prior to Illumina sequencing.
10. This step allows for the selective enrichment and PCR amplification of cDNA fragments with adapter molecules attached to both ends. The PCR is performed with two primers

provided by Illumina that anneal to the ends of the adapters. To avoid any skewing in the library representation PCR is limited to 18 cycles.

Acknowledgements

We thank Rongkun Shen, Henry Priest, Scott Givan, Chris Sullivan, and Doug Bryant for their contributions to our studies involving Illumina-based transcriptome sequencing. We would also like to thank Alice Barkan, Todd Castoe, Todd Michael, and Laurence von Kalm for their helpful comments on this chapter. This work was supported by Oregon State University and by grant ARF4435 from the Oregon Agricultural Research Foundation.

References

1. Schuster, S.C. (2008) Next-generation sequencing transforms today's biology. *Nat. Methods* **5**, 16–18.
2. Wold, B. and Myers, R.M. (2008) Sequence census methods for functional genomics. *Nat. Methods* **5**, 19–21.
3. Ronaghi, M., Karamohamed, S., Pettersson, B., Uhlén, M., and Nyrén, P. (1996) Real-time DNA sequencing using detection of pyrophosphate release. *Anal. Biochem.* **242**, 84–89.
4. Margulies, M., Egholm, M., Altman, W.E., Attiya, S., Bader, J.S., Bemben, L.A., Berka, J., Braverman, M.S., Chen, Y.J., Chen, Z., et al. (2005) Genome sequencing in microfabricated high-density picolitre reactors. *Nature* **437**, 376–380.
5. Metzker, M.L. (2005) Emerging technologies in DNA sequencing. *Genome Res.* **15**, 1767–1776.
6. Chaisson, M.J. (2008) Short read fragment assembly of bacterial genomes. *Genome Res.* **18**, 324–330.
7. Dohm, J.C., Lottaz, C., Borodina, T., and Himmelbauer, H. (2007) SHARCGS, a fast and highly accurate short-read assembly algorithm for de novo genomic sequencing. *Genome Res.* **17**, 1697–1706.
8. Adams, M.D., Celniker, S.E., Holt, R.A., Evans, C.A., Gocayne, J.D., Amanatides, P.G., Scherer, S.E., Li, P.W., Hoskins, R.A., Galle, R.F., et al. (2000) The genome sequence of *Drosophila melanogaster*. *Science* **287**, 2185–2195.
9. Venter, J.C., Adams, M.D., Myers, E.W., Li, P.W., Mural, R.J., Sutton, G.G., Smith, H.O., Yandell, M., Evans, C.A., Holt, R.A., et al. (2001) The sequence of the human genome. *Science* **291**, 1304–1351.
10. Velasco, R., Zharkikh, A., Troggio, M., Cartwright, D.A., Cestaro, A., Pruss, D., Pindo, M., Fitzgerald, L.M., Vezzulli, S., Reid, J., et al. (2007) A high quality draft consensus sequence of the genome of a heterozygous grapevine variety. *PLoS ONE* **2**, e1326.
11. Goldberg, S.M., Johnson, J., Busam, D., Feldblyum, T., Ferriera, S., Friedman, R., Halpern, A., Khouri, H., Kravitz, S.A., Lauro, F.M., et al. (2006) A Sanger/pyrosequencing hybrid approach for the generation of high-quality draft assemblies of marine microbial genomes. *Proc. Natl. Acad. Sci. U S A* **103**, 11240–11245.
12. Swaminathan, K., Varala, K., and Hudson, M.E. (2007) Global repeat discovery and estimation of genomic copy number in a large, complex genome using a high-throughput 454 sequence survey. *BMC Genomics* **8**, 132.
13. Macas, J., Neumann, P., and Navratilova, A. (2007) Repetitive DNA in the pea (*Pisum sativum* L.) genome: comprehensive characterization using 454 sequencing and comparison to soybean and *Medicago truncatula*. *BMC Genomics* **8**, 427.
14. Wicker, T., Schlagenhauf, E., Graner, A., Close, T.J., Keller, B., and Stein, N.

(2006) 454 sequencing put to the test using the complex genome of barley. *BMC Genomics* 7, 275.

15. Moore, M.J., Dhingra, A., Soltis, P.S., Shaw, R., Farmerie, W.G., Folta, K.M., and Soltis, D.E. (2006) Rapid and accurate pyrosequencing of angiosperm plastid genomes. *BMC Plant Biol.* **6**, 17.
16. Barbazuk, W.B., Emrich, S.J., Chen, H.D., Li, L., and Schnable, P.S. (2007) SNP discovery via 454 transcriptome sequencing. *Plant J.* **51**, 910–918.
17. Thomas, R.K., Nickerson, E., Simons, J.F., Jänne, P.A., Tengs, T., Yuza, Y., Garraway, L.A., LaFramboise, T., Lee, J.C., Shah, K., et al. (2006) Sensitive mutation detection in heterogeneous cancer specimens by massively parallel picoliter reactor sequencing. *Nat. Med.* **12**, 852–855.
18. van Orsouw, N.J., Hogers, R.C., Janssen, A., Yalcin, F., Snoeijers, S., Verstege, E., Schneiders, H., van der Poel, H., van Oeveren, J., Verstegen, H., et al. (2007) Complexity reduction of polymorphic sequences (CRoPStrade mark): a novel approach for large-scale polymorphism discovery in complex genomes. *PLoS ONE* **2**, e1172.
19. Velculescu, V.E., Zhang, L., Vogelstein, B., and Kinzler, K.W. (1995) Serial analysis of gene expression. *Science* **270**, 484–487.
20. Brenner, S., Johnson, M., Bridgham, J., Golda, G., Lloyd, D.H., Johnson, D., Luo, S., McCurdy, S., Foy, M., Ewan, M., et al. (2000) Gene expression analysis by massively parallel signature sequencing (MPSS) on microbead arrays. *Nat. Biotechnol.* **18**, 630–634.
21. Ge, X., Wu, Q., and Wang, S.M. (2006) SAGE detects microRNA precursors. *BMC Genomics* 7, 285.
22. Lu, C., Tej, S.S., Luo, S., Haudenschild, C.D., Meyers, B.C., and Green, P.J. (2005) Elucidation of the small RNA component of the transcriptome. *Science* **309**, 1567–1569.
23. Lu, C., Kulkarni, K., Souret, F.F., MuthuValliappan, R., Tej, S.S., Poethig, R.S., Henderson, I.R., Jacobsen, S.E., Wang, W., Green, P.J., et al. (2006) MicroRNAs and other small RNAs enriched in the *Arabidopsis* RNA-dependent RNA polymerase-2 mutant. *Genome Res.* **16**, 1276–1288.
24. Fahlgren, N., Howell, M.D., Kasschau, K.D., Chapman, E.J., Sullivan, C.M., Cumbie, J.S., Givan, S.A., Law, T.F., Grant, S.R., Dangl, J.L., et al. (2007) High-throughput sequencing of *Arabidopsis* microRNAs: evidence for frequent birth and death of MIRNA genes. *PLoS ONE* **2**, e219.
25. Kasschau, K.D., Fahlgren, N., Chapman, E.J., Sullivan, C.M., Cumbie, J.S., Givan, S.A., and Carrington, J.C. (2007) Genome-wide profiling and analysis of *Arabidopsis* siRNAs. *PLoS Biol.* **5**, e57.
26. Henderson, I.R., Zhang, X., Lu, C., Johnson, L., Meyers, B.C., Green, P.J., and Jacobsen, S.E. (2006) Dissecting *Arabidopsis thaliana* DICER function in small RNA processing, gene silencing and DNA methylation patterning. *Nat. Genet.* **38**, 721–725.
27. Rajagopalan, R., Vaucheret, H., Trejo, J., and Bartel, D.P. (2006) A diverse and evolutionarily fluid set of microRNAs in *Arabidopsis thaliana*. *Genes Dev.* **20**, 3407–3425.
28. Barakat, A., Wall, K., Diloretto, S., dePamphilis, C., and Carlson, C. (2007) Conservation and divergence of microRNAs in *Populus*. *BMC Genomics* **8**, 481.
29. Barakat, A., Wall, K., Leebens-Mack, J., Wang, Y.J., Carlson, J.E., and Depamphilis, C.W. (2007) Large-scale identification of microRNAs from a basal eudicot (*Eschscholzia californica*) and conservation in flowering plants. *Plant J.* **51**, 991–1003.
30. Yao, Y. and Ni, Z. (2007) Cloning and characterization of microRNAs from wheat (*Triticum aestivum* L.). *Genome Biol.* **8**, R96.
31. Mockler, T.C., Chan, S., Sundaresan, A., Chen, H., Jacobsen, S.E., and Ecker, J.R. (2005) Applications of DNA tiling arrays for whole-genome analysis. *Genomics* **85**, 1–15.
32. Mardis, E.R. (2007) ChIP-seq: welcome to the new frontier. *Nat. Methods* **4**, 613–614.
33. Johnson, D.S., Mortazavi, A., Myers, R.M., and Wold, B. (2007) Genome-wide mapping of in vivo protein–DNA interactions. *Science* **316**, 1497–1502.
34. Robertson, G., Hirst, M., Bainbridge, M., Bilenky, M., Zhao, Y., Zeng, T., Euskirchen, G., Bernier, B., Varhol, R., Delaney, A., et al. (2007) Genome-wide profiles of STAT1 DNA association using chromatin immunoprecipitation and massively parallel sequencing. *Nat. Methods* **4**, 651–657.
35. Mikkelsen, T.S., Ku, M., Jaffe, D.B., Issac, B., Lieberman, E., Giannoukos, G., Alvarez, P., Brockman, W., Kim, T.K., Koche, R.P., et al. (2007) Genome-wide maps of

chromatin state in pluripotent and lineage-committed cells. *Nature* **448**, 553–560.

36. Albert, I., Mavrich, T.N., Tomsho, L.P., Qi, J., Zanton, S.J., Schuster, S.C., and Pugh, B.F. (2007) Translational and rotational settings of H2A.Z nucleosomes across the *Saccharomyces cerevisiae* genome. *Nature* **446**, 572–576.
37. Barski, A., Cuddapah, S., Cui, K., Roh, T.Y., Schones, D.E., Wang, Z., Wei, G., Chepelev, I., and Zhao, K. (2007) High-resolution profiling of histone methylations in the human genome. *Cell* **129**, 823–837.
38. Taylor, K.H., Kramer, R.S., Davis, J.W., Guo, J., Duff, D.J., Xu, D., Caldwell, C.W., and Shi, H. (2007) Ultradeep bisulfite sequencing analysis of DNA methylation patterns in multiple gene promoters by 454 sequencing. *Cancer Res.* **67**, 8511–8518.
39. Angly, F.E., Felts, B., Breitbart, M., Salamon, P., Edwards, R.A., Carlson, C., Chan, A.M., Haynes, M., Kelley, S., Liu, H., et al. (2006) The marine viromes of four oceanic regions. *PLoS Biol.* **4**, e368.
40. Cox-Foster, D.L., Conlan, S., Holmes, E.C., Palacios, G., Evans, J.D., Moran, N.A., Quan, P.L., Briese, T., Hornig, M., Geiser, D.M., et al. (2007) A metagenomic survey of microbes in honey bee colony collapse disorder. *Science* **318**, 283–287.
41. Sogin, M.L., Morrison, H.G., Huber, J.A., Mark Welch, D., Huse, S.M., Neal, P.R., Arrieta, J.M., and Herndl, G.J. (2006) Microbial diversity in the deep sea and the underexplored "rare biosphere". *Proc. Natl. Acad. Sci. U S A* **103**, 12115–12120.
42. Gilbert, M.T., Binladen, J., Miller, W., Wiuf, C., Willerslev, E., Poinar, H., Carlson, J.E., Leebens-Mack, J.H., and Schuster, S.C. (2007) Recharacterization of ancient DNA miscoding lesions: insights in the era of sequencing-by-synthesis. *Nucleic Acids Res.* **35**, 1–10.
43. Poinar, H.N., Schwarz, C., Qi, J., Shapiro, B., Macphee, R.D., Buigues, B., Tikhonov, A., Huson, D.H., Tomsho, L.P., Auch, A., et al. (2006) Metagenomics to paleogenomics: large-scale sequencing of mammoth DNA. *Science* **311**, 392–394.
44. Green, R.E., Krause, J., Ptak, S.E., Briggs, A.W., Ronan, M.T., Simons, J.F., Du, L., Egholm, M., Rothberg, J.M., Paunovic, M., et al. (2006) Analysis of one million base pairs of Neanderthal DNA. *Nature* **444**, 330–336.
45. Noonan, J.P., Coop, G., Kudaravalli, S., Smith, D., Krause, J., Alessi, J., Chen, F., Platt, D., Pääbo, S., Pritchard, J.K., et al. (2006) Sequencing and analysis of Neanderthal genomic DNA. *Science* **314**, 1113–1118.
46. Hodges, E., Xuan, Z., Balija, V., Kramer, M., Molla, M.N., Smith, S.W., Middle, C.M., Rodesch, M.J., Albert, T.J., Hannon, G.J., et al. (2007) Genome-wide in situ exon capture for selective resequencing. *Nat. Genet.* **39**, 1522–1527.
47. Porreca, G.J., Zhang, K., Li, J.B., Xie, B., Austin, D., Vassallo, S.L., LeProust, E.M., Peck, B.J., Emig, C.J., Dahl, F., et al. (2007) Multiplex amplification of large sets of human exons. *Nat. Methods* **4**, 931–936.
48. Hene, L., Sreenu, V.B., Vuong, M.T., Abidi, S.H., Sutton, J.K., Rowland-Jones, S.L., Davis, S.J., and Evans, E.J. (2007) Deep analysis of cellular transcriptomes – LongSAGE versus classic MPSS. *BMC Genomics* **8**, 333.
49. Emrich, S.J., Barbazuk, W.B., Li, L., and Schnable, P.S. (2007) Gene discovery and annotation using LCM-454 transcriptome sequencing. *Genome Res.* **17**, 69–73.
50. Eveland, A.L., McCarty, D.R., and Koch, K.E. (2007) Transcript profiling by 3′UTR sequencing resolves expression of gene families. *Plant Physiol.* **146**, 32–44.
51. Gowda, M., Li, H., Alessi, J., Chen, F., Pratt, R., and Wang, G.L. (2006) Robust analysis of 5′-transcript ends (5′-RATE): a novel technique for transcriptome analysis and genome annotation. *Nucleic Acids Res.* **34**, e126.
52. Kim, J.B., Porreca, G.J., Song, L., Greenway, S.C., Gorham, J.M., Church, G.M., Seidman, C.E., and Seidman, J.G. (2007) Polony multiplex analysis of gene expression (PMAGE) in mouse hypertrophic cardiomyopathy. *Science* **316**, 1481–1484.
53. Shendure, J., Porreca, G.J., Reppas, N.B., Lin, X., McCutcheon, J.P., Rosenbaum, A.M., Wang, M.D., Zhang, K., Mitra, R.D., and Church, G.M. (2005) Accurate multiplex polony sequencing of an evolved bacterial genome. *Science* **309**, 1728–1732.
54. Bainbridge, M.N., Warren, R.L., Hirst, M., Romanuik, T., Zeng, T., Go, A., Delaney, A., Griffith, M., Hickenbotham, M., Magrini, V., et al. (2006) Analysis of the prostate cancer cell line LNCaP transcriptome using a sequencing-by-synthesis approach. *BMC Genomics* **7**, 246.
55. Torres, T.T., Metta, M., Ottenwälder, B., and Schlötterer, C. (2008) Gene expression

profiling by massively parallel sequencing. *Genome Res.* **18**, 172–177.

56. Toth, A.L., Varala, K., Newman, T.C., Miguez, F.E., Hutchison, S.K., Willoughby, D.A., Simons, J.F., Egholm, M., Hunt, J.H., Hudson, M.E., et al. (2007) Wasp gene expression supports an evolutionary link between maternal behavior and eusociality. *Science* **318**, 441–444.
57. Weber, A.P., Weber, K.L., Carr, K., Wilkerson, C., and Ohlrogge, J.B. (2007) Sampling the *Arabidopsis* transcriptome with massively parallel pyrosequencing. *Plant Physiol.* **144**, 32–42.
58. Cheung, F., Haas, B.J., Goldberg, S.M., May, G.D., Xiao, Y., and Town, C.D. (2006) Sequencing *Medicago truncatula* expressed sequenced tags using 454 Life Sciences technology. *BMC Genomics* **7**, 272.
59. Michael, T.P., Mockler, T.C., Breton, G., McEntee, C., Byer, A., Trout, J.D., Hazen, S.P., Priest, H.D., Sullivan, C.M., Shen, R., et al. (2007) Network discovery pipeline elucidates conserved time of day specific cis-regulatory modules. *PLoS Genetics* **2(8)**, e 795.
60. Ewing, B. and Green, P. (1998) Base-calling of automated sequencer traces using phred. II. Error probabilities. *Genome Res.* **8**, 186–194.
61. Myers, E.W., Sutton, G.G., Delcher, A.L., Dew, I.M., Fasulo, D.P., Flanigan, M.J., Kravitz, S.A., Mobarry, C.M., Reinert, K.H., Remington, K.A., et al. (2000) A whole-genome assembly of *Drosophila*. *Science* **287**, 2196–2204.
62. Sutton, G., White, O., Adams, M., and Kerlavage, A. (1995) TIGR Assembler: a new tool for assembling large shotgun sequencing projects. *Genome Sci. Technol.* **1**, 9–19.
63. Jaffe, D.B., Butler, J., Gnerre, S., Mauceli, E., Lindblad-Toh, K., Mesirov, J.P., Zody, M.C., and Lander, E.S. (2003) Whole-genome sequence assembly for mammalian genomes: Arachne 2. *Genome Res.* **13**, 91–96.
64. Kent, W.J. (2002) BLAT—the BLAST-like alignment tool. *Genome Res.* **12**, 656–664.
65. Warren, R.L., Sutton, G.G., Jones, S.J., and Holt, R.A. (2007) Assembling millions of short DNA sequences using SSAKE. *Bioinformatics* **23**, 500–501.
66. Jeck, W.R., Reinhardt, J.A., Baltrus, D.A., Hickenbotham, M.T., Magrini, V., Mardis, E.R., Dangl, J.L., and Jones, C.D. (2007) Extending assembly of short DNA sequences to handle error. *Bioinformatics* **23**, 2942–2944.
67. Schatz, M.C. and Trapnell, C. (2007) High-throughput sequence alignment using Graphics Processing Units. *BMC Bioinformatics* **8**, 474.
68. Zhu, Y., Machleder, E., Chenchik, A., and Siebert, P. (2001) Reverse transcriptase template switching: a SMART approach for full-length cDNA library construction. *Biotechniques* **30**, 892–897.
69. Zhulidov, P.A., Bogdanova, E.A., Shcheglov, A.S., Vagner, L.L., Khaspekov, G.L., Kozhemyako, V.B., Matz, M.V., Meleshkevitch, E., Moroz, L.L., Lukyanov, S.A., et al. (2004) Simple cDNA normalization using kamchatka crab duplex-specific nuclease. *Nucleic Acids Res.* **32**, e37.

Chapter 6

Isolation of Plant Polysomal mRNA by Differential Centrifugation and Ribosome Immunopurification Methods

Angelika Mustroph, Piyada Juntawong, and Julia Bailey-Serres

Abstract

Polyribosomes (polysomes) form as multiple ribosomes engage in translation on a single mRNA. This process is regulated for individual mRNAs by both development and the environment. To evaluate the translation state of an mRNA, ribosomal subunits, ribosomes, and polysomes can be isolated from detergent-treated cell extracts by high-speed differential centrifugation. These ribonucleoprotein complexes can be further purified by centrifugation through sucrose density gradients. By fractionation of the gradient the amount of an individual mRNA in a sub-population of polysomes can be quantitatively determined. Here, we describe methods for the isolation and quantification of polysome complexes from plant tissues. The mRNA obtained can be further analyzed by methods that evaluate polysomal mRNA abundance at the individual transcript or global level. A modification of the conventional polysome isolation procedure is described for transgenic *Arabidopsis thaliana* that express an epitope-tagged version of ribosomal protein L18 (RPL18) that facilitates capture of ribosomes from crude cell extracts by a one-step immunoprecipitation method.

Key words: Polyribosomes, ribosome, translational regulation, immunoprecipitation, mRNA translation, sucrose density gradient, epitope-tagged ribosome, ribosomal protein L18, microarray, microgenomics.

1. Introduction

The control of mRNA translation is a critical component of gene expression in higher plants (1–3). There are three phases of translation: initiation, elongation, and termination. The initiation phase is a multi-step process whereby the 5'-capped and 3'-polyadenylated mRNA recruits first the 40S and then the 60S ribosomal subunit so that peptidyl chain elongation can commence (4). Polyribosomes

Dmitry A. Belostotsky (ed.), *Plant Systems Biology, vol. 553*

DOI 10.1007/978-1-60327-563-7_6 Springerprotocols.com

(polysomes) form as a result of sequential initiation of ribosomes on an mRNA and proceed in the elongation phase of polypeptide synthesis. Once the translating ribosome reaches a termination codon, the polypeptide is released and the 80S ribosome runs off the mRNA. It is well established that the initiation phase is rate-limiting. In most cases, the level of an mRNA in polysomes is well correlated with its translation. A reduction in the number of mRNA molecules in polysome complexes and the number of ribosomes per mRNA is diagnostic of a restriction in initiation of translation. mRNAs that are not undergoing translation are sequestered in cytosolic messenger RNA ribonucleoprotein (mRNP) complexes where they may be stabilized or degraded (5, 6).

To evaluate the translational regulation of an individual transcript or a population of mRNAs in an organ, it is necessary to measure the amount of the transcript in polysome complexes, relative to the total amount of transcript. Further insight of translational regulation is gained by assessment of the relative amount of transcript in small to large polysome complexes. Polysomes can be isolated from frozen plant material (whole plant, organ, dissected region, or mature pollen grains). The tissue must be rapidly frozen in liquid nitrogen upon harvest and pulverized to a fine powder. The pulverized tissue is then thawed in an extraction buffer under conditions that inhibit the activity of RNases. This is routinely accomplished by use of a buffer with a high pH and if necessary the addition of the RNase inhibitor heparin. The buffer must also contain magnesium chloride to stabilize the two-subunit ribosome complex and the translational inhibitors cycloheximide and chloramphenicol to block further translocation of the cytosolic and organellar ribosomes, respectively. Ionic and non-ionic detergents are typically included to disrupt ribosome association with the endoplasmic reticulum and cytoskeleton. The salt concentration can be adjusted to maintain monosomes that lack an mRNA (0.2 M KCl) or only monosomes that are associated with mRNA (0.8 M KCl) (7). Following centrifugation of the extract to remove cell debris (16,000–30,000 $\times g$), the supernatant is centrifuged at high speed (170,000 $\times g$) through a 2 M sucrose cushion to obtain a pellet fraction that is enriched in ribosome subunits, ribosomes, and polysomes. The pellet is resuspended in a buffer formulated to maintain ribosome complexes, briefly centrifuged at low speed to pellet insoluble material, and the ribosome complexes are further fractionated by centrifugation through a continuous sucrose gradient. The gradient is pumped through a UV detector and the ultraviolet absorbance at 254 nm is recorded; the gradient is fractionated into 12–18 fractions of equal volume. These can be used to evaluate the co-fractionation of specific mRNA of interest in fractions that contain polysomes, 80S monosomes, or less dense complexes. The fractions can also be used to evaluate the components of ribosomes (rRNA and proteins) and other mRNPs. The methods described here have proven successful in the evaluation of translational regulation in maize (seedling, mature leaf, endosperm, embryo, pollen grains), tobacco (mature leaf), tomato (mature leaf), and *Arabidopsis* (seedling organs, mature leaf) ((8–13); Bailey-Serres, unpublished).

The implementation of DNA microarray technology to studies of translational regulation has revealed that only a portion of the mRNAs of an individual gene co-purifies with polysomes (11, 13–15). Because of this, the genomic-level profiling of mRNAs associated with polysomes can illuminate the aspects of gene expression that cannot be visualized using conventional profiling of total cellular mRNA. To facilitate the evaluation of polysomal mRNA in high throughput studies, we developed a method for rapid purification of endogenous polysome complexes for DNA microarray studies (16). This was accomplished by construction of transgenic *Arabidopsis thaliana* that expresses ribosomal protein L18 (RPL18) with a FLAG epitope tag at the amino terminus. Transgenic lines were produced in which this chimeric ribosomal protein gene is driven by the near-constitutive cauliflower mosaic virus 35S promoter or cell-type-specific promoters ((16); Mustroph, Zanetti and Bailey-Serres, unpublished). Plants with the tagged RPL18 can be used, for example, to monitor dynamics in polysomal mRNA populations in response to an environmental stimulus or chemical compound. Frozen tissue can be pulverized and lysed in a buffer optimized for immunopurification of polysomes that possess the tagged RPL18 by absorption to anti-FLAG M2 agarose. Following the careful washing of the agarose matrix, the complexes are released from the agarose with an excess of a $FLAG_3$ peptide. RNA extracted from the complexes includes rRNA and intact mRNAs. The comparison of the UV absorbance profiles of ribosomes isolated by conventional differential centrifugation versus immunoprecipitation revealed that similar proportions of small and large polysomes were obtained by both methods (16). Of particular importance, the immunopurified polysomes included complexes of 1 to >20 ribosomes, as well as large (>2.5 kb) and low abundance mRNAs. However, the ribosomes of mitochondria and plastids are excluded from the immunoprecipitate. We have performed DNA microarray hybridization analyses with biological replicate samples using polysomes immunoprecipitated from lines with *p35S:HF-RPL18* as well as other *promoter:HF-RPL18* constructs. The results confirm that immunopurification of polysomes is biologically and technically reproducible. Thus, polysome analysis can be accomplished using traditional differential centrifugation methods or transgenic lines designed to efficiently isolate a sub-population of mRNA complexes from specific cell types.

2. Materials

1. All solutions and equipment used in this protocol need to be free of RNase. Glassware, Miracloth, pipette tips, tubes, and solutions must be sterilized by autoclaving for 15 min.
2. All steps are carried out on ice or at 4°C.
3. Unless otherwise stated, all solutions are prepared with sterile deionized water.
4. To prepare the plant material, tissue must be harvested directly into liquid nitrogen, ground to a fine powder using sufficient liquid nitrogen to maintain a frozen state. Pulverization can be accomplished with a porcelain mortar and pestle or with a coffee bean grinder. The pulverization of pollen grains is improved when a small amount of sterile diatomaceous earth is added to the mortar. The pulverized tissue can be stored at –80°C until use.

2.1. Conventional Isolation of Polysomes by Differential Centrifugation

2.1.1. Equipment

1. Preparative centrifuge with fixed angle or swinging bucket rotor accommodating 30 mL tubes (i.e., Beckman J2-21 high-speed centrifuge and JA-20 rotor, fitted with rubber inserts to accommodate 15 or 30 mL Corex tubes)
2. Ultracentrifuge with fixed angle rotor accommodating 30 mL tubes (i.e., Beckman L8-M ultracentrifuge and TY 70Ti rotor)
3. Thick-walled polycarbonate tubes (i.e., Beckman centrifuge tubes #355654), washed with 2.5% (v/v) hydrogen peroxide, rinsed twice with autoclaved water, and dried prior to use
4. Eppendorf or other microcentrifuge capable of centrifugation at 16,000 × *g*

2.1.2. Solutions and Chemicals

1. Sucrose (Ultracentrifuge grade, Fisher)
2. Heparin (Sigma-Aldrich) (*see* **Note 1**)

The following stock solutions are autoclaved and stored at room temperature

3. 2 M Tris, adjust to pH 9.0 with HCl
4. 2 M KCl
5. 0.5 M Ethylene glycol-bis(2-aminoethylether)-*N*,*N*,*N′*,*N′*-tetraacetic acid (EGTA), adjust to pH 8.0 with 10 M NaOH (*see* **Note 2**)

6. 1 M $MgCl_2$
7. 20% (v/v) Polyoxyethylene 10 tridecyl ether (PTE) (*see* **Note 3**)
8. 10% Sodium deoxycholate (DOC) (*see* **Note 4**)
9. 20% Detergent mix (*see* **Note 5**): 20% (w/v) polyoxyethylene(23)lauryl ether (Brij-35), 20% (v/v) Triton X-100, 20% (v/v) octylphenyl-polyethylene glycol (Igepal CA 630), 20% (v/v) polyoxyethylene sorbitan monolaurate 20 (Tween 20)

The following solutions should *not* be autoclaved, stored at –20°C in aliquots

10. 0.5 M Dithiothreitol (DTT)
11. 50 mg/mL Cycloheximide, dissolved in ethanol
12. 50 mg/mL Chloramphenicol, dissolved in ethanol
13. 0.5 M Phenylmethylsulfonyl fluoride (PMSF), dissolved in isopropanol

2.1.3. Buffers

1. **Polysome extraction buffer (PEB):** prepared on the day of each experiment and kept on ice

Final concentration		Amount of stock solution for 50 mL
0.2 M	Tris, pH 9.0	5 mL
0.2 M	KCl	5 mL
0.025 M	EGTA	2.5 mL
0.035 M	$MgCl_2$	1.75 mL
1%	Detergent mix (*see* **Notes 6, 7**)	2.5 mL
1%	DOC (*see* **Note 8**)	5 mL
1%	PTE	2.5 mL
5 mM	DTT	0.5 mL
1 mM	PMSF	0.1 mL
50 μg/mL	Cycloheximide	50 μL
50 μg/mL	Chloramphenicol	50 μL
0.5 mg/mL	Heparin (*see* **Note 1**)	

2. **Sucrose cushion solution**: keep at 4°C for a maximum of 12 weeks, add the last three compounds fresh for each experiment.

Final concentration		Amount of stock solution for 50 mL
0.4 M	Tris, pH 9.0	10 mL
0.2 M	KCl	5 mL
0.005 M	EGTA	0.5 mL
0.035 M	$MgCl_2$	1.75 mL
1.75 M	Sucrose (*see* **Note 9**)	30 g
Dissolve while heating to about 60°C, adjust to desired volume, autoclave for no longer than 15 min		
Add the following just before use		
5 mM	DTT	0.5 mL
50 μg/mL	Cycloheximide	50 μL
50 μg/mL	Chloramphenicol	50 μL

3. **Resuspension buffer**: prepared on the day of each experiment and kept on ice

Final concentration		Amount of stock solution for 10 mL
0.2 M	Tris, pH 9.0	1 mL
0.2 M	KCl	1 mL
0.025 M	EGTA	0.5 mL
0.035 M	$MgCl_2$	0.35 mL
5 mM	DTT	0.1 mL
50 μg/mL	Cycloheximide	10 μL
50 μg/mL	Chloramphenicol	10 μL

2.2. Immunoprecipitation of Polysomes from Lines Expressing FLAG-Tagged RPL18

2.2.1. Equipment

1. This technique is based on the usage of transgenic *A. thaliana* or other plants expressing a FLAG-tagged ribosomal protein (16). These stable transgenic lines are essential for this protocol.
2. Preparative centrifuge with fixed angle or swinging bucket rotor accommodating 30 mL tubes (i.e., Beckman J2-21 high-speed centrifuge and JA-20 rotor, fitted with rubber inserts to accommodate 15 or 30 mL Corex tubes)
3. Low-speed benchtop centrifuge with swinging buckets for 15 or 50 mL Falcon tubes (required speed 8,200 × *g*)
4. Rocking shaker, capable of shaking at about 60 rpm/min

2.2.2. Solutions and Chemicals

1. The same stock solutions are used as described in **Section 2.1.2**
2. α-FLAG agarose beads (Sigma, product number A 2220)
3. $FLAG_3$ peptide (Sigma, product number F 4799)
4. RNAsin (40 U/μL, Promega) (*see* **Note 1**)
5. Qiagen RNeasy kit (Catalog # 74904)
6. 8 M Guanidine–HCl, autoclaved
7. 99% (v/v) Ethanol

2.2.3. Buffers

1. **Polysome extraction buffer:** *see* **Section 2.1.3, Step 1**
2. **Wash buffer:** prepared on the day of each experiment and kept on ice

Final concentration		Amount of stock solution for 100 mL
0.2 M	Tris, pH 9.0	10 mL
0.2 M	KCl	10 mL
0.025 M	EGTA	5 mL
0.035 M	$MgCl_2$	3.5 mL
5 mM	DTT	1 mL
1 mM	PMSF	0.2 mL
50 μg/mL	Cycloheximide	100 μL
50 μg/mL	Chloramphenicol	100 μL
20 U/mL	RNAsin (*see* **Note 1**)	50 μL

2.3. Analysis of Sucrose Gradient Fractionated Polysomes

2.3.1. Equipment

1. Ultracentrifuge and swinging bucket rotor capable of 237,000 × *g* (i.e., Beckman L8-M Ultracentrifuge with rotor SW55.1)
2. Polyallomer tubes for gradients (Beckman, Catalog # 326819)
3. ISCO UA-5 UV detector, 185 Gradient Fractionator (ISCO Lincoln, NE)
4. Optional: A computer with a DAS-8 compatible data acquisition card connected to the data integrator output devise of the UA-5 detector unit (http://cepceb.ucr.edu/resources/protocol.htm#arab).
5. Florinert displacement fluid (i.e., Perfluoro-compound FC-40 (PC-FC40), ACROS Organics, Belgium)
6. Icruncher 2.1 program for normalization of polysome profiles (http://www.cepceb.ucr.edu/resources/Protocols/Downloads/ICruncher-2.1.xls)

2.3.2. Solutions

1. 10 × sucrose salts, autoclaved for 15 min, stored at –20°C: 0.4 M Tris–HCl pH 8.4, 0.2 M KCl, 0.1 M $MgCl_2$
2. 2 M sucrose (*see* **Note 9**), autoclave for 15 min

Autoclaved deionized water. We typically do not treat the water with diethylpyrocarbonate to destroy RNases.

2.3.3. Preparation of Sucrose Gradients

Prepare sucrose layers according to the following overview:

Sucrose (%)	2 M Sucrose (mL)	10 × Sucrose Salts (mL)	Sterile water (mL)	Chloramphenicol and Cycloheximide (μL)	Volume per gradient (mL)
60	44	5	1	5	0.75
45	49.5	7.5	18	7.5	1.5
30	33	7.5	34.5	7.5	1.5
15	11	5	34	5	0.75

1. Place the 5 mL ultracentrifuge tubes into a rack that can withstand –80°C
2. Starting with the 60% sucrose layer, pipette 0.75 mL into a 5-mL polyallomer centrifuge tube avoiding any air bubbles, and freeze for 1 h at –80°C
3. Add the next gradient layer, freeze again, and continue with the last two layers.
4. Store gradients at –80°C
5. The day of use, remove the gradients to be used from the freezer, thaw them in a 37°C incubator for exactly 1 h, and then cool them at 4°C for 1–1.5 h.
6. Important: Do not shake or drop thawed gradients at any time. For reproducible results, the gradients should be thawed in exactly the manner described.

3. Methods

3.1. Conventional Isolation of Polysomes (see Note 10)

1. Switch on the preparative centrifuge to cool down
2. Switch on ultracentrifuge to cool down
3. Estimate volume of pulverized tissue powder
4. Place pulverized tissue in a sterile beaker or mortar and add two times the volume of freshly prepared polysome extraction

buffer (*see* **Section 2.1.3, Step 1**; *see* **Table 6.1** for guidelines of tissue amounts to be used). Examples for estimated polysomal RNA yields are given in **Table 6.2**.

5. Let the mixture thaw on ice with occasional mixing
6. For small volumes, homogenize the mixture by use of a glass homogenizer
7. For larger volumes, the mixture is filtered through four layers of sterile cheesecloth and two layers of sterile Miracloth (Calbiochem, La Jolla, CA) into a beaker
8. Let the mixture stand on ice for 10 min (or until all samples are prepared)
9. For sample volume less than 1.5 mL, centrifuge samples at 4°C, 16,000 × *g*, for 15 min in a microcentrifuge. For larger sample volumes, centrifuge at 4°C, 16,000 × *g* for 15 min (i.e., using a Beckman J2-21 high-speed centrifuge fitted with a JA-20 rotor, run at 11,500 rpm)
10. Pour the supernatant into a new tube, using Miracloth to filter. Repeat centrifugation step to ensure removal of material that pellets at 16,000 × *g*

Table 6.1
Overview of amounts of plant material used in the described methods

Method of polysome isolation	Purpose	Plant material	Amount (OD_{260} units)	Packed volume of pulverized tissue (mL)
Sucrose cushion concentration	Analysis of polysome and monosome levels	Mature leaves	400	0.5–1
		Seedlings	400	1
	RNA isolation from gradient fractions	Mature leaves	1,000–4,000	2–3
		Seedlings	1,000–4,000	5
Immunoprecipitation	RNA for microarray analysis	Seedlings		3
		Cell-specific promoter (ubiquitous)		3
		Cell-specific promoter (limited cell number)		5

Table 6.2
Yields of polysomal RNA using different polysome isolation methods

Method of polysome isolation	Tissue	Expected RNA yield per mL of ground tissue
Sucrose cushion concentration	Mature leaves	30– 50 μg RNA/mL tissue
	Whole seedlings	10–20 μg RNA/mL tissue
	Shoots of seedlings	2–6 μg RNA/mL tissue
	Roots of seedlings	1–2 μg RNA/mL tissue
Immunoprecipitation	Whole seedlings	1–1.5 μg RNA/mL tissue
	Shoots of seedlings	500–1,000 ng RNA/mL tissue
	Roots of seedlings	300–500 ng RNA/mL tissue
	Cell-specific promoter (limited cell number)	30–100 ng RNA/mL tissue

11. If desired, save 10% volume of the clarified extract to isolate total RNA.
12. Arrange thick-walled polycarbonate tube in a diagonal tube rack and put 8 mL of sucrose cushion solution (*see* **Section 2.1.3, Step 2**) into each tube
13. Pour gently and slowly the clarified extract (above) on top of this solution, avoid mixing of the sample and sucrose solution (one can use a plastic Pasteur pipette to transfer the solution)
14. Balance the weight of the tubes with the two unit cap within 0.05 g. Install the two unit cap on each tube, set them on ice if the ultracentrifuge is not yet at 4°C
15. Centrifuge samples at 4°C, 170,000 × *g* for 3 h (50,000 rpm, TY 70Ti rotor). An alternative is to centrifuge at 116,000 × *g* (35,000 rpm, TY 70Ti rotor) overnight (approximately 18 h); this yields the same amount of monosomes and polysomes but more 40S and 60S subunits
16. After centrifugation, transfer tubes to ice, mark the pellet side on the tube
17. Carefully remove the supernatant and then the sucrose cushion, taking care not to disturb the pellet. The polysome pellet (P170) should be clear and sticky, with a light brown color
18. Wash the tube walls with sterile water gently, avoiding the pellet, and again remove the liquid

19. Resuspend the pellet in ice cold resuspension buffer (*see* **Section 2.1.3, Step 3**) by gently pipetting the solution up and down near the marked pellet region
20. Let sit on ice for 30 min
21. Transfer the resuspended sample to a 1.5 mL microfuge tube and briefly centrifuge at 4°C, transfer the supernatant to a new sterile microfuge tube and discard the pellet
22. Recommended step – measure the OD_{260} of the sample to estimate the RNA concentration and yield; this can only be accomplished if Heparin is not used
23. The suspension contains ribosomal subunits, ribosomes, and polysome complexes, and can be either used to perform polysomal profiles (*see* **Section 3.4**), or to directly isolate RNA (*see* **Section 3.2.4**) or proteins (*see* **Section 3.4**).

3.2. Isolation of Epitope-Tagged Polysomes by Immunopurification (*see* Note 11)

3.2.1. Tissue Extraction

1. Estimate volume of pulverized tissue powder, and add two times the volume of freshly prepared polysome extraction buffer (PEB, *see* **Section 2.1.3, Step 1**). For preparative immunoprecipitation use at least 2.5 mL of packed leaf tissue and 5 mL of PEB. Preparative immunoprecipitation from seedlings requires more tissue than for leaves (*see* **Table 6.1** for guidelines of tissue amounts to be used). Examples for estimated polysomal RNA yields are given in **Table 6.2**.
2. Let the mixture thaw on ice
3. Homogenize the mixture by use of a glass homogenizer
4. Let the mixture stand on ice for 10 min (or until all samples are prepared)
5. Centrifuge the samples at 4°C, 16,000 × *g*, for 15 min in a microcentrifuge
6. Pour supernatant into a new, sterile tube, using sterile Miracloth to filter. Repeat the centrifugation step to ensure removal of material that pellets at 16,000 × *g*
7. If desired, save 10% of the clarified extract to isolate total RNA
8. Recommended step – measure the OD_{260} of the sample to estimate the RNA concentration and yield

3.2.2. Preparation of the α-FLAG M2 Agarose Beads

1. Thoroughly suspend the α-FLAG M2 agarose gel in the reagent vial to make a uniform suspension of the resin. Transfer 100 μL of the beads to a new 1.5 mL tube. Use cut pipette tips for easier transfer.
2. Centrifuge at 8,200 × *g* for 60 s
3. Remove the supernatant with a Pasteur pipette, add 1.5 mL of wash buffer (*see* **Section 2.2.3, Step 2**), and resuspend beads

4. Centrifuge at 8,200 × *g* for 60 s
5. Remove the supernatant with a pipette and wash one more time with 1.5 mL of wash buffer before continuing with the immunoprecipitation

3.2.3. Immunoprecipitation of Polysomes

1. Mix 250–300 units of A_{260} of the clarified extract (*see* **Section 3.2.1**) with 100 μL of washed α-FLAG M2 agarose beads (*see* **Section 3.2.2**) in a 15 mL plastic Falcon tube. Bring volume to 5 mL with PEB (*see* **Note 12**)
2. To bind the epitope-tagged ribosomes to the affinity matrix, incubate for 2 h at 4°C with gentle back-and-forth shaking on a rocking platform
3. Centrifuge for 60 s at 8,200 × *g* at 4°C
4. Transfer the supernatant to a new tube. This is the supernatant of the immunoprecipitation or unbound fraction
5. Add 6 mL of PEB to the beads, mix by gently inverting the tube, incubate at 4°C for 5 min with gentle shaking on a rocking platform and centrifuge for 60 s at 8,200 × *g* at 4°C (first wash)
6. Remove the supernatant with pipette and add 6 mL of wash buffer (*see* **Section 2.2.3, Step 2**). Incubate at 4°C for 5 min with gentle shaking (second wash)
7. Centrifuge for 60 s at 8,200 × *g* at 4°C
8. Remove the supernatant with pipette and add 6 mL of wash buffer. Incubate at 4°C for 5 min with shaking (third wash)
9. Centrifuge for 60 s at 8,200 × *g* at 4°C
10. Repeat wash again for a total of four washes
11. Remove the supernatant. To elute the affinity-purified ribosomes, use a fine tipped pipette to remove as much of the supernatant as possible. Add to the beads 300 μL of wash buffer containing 200 ng/μL of $FLAG_3$ peptide, and 20 U/mL RNAsin (*see* **Note 1**). Incubate for 30 min at 4°C with shaking on a rocking platform.
12. Centrifuge for 60 s at 8,200 × *g* at 4°C. Transfer the supernatant to a new tube. If the supernatant still contains the beads (white or red particles), centrifuge again at 13,000 × *g* for 2 min at 4°C, and transfer to a new tube. It is extremely important to remove all beads.
13. The resulting solution is the eluate of the immunoprecipitation that contains released FLAG-tagged polysomes including the associated proteins and RNAs. This can be used to isolate RNA (*see* **Section 3.2.4**), proteins (*see* **Section 3.4**), or further fractionated on sucrose gradients (*see* **Section 3.4**) to assess the size distribution of the purified ribosomal subunits, monosomes, and polysomes

3.2.4. RNA Extraction

For extraction of RNA from the eluate, use the Qiagen RNeasy kit (*see* **Note 13**).

1. Add 2 volumes of 8 M guanidine-HCl to the eluate of the immunoprecipitation and vortex for 1 min
2. Add 3 volumes of 99% ethanol and vortex for 1 min
3. Precipitate the RNA at –20°C overnight
4. Centrifuge at 16,000 × *g* for 45 min
5. Remove supernatant and let the pellet dry for 20 min
6. Prepare extraction buffer adding 10 μL of β-mercaptoethanol to 1 mL of Qiagen RLT buffer (provided with the RNeasy kit, contains guanidine thiocyanate)
7. Resuspend the pellet in 450 μL of RLT buffer and vortex for 1 min
8. Add 250 μL of 99% ethanol and mix by inverting the tube. Do not vortex
9. Apply the sample into an RNeasy mini spin column. Incubate for 3 min
10. Centrifuge for 15 s at 16,000 × *g*
11. Add 700 μL of RW1 buffer (provided with the RNeasy kit, contains guanidine thiocyanate) and centrifuge for 15 s at 9,000 × *g*. Discard the flow through
12. Add 500 μL of Qiagen RPE buffer (provided with the RNeasy kit, 4 volumes of ethanol is added to RPE buffer before usage according to the manual) and centrifuge for 15 s at 9,000 × *g*. Discard the flow through
13. Add 500 μL of RPE buffer to the column and centrifuge for 2 min at 9,000 × *g*
14. Transfer the column to a new 2 mL microtube; centrifuge for 1 min at 16,000 × *g* to remove remaining ethanol
15. Transfer the column to a new 1.5 mL microfuge tube and add 50 μL of RNAse-free water. Incubate for 5 min
16. Elute RNA centrifuging for 1 min at 16,000 × *g*.
17. RNA can now be used for further analysis (i.e., cDNA synthesis).

3.3. Preparation of Crude Extracts for Sucrose Gradient Fractionation

Polysomes are sufficiently abundant in crude extracts from some tissues so that concentration by centrifugation through a sucrose cushion is unnecessary. We have successfully fractionated polysomes over sucrose gradients from crude extracts prepared from mature leaves of tobacco and *Arabidopsis* and developing endosperm of maize. The preparation is the same as described in **Section 3.2.1** for small tissue samples. Following centrifugation of the sample at 4°C, 16,000 × *g*, for 15 min, 750 μL of the clarified supernatant is loaded onto a 4.5 mL (20–60% w/v) sucrose gradient for fractionation of polysomes, as described in **Section 3.4**.

3.4. Polysome Absorbance Profile Analysis

Ribosome complexes obtained by pelleting through a sucrose cushion (*see* **Section 3.1**) or polysome immunopurification (*see* **Section 3.2**) can be further fractionated by ultracentrifugation through a sucrose gradient. This technique provides visual and quantitative information on relative levels of the 40S and 60S ribosomal subunits, 80S ribosomes (monosomes), and small to large polysomes. Therefore, it provides evidence of the integrity of the isolated polysomes. An example of an absorbance profile obtained from a sucrose cushion ribosome pellet from *Arabidopsis* seedlings is shown in **Fig. 6.1A**. A quantitative estimation of ribosomes in polysomal complexes can be obtained by integration of the area under the peaks of complexes of different masses.

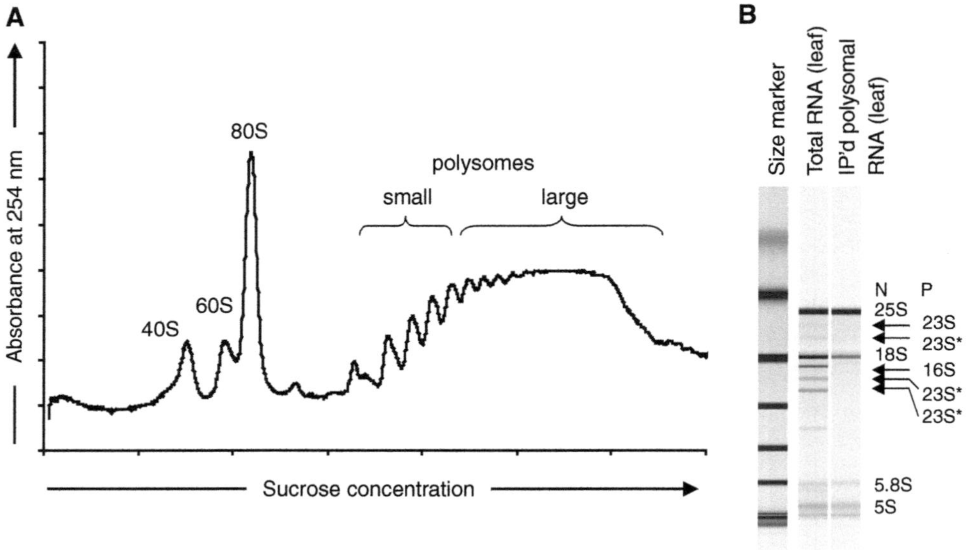

Fig. 6.1. **(A)** Example for a polysome profile of *Arabidopsis* seedlings, obtained after polysome isolation by differential centrifugation as described in **Section 3.1**, and sucrose gradient centrifugation as described in **Section 3.4**. The absorbance peaks represent single ribosome subunits (40S, 60S), monosomes (80S), and small to large polysomes. **(B)** RNA gel of total RNA and IP'd polysomal RNA from leaves of *Arabidopsis* seedlings expressing *p35S:HF-RPL18*. Polysomes were immunopurified and RNA was isolated as described in **Section 3.3**. N, nuclear rRNAs; P, plastid rRNAs (23S, 16S) and their degradation products (23S*).

Polysome DNA microarray analysis has been implemented as a tool for genome-wide study of translational regulation (9, 11, 13, 15). Generally, polysome-associated RNAs are used as templates for cRNA synthesis and hybridized to a DNA microarray platform. In each hybridization reaction, an equal amount of cRNA is used, even though the cellular level of polysomes may differ between samples (i.e., due to use of different growth conditions, genotypes, or organ samples). To accurately quantify the amount of individual mRNAs in polysomes, it is necessary to normalize the signal values obtained in each polysome RNA hybridization (9, 11, 13). This can be accomplished by quantifying the amount of polysomes in the fraction used to obtain mRNA for the hybridization, relative to the total amount of ribosomal complexes in the sample (9, 11, 13).

3.4.1. Procedure

1. Isolate polysomes as described in **Section 3.1** or **3.2**, or crude extracts as described in **Section 3.3**
2. Thaw sucrose gradients and equilibrate as described in **Section 2.3.3**
3. For analysis of the absorbance profile of polysomes, load 400 A_{260} units of the resuspended polysome pellet on top of each gradient. For preparation of ribosomal complexes fractionated into aliquots, this amount can be increased to 1,000–2,000 units
4. Balance tubes to within 0.05 g
5. Perform ultracentrifugation at 4°C, 237,000 × *g* (50,000 rpm, SW55.1 rotor) for 1.5 h. The run length can be increased or decreased by 10–15 min to alter the degree of separation of the ribosome complexes
6. While the gradient is spinning, prepare the ISCO absorbance detector (model # UA-5, ISCO, Lincoln, NE). Switch on 20 min prior to use to warm up the UV lamp. Assemble the peristaltic pump and gradient holder according to the manufacturer's instructions. Adjust the absorbance detector to 0.2 or 1.0 sensitivity for analytical and preparative runs, respectively. Use 150 cm/h chart speed
7. Run a sucrose gradient with the amount of sample buffer loaded on top of the gradients to establish the baseline absorbance profile. This gradient does not need to be centrifuged along with the experimental samples
8. After centrifugation, carefully remove the rotor from the centrifuge, place the buckets on ice, and remove the first gradient to be analyzed. Assemble the gradient in the UV detector holder, puncture the tube bottom, and run the displacement fluid into the tube at 0.75 mL/min flow rate. Record the A254 nm profile with chart recorder and using a data acquisition device if available ((9); *see* **Section 2.3.1**)

   While running the gradients, collect fractions (usually 12 fractions of 0.4 mL), if desired. Place them on ice immediately to avoid RNA degradation.
9. For RNA preparation: Immediately add 2 volumes of 8 M guanidine chloride and 3 volumes of 99% ethanol and mix well. Allow RNA precipitation at –20°C overnight. Pellet by centrifugation at 16,000 × *g*, 4°C for 45 min, followed by RNA extraction as described in **Section 3.2.4** (*see* **Note 13**).

10. For protein preparation: Add 2 volumes of 99% ethanol, mix well and allow to stand at 4°C overnight. Pellet proteins by centrifugation at 16,000 × *g*, 4°C for 15 min, and wash once with 70% ethanol. Pellets can be resuspended in 2X-SDS loading buffer, and loaded on an SDS polyacrylamide gel.
11. If absorbance profile data were electronically recorded, analyze polysome profile and calculate proportion of ribosomes, small and large polysomes (*(9)*; *see* **Section 2.3.1**)

4. Notes

1. Heparin and RNase inhibitor are only required for tissues with high RNase content, such as mature maize leaves. If there is evidence of rRNA or mRNA degradation in samples, the addition of heparin to the extraction buffer usually alleviates the problem. RNA degradation can result if samples are thawed prior to addition of the extraction buffer or the extraction is performed above 4°C.
2. EGTA dissolves only after adjusting the pH.
3. Shake bottle before pipetting the solution.
4. Use lung protection while weighing DOC.
5. Dissolve while heating to about 60°C.
6. These detergents can be omitted if only soluble ribosomes are to be isolated. However, the yield of ribosomes will be considerably lower.
7. Warm solution at 42°C before use; pipette with a 1,000 μL tip enlarged by cutting 0.5 cm from the end
8. This detergent is included to disrupt ribosome–cytoskeleton association. It can be omitted when working with *Arabidopsis* or maize seedlings, but should be included for mature leaves and seed endosperm.
9. Use high purity sucrose (i.e., Ultracentrifuge grade, Fisher), to ensure RNase-free conditions
10. This technique does not require special transgenic plants with a FLAG-tagged ribosomal protein. However, the centrifugation step might also result in pelleting of other RNA-binding protein complexes that are not translationally active.
11. This affinity-tag technique can be only used to enrich nuclear-encoded mRNAs. mRNAs encoded by and translated in mitochondria and plastids cannot be isolated with this method. **Figure 6.1B** shows a gel of total RNA in comparison to IP'd polysomal RNA. The organellar ribosomal RNAs are missing in the immunopurified d polysomal RNA sample.

12. For large-scale experiments, adjust the amount of beads according to the amount of extract. When using lines with cell-type-specific promoters expressed in a limited number of cells of an organ/tissue, use two–three times more tissue for the same amount of beads.
13. Alternatively, one can isolate RNA from fractions by other protocols such as that described in Fennoy and Bailey-Serres, 1995 (7), or by use of Trizol reagent, according to the supplemented protocol.

Acknowledgments

We thank all the individuals who have improved and expanded these polysome analyses methods, especially Cristina Branco-Price, Riki Kawaguchi, Reed Sorensen, Joanna Frazek-Warner, Alan Williams, and Maria Eugenia Zanetti. This work was supported by the U.S. National Science Foundation (DBI-0211857, IBN-0420152 and IOS-0750811).

References

1. Bailey-Serres, J. (1999) Selective translation of cytoplasmic mRNAs in plants. *Trends Plant Sci.* **4**(4), 142–148.
2. Kawaguchi, R. and Bailey-Serres, J. (2002) Regulation of translational initiation in plants. *Curr. Opin. Plant Biol.* **5**(5), 460–465.
3. Browning, K.S. (2004) Plant translation initiation factors: it is not easy to be green. *Biochem. Soc. Trans.* **32**, 589–591.
4. Proud, C.G. (2007) Signaling to translation: how signal transduction pathways control the protein synthetic machinery. *Biochem J.* **403**(2), 217–234.
5. Parker, R. and Sheth, U. (2007) P bodies and the control of mRNA translation and degradation. *Mol. Cell.* **25**(5), 635–646.
6. Hoyle, N.P., Castelli, L.M., Campbell, S.G., Holmes, L.E., and Ashe, M.P. (2007) Stress-dependent relocalization of translationally primed mRNPs to cytoplasmic granules that are kinetically and spatially distinct from P-bodies. *J. Cell Biol.* **179**(1), 65–74.
7. Fennoy, S.L. and Bailey-Serres, J. (1995) Post-transcriptional regulation of gene expression in oxygen-deprived roots of maize. *Plant J.* **7**, 287–295.
8. Fennoy, S.L., Nong, T., and Bailey-Serres, J. (1998). Transcriptional and post-transcriptional processes regulate gene expression in oxygen-deprived roots of maize. *Plant J.* **15**, 727–735.
9. Kawaguchi, R., Williams, A.J., Bray, E.A., and Bailey-Serres, J. (2003) Translational regulation in response to water deficit stress in *Nicotiana tobacum*. *Plant Cell Environ.* **26**, 221–229.
10. Williams, A.J., Werner-Fraczek, J., Chang, I.F., and Bailey-Serres, J. (2003) Regulated phosphorylation of 40S ribosomal protein S6 in root tips of maize. *Plant Physiol.* **132**(4), 2086–2097.
11. Kawaguchi, R., Girke, T., Bray, E.A., and Bailey-Serres, J. (2004) Differential mRNA translation contributes to gene regulation under non-stress and dehydration stress conditions in *Arabidopsis thaliana*. *Plant J.* **38**(5), 823–839.
12. Slaymaker, D.H. and Hoppey, C.M. (2006) Reduced polysome levels and preferential recruitment of a defense gene transcript into polysomes in soybean cells treated with the syringolide elicitor. *Plant Sci.* **170**, 54–60.
13. Branco-Price, C., Kawaguchi, R., Ferreira, R.B., and Bailey-Serres, J. (2005) Genome-wide analysis of transcript abundance and translation in *Arabidopsis* seedlings

subjected to oxygen deprivation. *Ann. Bot. (Lond)* **96**(4), 647–660.

14. Kawaguchi, R. and Bailey-Serres, J. (2005) mRNA sequence features that contribute to translational regulation in *Arabidopsis. Nucleic Acids Res.* **33**(3), 955–965.
15. Nicolaï, M., Roncato, M.A., Canoy, A.S., Rouquie, D., Sarda, X., Freyssinet, G., and Robaglia, C. (2006) Large-scale analysis of mRNA translation states during sucrose starvation in *Arabidopsis* cells identifies cell proliferation and chromatin structure as targets of translational control. *Plant Physiol.* **141**(2), 663–673.
16. Zanetti, M.E., Chang, I.F., Gong F., Galbraith D.W., and Bailey-Serres J. (2005) Immunopurification of polyribosomal complexes of *Arabidopsis* for global analysis of gene expression. *Plant Physiol.* **138**(2), 624–635.

Chapter 7

Chromatin Charting: Global Mapping of Epigenetic Effects

Chongyuan Luo and Eric Lam

Abstract

To tackle the question of how chromatin organization is involved in global regulation of genome-related processes such as transcription, we have recently created a collection of 277 transposon-tagged *Arabidopsis* lines comprised of a single insert with a common luciferase reporter cassette and a *LacO* repeat array for visual tracking of the tagged region via fluorescent protein fusion technology. Using this collection of plants, one can begin to map transgene position effects as well as global epigenetic control in response to developmental or externally applied cues. In this chapter, we will outline the approach and methods for deploying this novel resource for the study of global gene control, using *Arabidopsis* as a convenient model system.

Key words: Chromatin, transposon-tagged lines, *Arabidopsis*, epigenetics, luciferase, dsRNA suppression, position effects.

1. Introduction

1.1. Chromatin-Based Regulation of Gene Expression: Epigenetic vs. Genetic Mechanisms

Transcription control is a major output of genetic information that contributes to all phases of development in the life of an organism. Superimposed on the textbook model of gene control via *cis*-acting promoter elements and DNA-binding transcription factor is the growing appreciation for the importance of epigenetic mechanisms that act via chromatin modifications. Epigenetic mechanisms result in heritable changes in gene expression without alterations in the sequence of the gene(s) involved. One classic example from plants is paramutation, first discovered in maize more than 50 years ago (1). At the R locus, responsible for seed color, Brink observed that alteration of one allele (*R-r*) by another (*R-stippled*) is heritable (but reversible) even after its segregation from *R-stippled*. Paramutation has since been documented in

Dmitry A. Belostotsky (ed.), *Plant Systems Biology, vol. 553*

DOI 10.1007/978-1-60327-563-7_7 Springerprotocols.com

various plant species (including *Arabidopsis*) as well as animal systems (reviewed in (2)). Nucleolar dominance, in which rRNA genes from one parent are specifically silenced in a genetic hybrid, is another well-known epigenetic phenomenon that exists in many eukaryotes (3). Other epigenetic control mechanisms that have been described in eukaryotes include *trans*-inactivation by the *brown* (dominant) allele (*Br(D)*) of the respective bw(+) WT allele (importantly in the context of this chapter, this is correlated with its spatial repositioning in the nucleus to associate with centric heterochromatin); gene silencing at the centromeres of many organisms; and X-inactivation in mammals (summarized in (2, 4, 5)). The common mediator of these epigenetic silencing mechanisms is the presence or formation of more highly condensed heterochromatin and its propagation along the chromosome. Epigenetic mechanisms can profoundly affect plant development such as in the case of genomic imprinting during gametogenesis, control of flowering time by vernalization, and the regulation of meristem size and identity (summarized in (6, 7)).

1.2. Distinct Molecular Mechanisms of Epigenetic Control Exist in Eukaryotes

Two major regulatory pathways that are commonly involved in epigenetic control are DNA methylation at cytosine residues (8, 9) and specific types of covalent modification that include methylation and/or acetylation of specific lysine residues of histones H3 and H4. In general, cytosine hypermethylation and decreased acetylation of histones are correlated with gene silencing while specific methylation states of the amino terminus of histone H3 is either associated with activated (H3K4m) or suppressed (H3K9m) gene expression (10). Although the precise nature of the "histone code" that has been speculated to allow quantitative prediction of gene expression remains to be defined (11), it is clear that many histone modifications play a major role in epigenetic control (6, 12). More recently, DNA methylation and histone modification at heterochromatic regions neighboring centromeres and the nucleolus, and in transcriptional gene silencing (TGS) of transgenes, have been shown to depend on components in the RNA interference (RNAi) and small interfering RNA (siRNA) pathways (4, 13, 14). The discovery that an RNA-dependent RNA polymerase (RdRP), an enzyme involved in some double-stranded RNA (dsRNA)-initiated gene silencing pathways, is an important component of paramutation in maize suggests that similar mechanisms may be shared in different types of epigenetic phenomena (15). This is supported by genetic data which show that mutants at three different paramutation loci can all activate a transcriptionally silenced transgene in maize (16). In all three cases, alteration of the DNA methylation state at the transgene locus is correlated with reactivation of the transgenes. However, with two of the mutants, the transgene remained active after reintroduction of the wild-type allele which suggested that the chromatin state at the transgene locus has been altered in a heritable manner. These observations, as well as studies examining the effects of RNAi mutants on transposon and transgene silencing (17–19), suggest that there are multiple epigenetic silencing pathways with distinct characteristics. In addition, these and other genetic studies in maize and *Arabidopsis* demonstrate that transgenes can be convenient markers to monitor epigenetic mechanisms in plants (2, 4, 8).

1.3. Global Organization of Chromatin and Transcription Activity

To model epigenetic control mechanisms in detail, an understanding of the relationship between chromatin organization and gene expression is essential. Interphase chromatin of eukaryotes has been shown to exist in distinct subnuclear compartments called chromatin territories (CTs), with relatively little intermixing between neighboring chromosomes (20, 21). The interchromatin space (ICS) is envisioned to contain most of the protein complexes/factories that are involved in splicing, repair, and transcription. In this regional organization type of model, heterochromatic regions within each chromosome provide the backbone/anchor for maintaining the CT, while the perichromatin region (PR) observed in EM studies has been suggested to correspond to the portion of the euchromatin that is accessible for gene expression in the particular cell context (summarized in (21)). The make-up of the euchromatin in the PR can be dynamically altered by epigenetic mechanisms such as DNA methylation and histone deacetylation. This regional model of chromatin organization thus provides a structural basis to rationalize epigenetic regulation of gene expression. One prediction of the regional model for chromatin organization is the existence of "position effects" (PEs) that can quantitatively modulate the expression level of transgenes inserted into different parts of the genome. With support from the NSF Plant Genome Research Program (PGRP) in the past 7 years, we established a set of "Chromatin Charting" lines that consists of mapped single insertions dispersed throughout the genome of *Arabidopsis*, using transposon-mediated tagging techniques developed previously (22). By placing selected reporter genes to compare transcription activity and a *LacO* array to allow visual tracking of the inserted loci in nuclei of live plants (23), this collection provides a novel resource for the discovery of position effect loci (PELs) and facilitates their comprehensive study at the functional and physical levels. Since these elements of our inserts are common between individual lines, differences in their reporter gene activity and physical property thus likely reflect influences by the insertion neighborhood's characteristics, some of which can be epigenetic in origin. We refer to the relative level of reporter gene expression as "transcription potential", a term chosen to reflect and emphasize the relative degree that a common gene unit can be activated at different locations in the genome. Details of our project schemes and methodologies can be obtained from our web sites (http://Charting.cshl.org; http://aesop.rutgers.edu/

~lamlab/ccharting.html). Briefly, eight transgenic "launchpad" lines (CCP4 lines) with the Chromatin Charting construct (*pCCharting*; **Fig. 7.1A**) were crossed with six transgenic lines that overexpress the maize Ac transposase to mobilize the dormant

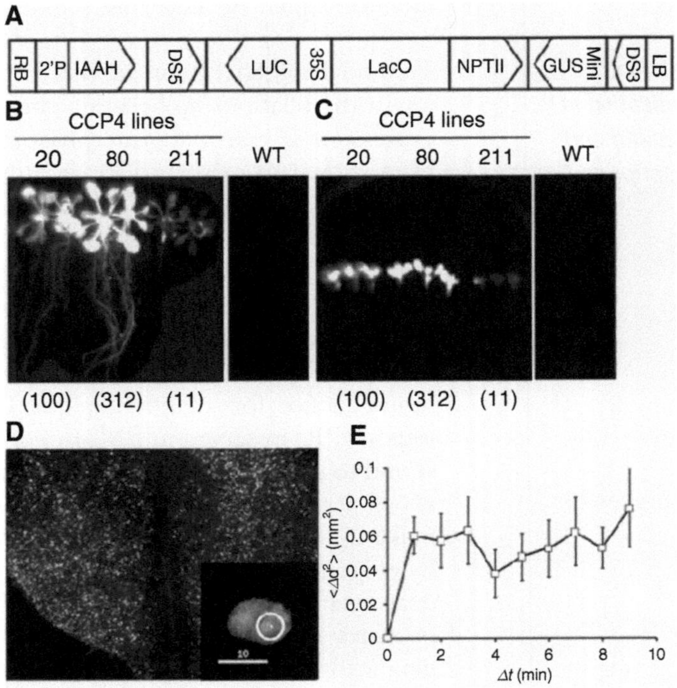

Fig. 7.1. Chromatin charting vector and its uses. **(A)** Relevant structure of the *pCCharting* construct (24). **RB** and **LB,** right and left border of T-DNA; **2′P**, promoter for the IAAH selection gene; **IAAH**, indole acetamide hydrolase gene; **DS5** and **DS3**, 5′ and 3′ border sequences for maize Ds element; **LUC**, firefly luciferase gene; **35S**, CaMV 35S promoter; ***LacO***, lac operator array; **NPTII**, neomycin phosphotransferase gene; **GUS**, β- glucuronidase gene; **Mini**, minimal CaMV 35S promoter (–46 to +8). Sizes of elements are not drawn to scale. In vivo imaging of luciferase activity in whole seedlings (**B**) and inflorescence of mature plants (**C**) from three different CCP4 lines and wild-type plants (WT). Two-week-old seedlings and floral tissue from 5-week-old plants are sprayed with in vivo luciferase assay solution and imaged with a Biophotonic camera system (Lumazone FA, MAG Biosystems). The numbers on the bottom of each set of plants shown in the panels indicate the relative luciferase activities measured by the standard in vitro assay. (**D**) Visualization of *LacO* array by transient expression of GFP-LacI-NLS protein. Transient expression was performed with a CCT71 (24) plant according to **Section 3.4.2**. The panel shows a rosette leaf that has been infiltrate with *Agrobacterium* containing the EL700 construct and induced with Dex as described in **Section 3**. Fluorescence image was collected with a Leica stereofluorescence microscope at relatively low magnification and shows nuclear fluorescence of the expressed GFP-LacI-NLS protein. Inset shows a Z-section of a single nucleus from the same leaf sample examined with the DeltaVision microscope system. The bright fluorescence spot, outlined with a *white circle*, indicates the decorated *LacO* array. The bar in the inset corresponds to 10 μm. (**E**) Mean-square change of 3-D distances between seven pairs of *LacO* spots is plotted against Δt. The averaged data from the seven sets of data points are shown with standard deviations.

Ds transposable element in our launchpad lines. Novel and stable transposants can then be recovered in the F2 generation using a combination of positive selection (Kan^r) for our *CCharting* Ds element and negative selection against the parental launchpad and Ac-expressing loci, both of which contain an IAAH marker that confers 1-naphthaleneacetamide (NAM) sensitivity (22). From screening 11,682 F2 families derived from about 2,000 independent crosses performed in the Lam and Martienssen labs, we recovered 611 stable transposants (CCT lines). We opted for using the more complicated and laborious transposon-mediated approach instead of random T-DNA integration via *Agrobacterium* since the latter is known to generate complex insertion events at a much higher frequency. To date, genomic locations for 271 CCT lines among this first collection has been determined by TAIL-PCR and validated using locus-specific primers. Together with the six mapped and confirmed CCP4 lines, this collection of 277 "Chromatin Charting" lines have been deposited into the Ohio State Stock Center (ABRC). Using this set of plant lines, one can now screen for regions in the genome that may display locus-specific silencing or activation of the common luciferase marker gene that is tissue-, developmental stage-, or signaling pathway-specific. As a proof-of-concept study, we have recently discovered a set of root-specific silencing loci at the north end of Chr. 2 adjacent to the NOR (24). Several screens for loci that respond to tissue- and growth condition-specific signals are underway in our lab. However, we believe that this resource can be deployed by various plant biology investigators to search for evidence of control at the chromatin level by their pathways of interest. This chapter thus aims to describe the materials and methods that we have established in our laboratory for this purpose.

2. Materials

2.1. Seed Germination

1. 50% Bleach
2. Solid Growth Medium: 0.5X Murashige and Skoog (MS) mineral salts, 1% sucrose, 0.25% Phytagel™ (Sigma), 50 μg/ml Kanamycin (optional)
3. 3 M Micropore™ surgical tape, ½ in.
4. SterilGARD Laminar Flow Hood (Baker) or a comparable aseptic environment

2.2. Luciferase Assays

1. Biotium Firefly Luciferase Assay Kit
2. D-Luciferin, potassium salt
3. Triton X-100

4. Biotek Synergy™ HT Multi-Detection Microplate Reader
5. Costar 96-well assay plate
6. Costar 96-well assay plate (White Plate, Clear Bottom with Lids)
7. Bio-Rad Quick Start Bradford Protein Assay Kit
8. Mettler AE200 balance or any other type with appropriate scale resolution
9. In vivo luciferase assay solution: 0.3 mg/ml D-Luciferin, 0.01% Triton X-100

2.3. Root and Shoot Tissue Extracts

1. Two- to four-week-old *Arabidopsis* plants grown vertically for at least 5 days
2. Wheaton Instruments Overhead Stirrer
3. 5X Firefly Luciferase Assay Lysis Buffer Kit (Biotium; catalog #30003-1)
4. Solid medium for vertical growth of *Arabidopsis*: 0.5X Murashige and Skoog (MS) mineral salts, 1% sucrose, 1% agar
5. 150 mm × 15 mm Petri dish

2.4. Microscopy with Chromatin Charting Visualization (CCV) Constructs by Transgenic or Transient Expression Approaches

1. Nikon TE200 microscope
2. Applied Precision DeltaVision image restoration microscope system Version 3.5
3. Nikon PlanApo 60X, 1.2 N.A/water-immersion objective lens
4. softWoRx 3.6.1 Suite software package included in DeltaVision system
5. *Agrobacterium tumefaciens* GV3101 strains containing CCV binary vectors EL700 or JM71 (23, 24)
6. *Agrobacterium* suspension buffer: 10 mM $MgCl_2$, 10 mM MES, pH 4.5, 200 μM 3′, 5′-dimethoxy-4′-hydroxy-acetophenone (Acetosyringone)
7. A 1 ml syringe without needle (Becton Dickinson and Co., Tuberculin slip tip)

3. Methods

3.1. Germination of Seeds for Arabidopsis *Lines*

1. Seeds of *Arabidopsis* CCP4 or CCT lines are vortexed with 50% bleach for 2–5 min, and then rinsed approximately five times with sterile water to remove trace of bleach. After washing, spread seeds on solid growth media under a SterilGARD laminar flow hood. To prevent false positives during antibiotic selection, enough spacing among seeds is required.

For lines showing reduced Kanamycin resistance due to silencing of the *Npt II* gene, germinate seeds on 0.5X solid MS plates without Kanamycin.

2. Seal plates with 3 M Micropore^{TM} surgical tape ½ in.
3. Synchronize seed germination at 4°C for 48 h before transferring plates to normal growth condition.

3.2. Luciferase Assay – Whole Tissues and Extract-Based Assays

To quantify the expression level of Firefly Luciferase gene, luciferase activity can be measured either in tissue homogenate (in vitro assay) or intact tissues (in vivo assay; **Fig. 7.1B** and **C**). In vitro luciferase assay is in general more sensitive than in vivo assay but also more time-consuming and needs additional attention and equipment. Like any other biochemical assays that involve protein extraction, the in vitro luciferase assay requires rapid sample handling and cooling to prevent protein degradation during extract preparation. To achieve maximum sensitivity and reproducibility, we suggest using commercial luciferase assay kit for the in vitro luciferase assay. Luciferase activity acquired from in vitro assays is normalized with total protein concentration of the tissue homogenate. Protein concentration is measured with Bio-Rad Quick Start Bradford Protein Assay Kit following the manufacturer's instruction. In vivo (with intact tissues) luciferase assay is less quantitative and sample variations in luciferin absorption, surface properties, and probe-to-sample distance can all potentially affect the observed light output. Without a soluble extract, in vivo luciferase activity is often normalized with fresh tissue weight. However, correlation between tissue weight and protein content of plants can be affected by developmental stage, tissue type, and age (*see* **Note 1**).

3.2.1. In Vitro Luciferase Assay

1. Harvest 20–100 mg *Arabidopsis* tissues in 1.5 ml micro-centrifuge tubes.
2. Briefly grind with Overhead Stirrer (Wheaton Instruments) for 5 s. Add 200 μl 1X luciferase lysis buffer and continue grinding for another 15 s. Refrain from using high speed, as the grinder may become overheated. Place samples onto ice immediately after grinding.
3. Clarify the samples by centrifugation at above 10,000 × *g* for 5 min at 4°C. Transfer the supernatant to a new pre-cooled micro-centrifuge tube.
4. Use 2–10 μl of tissue extracts for measuring luciferase assay. Signal might exceed the upper limit of the plate reader if too much extract is used. We suggest doing an exploratory assay before the experiment to find out the appropriate amount of extract to be used for the assay. Add 1X luciferase lysis buffer to wells to make the final volume 50 μl.

5. Use multi-channel pipettes to add 50 μl of luciferase assay solution provided in the luciferase assay kit to each well. Photon intensity generated by oxidation of luciferin will continue to decrease shortly after the reaction has started, so always try to minimize the intervals between adding luciferase assay solution to the different wells. Load the 96-well plate onto a plate reader and start reading as soon as possible.

3.2.2. Quantification of Protein Concentration

To standardize the luciferase activity of each sample, total protein concentration of the extract is measured after the luciferase assay. The detergent contained in the luciferase lysis buffer interferes with the Quick Start Bradford protein assay solution if more than 5 μl of extract is used. So using relatively small volumes of extracts (2–3 μl) for Bradford assay is important to get reliable data.

1. Prepare a standard curve with the bovine serum albumin (BSA) protein standard provided with Quick Start Bradford protein assay solution.
2. Mix 2 μl of cell extracts with 18 μl of distilled water in microcentrifuge tube.
3. Add 1 ml of Quick Start Bradford protein assay solution and shake to mix.
4. Incubate at room temperature for at least 5 min but shorter than 1 h to let the reaction complete.
5. Measure absorbance at 595 nm by a microplate reader or a spectrophotometer.
6. Calculate protein concentration by using the standard curve generated in Step 1.

3.2.3. Intact Tissue Luciferase Assay (In Vivo Assay)

1. Add 30 μl water to each of the wells in a 96-well assay plate to prevent over-drying of the leaf samples.
2. Snip a medium size rosette leaf from 3- to 4-week-old *Arabidopsis* plants. The leaf piece should be able to fit comfortably into the well of a Costar 96-well assay plate. Record the weight of each leaf. Transfer the leaf piece into a well of the assay plate with the abaxial side of the leaf on top since the luciferin solution can then be taken up more readily.
3. Keep the plate covered with a lid to minimize water loss from leaf samples during sample preparation.
4. Spread the plate with in vivo luciferase assay solution (0.3 mg/ml D-Luciferin, 0.01% Triton X-100). Leave the plate on a bench for 5 min to allow uptake of D-Luciferin into cells. At the same time set up the program for the plate reader.

5. Load assay plate onto the plate reader and read for at least 30 min.
6. Normalize the measured luciferase activity with the weight of the leaf sample.

3.3. Comparative Analysis of Root and Shoot Expression Between Lines

1. Germinate *Arabidopsis* seeds on 0.5X MS plates or selection plates if desired. Synchronize seed germination at 4°C for 48 h. Move plates to normal growth condition and let plants grow for 7 days.
2. Carefully transfer seedlings to MS medium containing 1% agar with sterilized forceps. Grow plants in vertically oriented plates for 1–2 weeks.
3. To quantify luciferase activity in root and shoot, cut *Arabidopsis* plants at the base of their hypocotyls and homogenize shoot and root tissues separately in Luciferase Lysis Buffer. We suggest using half the volume of the lysis buffer that was used for shoot to homogenize root tissue.
4. Follow **Sections 3.2.1** and **3.2.2** to measure luciferase activity by in vitro luciferase assay and total protein concentration of lysates for normalization.

3.4. Microscopy Assays

To visualize the tagged loci in living plants and measure physical parameters of chromatin behavior, two approaches can be carried out: (1) Cross CCP4 or CCT lines to stable transgenic CCV lines EL700 and JM71, which can express GFP-LacI-NLS protein after induction via dexamethasone (Dex) treatment or EYFP-LacI-NLS upon ethanol treatment, respectively (23, 24). Since a minimum of at least two spots are needed to quantify diffusion dynamics, endoreduplicated epidermal cells can be analyzed in the F1 generation while F2 plants homozygous at the tagged locus can be used to analyze diploid cells. (2) Transient expression of GFP-LacI-NLS protein in rosette leaf cells by *Agrobacterium* infiltration. Approach 1 is adopted when visualization in various tissues (root or shoot) or cell types (epidermal cell or mesophyll cells) is desired, while approach 2 is suitable for rapid observation of chromatin dynamics in many lines. The success of visualizing *LacO* arrays *in planta* highly depends on the background-to-signal ratio (free GFP-LacI-NLS protein in nuclei vs. GFP-LacI-NLS binding to the *LacO* arrays). Over-induction by either a high concentration of Dex or long induction times can lead to higher background GFP fluorescence and increased difficulty in detecting *LacO* arrays, whereas short induction may result in insufficient signal intensity. Based on our experience with EL700 transgenic plants (23), 8–16 h of induction with 0.3 μM Dex is suggested for *Arabidopsis* seedlings.

3.4.1. Visualization of LacO *Array-Tagged Loci by Inducible Expression of GFP-LacI-NLS Protein*

1. Cross CCT or CCP4 lines into a stably transformed EL700 line that can express GFP-LacI-NLS protein upon Dex induction.
2. Germinate Fl or F2 seeds on selection medium (50 μg/ml Kanamycin and 15 μg/ml Hygromycin). Use PCR approach to identify plants containing homozygous *pCCharting* locus.
3. One- to two-week-old whole seedlings or detached rosette leaves from 2- to 3-week-old plants are floated on 0.3 μM Dex solution. After 8–16 h, seedlings or leaves were placed between microscope slides and cover slips with water and mounted on the microscope stage for observation.

3.4.2. Transient Expression of GFP-LacI-NLS Protein for Visualization of the LacO *Array*

1. Germinate CCP4 or CCT lines on 50 μg/ml Kanamycin selection plates.
2. Transfer 7- to 10-day-old seedlings to soil. Plants should be covered during the first 2 days after transfer with plastic domes.
3. At the same time, inoculate an *Agrobacterium* overnight culture in 5 ml LB with 50 μg/ml Kanamycin, 50 μg/ml Gentamicin.
4. Adjust OD_{600} of the bacteria culture to 0.5 with LB. Spin down bacteria at 12,000 × *g* for 5 s and resuspend the pellet in *Agrobacterium* suspension buffer.
5. Leave the bacteria suspension at room temperature for at least 3 h to induce *vir* genes.
6. Infiltrate the under (abaxial) side of whole rosette leaf with a 1 ml syringe.
7. Infiltrated leaves can be detached 30 h after infiltration and floated on 0.3 μM Dex solution for 8–16 h to induce the expression of the GFP-LacI-NLS protein.
8. Place infiltrated leaves between microscope slides and cover slips with water and mount on microscope stage for imaging (*see* example in **Fig. 7.1D**).

3.4.3. Microscopy Analyses and Quantification of Intranuclear Dynamics via LacO *Array Tracking with GFP-LacI-NLS*

1. To avoid the difficulty of searching for cells with adequate fluorescent protein expression under high magnification, we normally screen the tissue/seedling samples by epifluorescence microscopy at 4–20 × magnification and mark the areas of potential interest for further imaging.
2. For most *Arabidopsis* cells, images from 40 to 60 layers with 0.2 μm *Z* steps are enough to cover the nuclei as well as providing reasonable resolution along the *Z*-axis. We use exposure times between 0.3 and 1 s for each *Z*-section. The best exposure time for each experiment can vary significantly depending on the dynamic range and sensitivity of the particular CCD camera used. Exploratory experiments are usually necessary in order to find the optimal exposure times that minimize both saturation of the camera and bleaching of the

fluorescent protein. Saturation of camera will compromise the quantitative analysis of signal intensities, whereas bleaching of fluorescent protein will reduce image quality during continuous imaging (*see* **Note 2**).

3. To perform chromatin dynamic analysis, we routinely track each *Arabidopsis* nucleus for 10 min.
4. After image collection, image stack files are deconvolved using softWoRx Suite software from Applied Precision Inc. to reconstruct the 3-D fluorescence images.
5. Two methods can be used to measure distances between two *LacO* array spots. If only a few spots are being analyzed, distances ***d*** between two *LacO* array spots can be obtained directly from softWoRx Suite. Alternatively, if distances between a relatively large number of spots within the same nucleus are desired, distances can be calculated from coordinates of each spots using the Pythagorean theorem. *X* and *Y* coordinates of a particular pixel can be read from the 3-D image using the softWoRx software while layer number is used to calculate distance between two spots on the *Z*-axis. Squared distance between two spots, $\boldsymbol{d}^2$, can be derived as $\boldsymbol{d}^2 = (x_1 - x_2)^2 + (y_1 - y_2)^2 + [(z_1 - z_2) \times s]^2$ (x_1, x_2, y_1, y_2: *x* and *y* coordinates for the two spots being analyzed; z_1, z_2: *Z* layer numbers for the two spots; *s*: layer step distance). In most cases, the signal for each *LacO* array spots can be detected in several consecutive layers. The layer that the *LacO* array spot shows the highest fluorescence intensity will be taken as where the spot localize to.
6. Compute changes of the squared distance (or mean-squared displacement as a function of time), $\Delta d^2 = < d(t_x) - d(t_0) >^2$, where $d(t_0)$ is the 3-D distance between two spots at the first time point measured and $d(t_x)$ is the 3-D distance between the two spots at a subsequent time point *x*. Mean Δd^2 of all *LacO* spot pairs are then plotted with $\Delta t = t_x - t_0$ to generate the curve. If any one of the two spots being measured is freely diffusing without restriction, Δd^2 will increase continuously along with increase of Δt. However, if the movement of both spots are constrained to a certain area, the increase of Δd^2 will reach a plateau and become independent of Δt (25).

4. Notes

1. In addition to difficulties in standardization, the accuracy of the in vivo assay is also affected by variable uptake rate of D-Luciferin and sub-cellular physiology, which can affect the

turnover rate of luciferin by altering the reaction environment of luciferin oxidation. Despite these complications, the in vivo luciferase assay is nevertheless suitable for large-scale screening applications because of the increased speed and lower cost that can significantly facilitate a higher throughput.

2. Photobleaching is a problem for any continuous in vivo imaging approach using fluorescence proteins. If the amount of work required to create an optimized visualization construct is acceptable, introducing a more stable fluorescent protein variant should be first considered. Without modifying the visualization construct, using a CCD camera with higher sensitivity is the most effective solution. The latter can reduce exposure time while acquiring images with the same or better quality. Besides purchasing a new camera, less image layers or sacrificing some image resolution can also reduce exposure times.

Acknowledgments

This work is supported by an NSF Plant Genome Research Program grant to E.L. (#0077617).

References

1. Brink, R.A. (1956) A genetic change associated with the R locus in maize which is directed and potentially reversible. *Genetics* **41**, 872–889.
2. Chandler, V. and Stam, M. (2004) Chromatin conversations: mechanisms and implications of paramutation. *Nat. Rev. Genet.* **5**, 532–544.
3. Grummt, I. and Pikaard, C.S. (2003) Epigenetic silencing of RNA polymerase I transcription. *Nat. Rev. Mol. Cell Biol.* **4**, 641–649.
4. Matzke, M.A. and Birchler, J.A. (2005) RNAi-mediated pathways in the nucleus. *Nat. Rev. Genet.* **6**, 24–35.
5. Wolffe, A.P. and Matzke, M.A. (1999) Epigenetics: regulation through repression. *Science* **286**, 481–486.
6. Hsieh, T.-F. and Fischer, R.L. (2005) Biology of chromatin dynamics. *Annu. Rev. Plant Biol.* **56**, 327–351.
7. Reyes, J.C. (2006) Chromatin modifiers that control plant development. *Curr. Opin. Plant Biol.* **9**, 21–27.
8. Bender, J. (2004) DNA methylation and epigenetics. *Annu. Rev. Plant Biol.* **55**, 41–68.
9. Chan, S.W.-L., Henderson, I.R., and Jacobsen, S.E. (2005) Gardening the genome: DNA methylation in *Arabidopsis thaliana*. *Nat. Rev. Genet.* **6**, 351–360.
10. Fuchs, J., Demidov, D., Houben, A., and Schubert, I. (2006) Chromosomal histone modification patterns – from conservation to diversity. *Trends Plant Sci.* **11**, 199–208.
11. Barrera, L.O. and Ren, B. (2006) The transcriptional regulatory code of eukaryotic cells – insights from genome-wide analysis of chromatin organization and transcription factor binding. *Curr. Opin. Cell Biol.* **18**, 291–298.
12. He, Y. and Amasino, R.M. (2004) Role of chromatin modification in flowering-time control. *Trends Plant Sci.* **10**, 30–35.
13. Lippman, Z. and Martienssen, R. (2004) The role of RNA interference in heterochromatic silencing. *Nature* **431**, 364–370.
14. Pontes, O., Li, C.F., Nunes, P.C., Haag, J., Ream, T., Vitins, A., Jacobsen, S.E., and Pikaard, C.S. (2006) The *Arabidopsis* chromatin-modifying nuclear siRNA pathway involves a nucleolar RNA processing center. *Cell* **126**, 79–92.

15. Alleman, M., Sidorenko, L., McGinnis, K., Seshadri, V., Dorweiler, J.E., White, J., Sikkink, K., and Chandler, V.L. (2006) An RNA-dependent RNA polymerase is required for paramutation in maize. *Nature* **442**, 295–298.
16. McGinnis, K.M., Springer, C., Lin, Y., Carey, C.C., and Chandler, V. (2006) Transcriptionally silenced transgenes in maize are activated by three mutations defective in paramutation. *Genetics* **173**, 1637–1647.
17. Elmayan, T., Proux, F., and Vaucheret, H. (2005) *Arabidopsis RPA2*: a genetic link among transcriptional gene silencing, DNA repair, and DNA replication. *Curr. Biol.* **15**, 1919–1925.
18. Lippman, Z., *et al.* (2004) Role of transposable elements in heterochromatin and epigenetic control. *Nature* **430**, 471–476.
19. Probst, A.V., Fransz, P.F., Paszkowski, J., and Mittelsten-Scheid, O. (2003) Two means of transcriptional reactivation within heterochromatin. *Plant J.* **33**, 743–749.
20. Cremer, T., Cremer, M., Dietzel, S., Muller, S., Solovei, I., and Fakan, S. (2006) Chromosome territories – a functional nuclear landscape. *Curr. Opin. Cell Biol.* **18**, 307–316.
21. Kosak, S.T. and Groudine, M. (2004) Gene order and dynamic domains. *Science* **306**, 644–647.
22. Sundaresan, V., Springer, P., Volpe, T., Howard, S., Jones, J.D.G., Dean, C., Ma, H., and Martienssen, R. (1995) Patterns of gene action in plant development revealed by enhancer trap and gene trap transposable elements. *Genes Dev.* **9**, 1797–1810.
23. Kato, N. and Lam, E. (2001) Detection of chromosomes tagged with green fluorescent protein in live *Arabidopsis thaliana* plants. *Genome Biol.* **2**, research 0045.1–0045.10.
24. Rosin, R., Watanabe, N., Cacas, J.-L., Kato, N., Arroyo, J.M., Fang, Y., May, B., Vaughn, M., Simorowski, J., Ramu, U., McCombie, R.W., Spector, D.L., Martienssen, R.A., and Lam, E. (2008) Genome-wide transposon tagging reveals location-dependent effects on transcription and chromatin organization in *Arabidopsis. Plant J.* **55(3)**: 514–525.
25. Kato, N. and Lam, E. (2003) Chromatin of endoreduplicated pavement cells has greater range of movement than that of diploid guard cells in *Arabidopsis thaliana*. *J. Cell Sci.* **116**, 2195–2201.

Chapter 8

Clone-Based Functional Genomics

Annick Bleys, Mansour Karimi, and Pierre Hilson

Abstract

Annotated genomes have provided a wealth of information about gene structure and gene catalogs in a wide range of species. Taking advantage of these developments, novel techniques have been implemented to investigate systematically diverse aspects of gene and protein functions underpinning biology processes. Here, we review functional genomics applications that require the mass production of cloned sequence repertoires, including ORFeomes and silencing tag collections. We discuss the techniques employed in large-scale cloning projects and we provide an up-to-date overview of the clone resources available for model plant species and of the current applications that may be scaled up for systematic plant gene studies.

Key words: Functional genomics, recombinational cloning, clone collections, ORFeome, hairpin RNA, artificial microRNA.

1. Introduction

A decade ago, Hieter and Boguski (1) proposed to divide the term genomics into two disciplines: structural genomics and functional genomics. Structural genomics referred to the initial characterization of genome sequences and it started with the publication of the chromosome sequence of the bacterium *Haemophilus influenzae* in 1995 (2). The first eukaryotic genome sequences were released between 1996 and 2000 – budding yeast (*Saccharomyces cerevisiae*) (3), roundworm (*Caenorhabditis elegans*) (4), and fruit fly (*Drosophila melanogaster*) (5) – followed by the initial draft of the human genome sequence in 2001 (6). The first genome sequence of a flowering plant, the weed *Arabidopsis thaliana* (7), was finalized in 2000. Such scientific achievements enable the comparative analysis of organisms from diverse phyla and the identification of

Dmitry A. Belostotsky (ed.), *Plant Systems Biology, vol. 553*

DOI 10.1007/978-1-60327-563-7_8 Springerprotocols.com

life processes uniting or distinguishing them at the molecular level. Rice (*Oryza sativa*), the first crop genome to be sequenced (8), is particularly interesting because it shares common sets of genes with major food and feed monocotyledonous crops such as corn (*Zea mays*), wheat (*Triticum aestivum*), rye (*Secale cereale*), and barley (*Hordeum vulgare*). It was followed by poplar (*Populus trichocarpa*) (9) and grape vine (*Vitis vinifera*) (10). Collectively, these annotated genomes provide information about hundreds of thousands of plant genes. In most cases, their functions are inferred indirectly by sequence homology across species, linking novel genes with others previously characterized or with encoded protein domains of known biochemical activity. However the value of homologous relationships remains limited. For example, only 9.5% of the 27,589 structurally annotated *Arabidopsis* genes have been shown to be involved in known biological processes and for 6% of them the molecular function has been determined experimentally (www.arabidopsis.org/portals/masc/2007_MASC_Report.pdf).

The second discipline, functional genomics, takes advantage of the resources accrued in structural genomics projects but involves additional approaches to identify gene functions at a large scale, including technologies to profile comprehensively the molecular components of biological systems, such as their transcriptome, proteome, and metabolome. Other approaches are designed to determine which molecules interact with proteins, in which subcellular compartments proteins are localized, which is their biochemical activity and which are the effects of loss-of-function or gain-of-function genetic perturbations. The latter can be grouped under the denomination "clone-based functional genomics" because they all require the isolation and manipulation of specific fragments of the genomes under investigation: for example, open reading frames (ORFs) for protein characterization, or gene-specific tags used to target transcripts for degradation by RNA interference (RNAi).

2. Milestone Studies in Yeast and Animals

The phenotypic analysis of mutant series lacking particular genes can now be achieved systematically by taking advantage of genome sequence information. Large-scale reverse genetic screens in budding yeast elegantly demonstrate the power of such approaches. In the pre-genome era, random insertional mutations were generated en masse in yeast cell populations by transient transposition of a marked Tyl transposable element (11, 12). Later, the "Saccharomyces Genome Deletion Project" consortium exploited the annotated genome sequence to create a unique collection of yeast knock-out strains via double homologous recombination (13). Samples of this population were grown under different conditions and the fitness of numerous mutants was scored by analyzing the genetic footprint or pattern of polymerase chain reaction (PCR) products, specific to each gene insertion. In each deletion strain, one ORF was replaced by a cassette carrying specific 20-base TAG sequences serving as unique "molecular bar codes". The fitness contribution of each gene was measured after growing the pooled deletion strains under challenging conditions and by hybridizing DNA extracted from the mixed cell cultures to microarrays of complementary TAG sequences to track the presence or disappearance of each individual strain (14). Similar yeast deletion strain collections could also be used to investigate genetic interactions, such as synthetic lethality, at the genome scale. In this case, yeast double mutants were created by mating a mutated gene of interest into an array of viable gene deletion mutants to study the effect of the second mutation, possibly suppressing or enhancing the original phenotype (15, 16).

Unfortunately, not all eukaryotic genomes can be easily modified via homologous recombination. Therefore RNAi rapidly became the method of choice to knock down large sets of genes, once the basic elements involved in gene silencing had been identified (17–19). Briefly, double-stranded RNA (dsRNA) can mediate post-transcriptional splicing or translational arrest of the homologous mRNA, preventing the production of the cognate protein. Although the loss of function triggered by RNAi may be partial in some cases, many high-throughput RNAi screens have been carried out successfully in worm, as well as in mammalian and fly cells (18, 20–24). A number of genome-scale RNAi libraries have been created for that purpose, relying on different intermediates, such as long dsRNAs or hairpin RNAs, short hairpin RNAs, in vitro diced small-interfering RNAs (siRNAs), synthetic siRNAs, and artificial microRNAs (amiRNAs), which all have specific advantages and disadvantages.

The association between proteins is fundamental to most cellular processes, and the properties of a biological system are dictated by the topology of protein–protein interaction (PPI) networks (interactomes). PPIs are routinely mined to infer potential functional relationships, because proteins that physically associate are probably involved in related processes. In recent years, several large initiatives have focused on building the initial draft of global interactomes. The most commonly used method for high-throughput mapping of pairwise PPIs is the yeast two-hybrid (Y2H) system (25) that is based on the translational fusion of protein pairs either to the DNA-binding or transcriptional activation domains of a transcription factor (TF) that are inactive when separated. The reconstitution of the TF activity is only possible when two tested proteins interact and bring the two TF domains together. This event is detected by monitoring the transcription and activity of a reporter gene, generally conferring viability to a specially designed yeast strain. The 6,000 proteins of yeast have been assayed for pairwise interaction (26, 27, *27a*) and interactome drafts have been created for fly (28), worm (29), and human (30–32).

The Y2H system requires the reconstitution of a TF in the nucleus, which is an important limitation. Therefore, several alternative protein fragment complementation assays have been designed, including the reconstitution of the protein activity of β-galactosidase (33, 34), β-lactamase (35), green fluorescent protein (GFP) (36), dihydrofolate reductase (37, 38), and ubiquitin. The split-ubiquitin membrane Y2H system is particularly interesting because it can detect interactions between pairs of proteins associated with membranes (39, 40).

Another technique, called tandem affinity purification (TAP), was developed to isolate protein complexes and facilitate the analysis of their components by mass spectrometry (MS). It also requires the isolation of gene-specific sequences either to clone the ORFs of interest in phase with a TAP tag contained in an expression vector or to insert – by homologous recombination into chromosomes – the TAP cassette at the 3′ end of the genes under study. The TAP/MS analysis of almost 500 protein complexes isolated from yeast cells has already been completed based on the TAP tagging of most annotated ORFs (41, 42).

Nevertheless, there is more to protein functions than their physical association. Screens based on cloned genes have also been designed to search for enzymatic activity, protein–ligand binding, and DNA motif recognition. To streamline these different proteomic approaches, versatile ORFeome libraries were constructed that contain almost all ORFs annotated in yeast and bacterial genomes or significant fractions of the protein-coding genes identified in animal species (43).

The few examples listed above illustrate how basic elements embedded in the chromosomes of eukaryotic species can now be mass-produced in standardized formats for a range of applications. But such feats are only recently possible and rely on techniques developed to capture and transfer fragments efficiently and reliably between DNA molecules.

3. High-Throughput Cloning Techniques

Conventional restriction/ligation cloning methods are cumbersome for large-scale cloning efforts because they require sequence analysis and search for compatible restriction sites and involve multiple successive DNA manipulations. To streamline the process, several techniques have been developed for the transfer, in a single step, of particular segments between different dsDNA molecules.

Ligation-independent cloning relies on the generation of sticky ends in the DNA fragments (PCR products) and vectors by the exonuclease activity of T4 DNA polymerase (44). Hybridization of the complementary 5′ tails results in the formation of recombinant circular molecules that are repaired through in vivo ligation after efficient bacterial transformation. Recently, this technique has been used to allow high-throughput cloning of expressed sequence tags in an improved virus-induced gene silencing vector derived from the tobacco rattle virus (45). Similarly, the Uracil-Specific Excision Reagent (USER^{TM}; New England Biolabs, www.neb.com) DNA engineering method enables assembly of recombinant molecules from multiple PCR products without the need for in vitro ligation (46, 47). The procedure is based on PCR primers with a single deoxyuridine residue near their 5′ end and treatment of the resulting PCR products with a commercial USER enzyme mix (48), leading to the formation of abasic sites that destabilize base-pairing in dsDNA. After dissociation of the oligonucleotides lying upstream from the cleavage site, the PCR fragments are flanked by long 3′ overhangs, allowing directional assembly of the vector and USER-treated PCR fragment(s) into a single recombinant molecule. Despite the various modifications to the original method, uracil excision-based cloning has remained largely unused, most probably because of its incompatibility with proofreading DNA polymerases that stall at deoxyuridines present in DNA templates. However, this drawback has been resolved by the PfuCx DNA polymerase (46). Recently, a yeast expression vector has been redesigned to enable high-throughput USER cloning of plant P450 cytochromes (49).

Yeast-based recombineering, relying on in vivo homologous recombination, has been used to clone (multiple) DNA fragments into plant binary T-DNA vectors by one-step transformation (50). Other methods rely on in vitro site-specific recombination: the Univector plasmid-fusion (51), In-Fusion^{TM} cloning (52), and the Gateway^{TM} recombinational cloning systems. The Gateway system is commercialized by the Invitrogen Corporation (www.invitrogen.com) and is arguably the most popular nowadays. It is based on the enzymes that catalyze the insertion and excision of the λ phage genome into and from the *Escherichia coli* chromosome and on the DNA attachment sites involved in these reactions. The original sites have been modified for in vitro site-specific recombination to capture DNA segments into so-called entry clones and, from these, into various destination vectors (53) (**Fig. 8.1**). The BP reaction is catalyzed by the BP Clonase enzyme mix that transfers a DNA fragment of interest - for example, a PCR product - flanked by two *att*B sites into a donor vector carrying

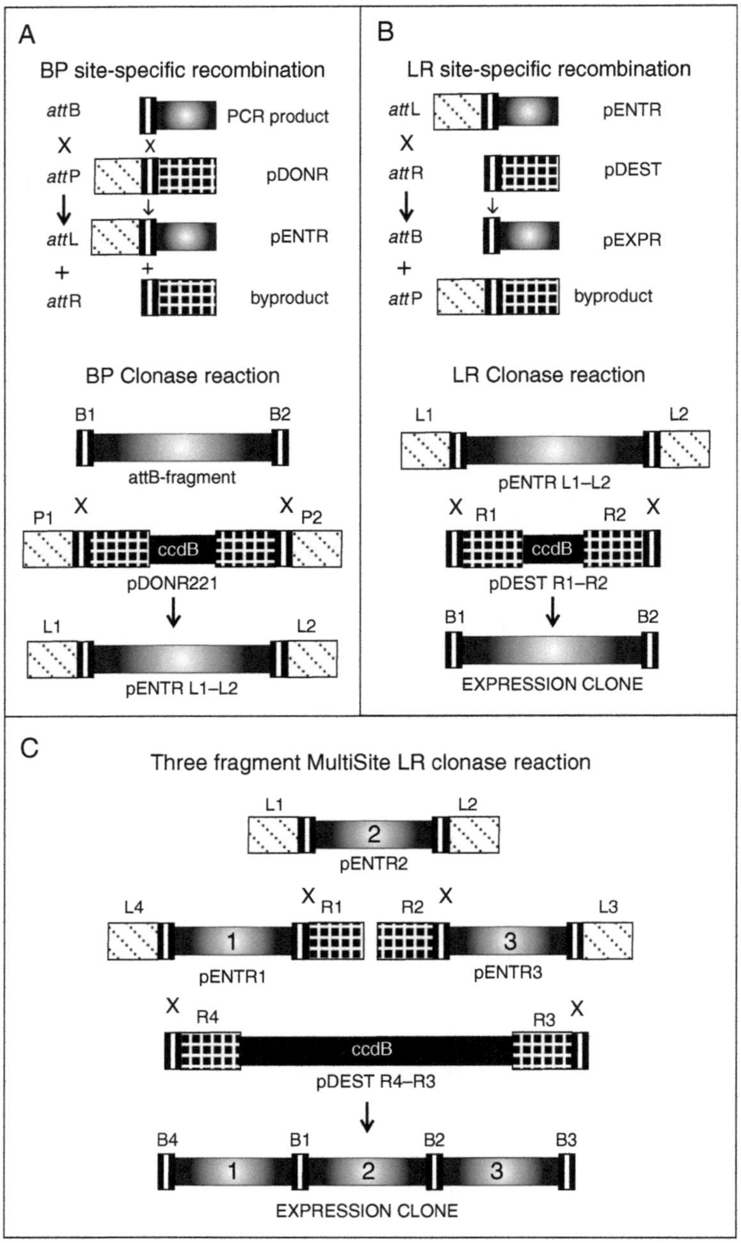

Fig. 8.1. Schematic representation of *att* sites and Gateway recombination reactions. (**A**) In a BP clonase reaction, *att*B sites (in a PCR product or plasmid) recombine with the matching *att*P sites of a donor vector (pDONR) to yield *att*L sites in a novel entry vector (pENTR) and *att*R sites in a byproduct. (**B**) In an LR clonase reaction, *att*L sites in an entry vector (pENTR) recombine with the matching *att*R sites of a destination vector (pDEST) to yield *att*B sites in a novel expression vector (pEXPR) and *att*P sites in a byproduct. (**C**) In a single MultiSite LR clonase reaction, the compatible *att* sites carried by entry clones originating from independent BP clonase reactions and by a MultiSite destination vector recombine to yield a single contig in which the DNA fragments of interest are separated by short *att*B sites. Inside-out gradient box, DNA fragment of interest assembled in BP and LR clonase reactions; black box with vertical white stripe, *att*B sites, also at the core of the *att*P, *att*L, and *att*R sites; diagonally lined box, portion of the *att*P and *att*L sites; lattice box, portion of the *att*P and *att*R sites. B1 to B4, *att*B1 to *att*B4 sites; L1 to L4, *att*L1 to *att*L4; R1 to R4, *att*R1 to *att*R4. Ref. 61; reproduced with permission from the American Society of Plant Biologists ©.

two matching *att*P sites, resulting in an entry clone in which the fragment is flanked by two *att*L sites. This fragment can then be transferred into different destination vectors carrying two *att*R sites through the LR reaction catalyzed by the LR Clonase enzyme mix. Following recombination of the matching *att*L and *att*R sites, the DNA fragment of interest is inserted and again flanked by *att*B sites, resulting in a novel expression clone. To enable directional cloning, variants of the original *att*B, *att*P, *att*L, and *att*R sites have been engineered so that *att*B1 will react specifically with *att*P1, but not with *att*P2, *att*P3, etc. (53–55). Moreover, the *att*B site sequences always maintain the frame register, which is necessary for translational fusions with the N- or C-terminus of the protein encoded in the cloned ORF.

4. Gateway Vectors for Plant Cell Transformation

Binary T-DNA vectors used for *Agrobacterium tumefaciens*-mediated transformation of plant cells are large plasmids that can be cumbersome to manipulate in classical restriction/ligation schemes. Therefore, several research teams have recently developed Gateway versions of such vectors to streamline ectopic gene expression, gene silencing, and promoter studies in transgenic plants (56–58). These original vector sets have later been complemented with constructs that express protein fusions carrying fluorescent, purification, or epitope tags (59–61) (**Table 8.1**).

Typically, the constructs are created by in vitro recombination, transformed in *E. coli* strains, and segregated from other reaction byproducts and input vectors through appropriate antibiotic selection and *ccdB* counterselection (53). Since small high-copy *E. coli* plasmids are routinely introduced into plant cells or protoplasts via methods that do not require *Agrobacterium*-mediated delivery, such as particle bombardment, polyethylene glycol/Ca^{2+} transfection, or electroporation, these plasmids have also been adapted for plant transgene construction via Gateway recombinational cloning (61) (**Table 8.1**).

5. A Unified Framework for the Modular Assembly of Plant Transgenes

An important asset of the Gateway system is the versatility of its entry clones. Once such an entry clone, for example an ORF clone, has been constructed and the sequence of its insert has been validated, the same plasmid can be recombined reliably with

Table 8.1
Gateway vectors for functional assays in plant cells

Name[a]	Vector backbone	BSM[c]	Gateway Cassette Rf[d]	T-DNA marker[e]	Applications	References
Binary T-DNA vectors						
p*2GW7	pPZP200	Sp/Sm	attR1-attR2	K,H,B	Overexpression or antisense (35S pro)	(57)
p*2WG7	pPZP200	Sp/Sm	attR1-attR2	K,H,B	Overexpression or antisense (35S pro)	(57)
p*7WG2	pPZP200	Sp/Sm	attR1-attR2	K,H,B	Overexpression or antisense (35S pro)	(57)
p*7WG2D	pPZP200	Sp/Sm	attR1-attR2	K,H,B	Overexpression together with a visible marker (35S pro)	(57)
pMDC32	pCAMBIA	Km	attR1-attR2	H	Overexpression (35S pro)	(56)
pEarleyGate100	pCAMBIA	Km	attR1-attR2	B	Overexpression (35S pro)	(59)
pIPK001		Sp/Sm	attR1-attR2	H	Overexpression (no pro)	(112)
pIPK002		Sp/Sm	attR1-attR2	H	Overexpression (ZmUbi1 pro)	(112)
pIPK003		Sp/Sm	attR1-attR2	H	Overexpression (OsAct1 pro)	(112)
pIPK004		Sp/Sm	attR1-attR2	H	Overexpression (d35S pro)	(112)

pIPK005		Sp/Sm	attR1-attR2	H	Overexpression (TaGstA1 pro)	(112)
pK7m34GW2-8WG3	pPZP200	Sp/Sm	attR1-attR2, attR4-attR3,	K	Overexpression from two independent cassettes	(61)
pK7m34GW2-8WG3-9m56GW4	pPZP200	Sp/Sm	attR1-attR2, attR4-attR3, attR5-attR5	K	Overexpression from three independent cassettes	(61)
p*GWL7	pPZP200	Sp/Sm	attR1-attR2, RfA	K,H,B,NM	Promoter analysis (LUC)	(57)
p*GWFS7	pPZP200	Sp/Sm	attR1-attR2, RfA	K,H,B,NM	Promoter analysis (GFP-GUS)	(57)
pMDC107	pCAMBIA	Km	attR1-attR2, RfA	H	Promoter analysis (GFP-6xHis tag)	(56)
pMDC111	pCAMBIA	Km	attR1-attR2, RfB	H	Promoter analysis (GFP-6xHis tag)	(56)
pMDC110	pCAMBIA	Km	attR1-attR2, RfC	H	Promoter analysis (GFP-6xHis tag)	(56)
pMDC162	pCAMBIA	Km	attR1-attR2, RfA	H	Promoter analysis (GUS)	(56)
pMDC163	pCAMBIA	Km	attR1-attR2, RfB	H	Promoter analysis (GUS)	(56)
pMDC164	pCAMBIA	Km	attR1-attR2, RfC	H	Promoter analysis (GUS)	(56)
pEarleyGate301	pCAMBIA	Km	attR1-attR2, RfB	B	Promoter analysis (HA tag)	(59)
pEarleyGate302	pCAMBIA	Km	attR1-attR2, RfB	B	Promoter analysis (FLAG tag)	(59)

(continued)

Table 8.1 (continued)

Name[a]	Vector backbone	BSM[c]	Gateway Cassette Rf[d]	T-DNA marker[e]	Applications	References
pEarleyGate301	pCAMBIA	Km	attR1-attR2, RfB	B	Promoter analysis (Myc tag)	(59)
pEarleyGate301	pCAMBIA	Km	attR1-attR2, RfB	B	Promoter analysis (AcV5 tag)	(59)
pK7S-NFm14GW	pPZP200	Sp/Sm	attR4-attL1, RfB	K	Promoter analysis (NLS-GFP-GUS)	(61)
p*7GWIWG2(I)	pPZP200	Sp/Sm	attR1-attR2, RfA	K,H,B	Hairpin RNA expression (35S pro)	(57)
p*7GWIWG2(II)	pPZP200	Sp/Sm	attR1-attR2, RfA	K,H,B	Hairpin RNA expression (35S pro)	(57)
pHELLSGATE12	pART27	Sp/Sm	attR1-attR2	K	Hairpin RNA expression (35S pro)	(78)
pAGRIKOLA	pGreen	Km	attR1-attR2	B	Hairpin RNA expression (35S pro)	(79)
pSTARGATE	pART27	Sp/Sm	attR1-attR2	H	Hairpin RNA expression (ubiquitin pro)	[f]
pWATERGATE	pART27	Sp/Sm	attR1-attR2	K	Hairpin RNA expression (At RbcS pro)	[f]
pIPK006		Sp/Sm	attR1-attR2	H	Hairpin RNA expression (no pro)	(112)
pIPK007		Sp/Sm	attR1-attR2	H	Hairpin RNA expression (ZmUbi1 pro)	(112)
pIPK008		Sp/Sm	attR1-attR2	H	Hairpin RNA expression (OsAct1 pro)	(112)

pIPK009		Sp/Sm	attR1-attR2	H	Hairpin RNA expression (d35S pro)	(112)
pIPK010		Sp/Sm	attR1-attR2	H	Hairpin RNA expression (TaGstA1 pro)	(112)
pOpOff	pART27	Sp/Sm	attR1-attR2	K	DEX-inducible hairpin RNA expression	(83) f
p*7WGC2	pPZP200	Sp/Sm	attR1-attR2, RfA	K,H,B,NM	CFP tag at N-terminus (35S pro)	(57)
p*7WGF2	pPZP200	Sp/Sm	attR1-attR2, RfA	K,H,B,NM	GFP tag at N-terminus (35S pro)	(57)
p*7WGR2	pPZP200	Sp/Sm	attR1-attR2, RfA	K,H,B,NM	RFP tag at N-terminus (35S pro)	(57)
p*7WGY2	pPZP200	Sp/Sm	attR1-attR2, RfA	K,H,B,NM	YFP tag at N-terminus (35S pro)	(57)
hRLUC-attR[b]	pPZP222	Sp/Sm	attR1-attR2	B	LUC tag at N-terminus (35S pro)	[g]
YFP-attR	pBin19	Km	attR1-attR2	K	YFP tag at N-terminus (35S pro)	[g]
pMDC45	pCAMBIA	Km	attR1-attR2, RfA	H	GFP tag at N-terminus (35S pro)	(56)
pMDC44	pCAMBIA	Km	attR1-attR2, RfB	H	GFP tag at N-terminus (35S pro)	(56)
pMDC43	pCAMBIA	Km	attR1-attR2, RfC	H	GFP tag at N-terminus (35S pro)	(56)
pEarleyGate104	pCAMBIA	Km	attR1-attR2, RfB	B	YFP tag at N-terminus (35S pro)	(59)

(continued)

Table 8.1 (continued)

Name[a]	Vector backbone	BSM[c]	Gateway Cassette Rf[d]	T-DNA marker[e]	Applications	References
pEarleyGate201	pCAMBIA	Km	attR1-attR2, RfB	B	HA tag at N-terminus (35S pro)	(59)
pEarleyGate202	pCAMBIA	Km	attR1-attR2, RfB	B	FLAG tag at N-terminus (35S pro)	(59)
pEarleyGate203	pCAMBIA	Km	attR1-attR2, RfB	B	Myc tag at N-terminus (35S pro)	(59)
pEarleyGate204	pCAMBIA	Km	attR1-attR2, RfB	B	AcV5 tag at N-terminus (35S pro)	(59)
pEarleyGate205	pCAMBIA	Km	attR1-attR2, RfB	B	TAP tag at N-terminus (35S pro)	(59)
pEarleyGate 401	pBin19	Km	attR1-attR2	B	YFP tag at N-terminus (35S pro) (Monocot)	(59)
pEarleyGate 402	pBin19	Km	attR1-attR2	B	FLAG tag at N-terminus (35S pro) (Monocot)	(59)
pEarleyGate 403	pBin19	Km	attR1-attR2	B	HA tag at N-terminus (35S pro) (Monocot)	(59)
attR-YFP	pBin19	Km	attR1-attR2	K	YFP tag at C-terminus (35S pro)	[g]
attR-hRLUC[b]	pPZP222	Sp/Sm	attR1-attR2	B	LUC tag at C-terminus (35S pro)	[g]
p*7CWG2	pPZP200	Sp/Sm	attR1-attR2, RfA	K,H,B,NM	CFP tag at C-terminus (35S pro)	(57)
p*7FWG2	pPZP200	Sp/Sm	attR1-attR2, RfA	K,H,B,NM	GFP tag at C-terminus (35S pro)	(57)

p*7RWG2	pPZP200	Sp/Sm	attR1-attR2, RfA	K,H,B,NM	RFP tag at C-terminus (35S pro)	(57)
p*7YWG2	pPZP200	Sp/Sm	attR1-attR2, RfA	K,H,B,NM	YFP tag at C-terminus (35S pro)	(57)
pK7Fm24GW	pPZP200	Sp/Sm	attR4-attR2	K	GFP tag at C-terminus (genomic fragment)	(61)
pMDC83	pCAMBIA	Km	attR1-attR2, RfA	H	GFP-6xHis tag at C-terminus (35S pro)	(56)
pMDC84	pCAMBIA	Km	attR1-attR2, RfB	H	GFP-6xHis tag at C-terminus (35S pro)	(56)
pMDC85	pCAMBIA	Km	attR1-attR2, RfC	H	GFP-6xHis tag at C-terminus (35S pro)	(56)
pMDC139	pCAMBIA	Km	attR1-attR2, RfA	H	GUS tag at C-terminus (35S pro)	(56)
pMDC140	pCAMBIA	Km	attR1-attR2, RfB	H	GUS tag at C-terminus (35S pro)	(56)
pMDC141	pCAMBIA	Km	attR1-attR2, RfC	H	GUS tag at C-terminus (35S pro)	(56)
pEarleyGate101	pCAMBIA	Km	attR1-attR2, RfB	B	YFP-HA tag at C-terminus (35S pro)	(59)
pEarleyGate102	pCAMBIA	Km	attR1-attR2, RfB	B	CFP-HA tag at C-terminus (35S pro)	(59)
pEarleyGate103	pCAMBIA	Km	attR1-attR2, RfB	B	GFP-His tag at C-terminus (35S pro)	(59)
pK7FWGF2	pPZP200	Sp/Sm	attR1-attR2, RfA	K	GFP tag both at C- and N-terminus (35S pro)	(57)

(continued)

Table 8.1 (continued)

Name[a]	Vector backbone	BSM[c]	Gateway Cassette Rf[d]	T-DNA marker[e]	Applications	References
pMDC30	pCAMBIA	Km	attR1-attR2, RfC	H	Inducible gene expression	(56)
pMDC7	pER8	Sp/Sm	attR1-attR2, RfB	H	Inducible gene expression	(56)
pJCGLOX	pCAMBIA	Cm	attR1-attR2	K	Inducible gene expression	(113)
pJLOX	pCAMBIA	Cm	attR1-attR2	H	Inducible gene expression	(73)
pMDC150	pMoa	Km	attR1-attR2	B	Inducible gene expression (activator)	(114)
pMDC160	pMoa	Km	attR1-attR2	B	Inducible gene expression (responder)	(114)
pMDC220	pMoa	Km	attR1-attR2	H	Inducible gene expression (responder)	(114)
pMDC221	pMoa	Km	attR1-attR2	H	Inducible gene expression (responder)	(114)
pLB12	pMoa	Km	attR1-attR2	K	Inducible gene expression (activator/ responder)	(114)
N TAPi	pPZP200	Sp/Sm	attR1-attR2	B	TAP technology (35S pro)	(115)

CTAPi	pPZP200	Sp/Sm	attR1-attR2	B	TAP technology (35S pro)	(115)
Ubi-NTAP-1300	pCAMBIA	Km	attR1-attR2	K	TAP technology (ubiquitin pro)	(116)
pKCTAP	pPZP200	Sp/Sm	attR4-attR3	K	TAP technology (35S pro)	(64)
pKNTAP	pPZP200	Sp/Sm	attR4-attR2	K	TAP technology (35S pro)	(64)
pTRV2-attR1-attR2	pCAMBIA	Km	attR1-attR2		Virus-induced gene silencing (TRV)	(117)
p*GW	pPZP200	Sp/Sm	attR1-attR2, RfA	K,H,B	Single fragment recombination	(57)
p*GWD,0	pPZP200	Sp/Sm	attR1-attR2, RfA	K,B	Single fragment recombination	h
pMDC99	pCAMBIA	Km	attR1-attR2, RfC	H	Single fragment recombination	(56)
pMDC100	pCAMBIA	Km	attR1-attR2, RfC	K	Single fragment recombination	(56)
pMDC123	pCAMBIA	Km	attR1-attR2, RfC	B	Single fragment recombination	(56)
p*m42GW,3	pPZP200	Sp/Sm	attR4-attR2	K,H,B,NM	MultiSite (two-fragment recombination without terminator)	(62)
p*m43GW	pPZP200	Sp/Sm	attR4-attR3	K,H,B,NM	MultiSite (three-fragment recombination without terminator)	(62)

(continued)

Table 8.1 (continued)

Name[a]	Vector backbone	BSM[c]	Gateway Cassette Rf[d]	T-DNA marker[e]	Applications	References
p*7m24GW,3	pPZP200	Sp/Sm	attR4-attR2	K,H,B,NM	MultiSite (two-fragment recombination with terminator)	(62)
p*7m34GW	pPZP200	Sp/Sm	attR4-attR3	K,H,B,NM	MultiSite (three-fragment recombination with terminator)	(62)
pHSC	pPZP200	Sp/Sm	attR1-attR2	H	Conditional expression of CRE	(118)
High-copy vectors						
p2GW7,0		Am	attR1-attR2, RfA	na	Overexpression or antisense (35S pro)	(57)
pSAT6-NP-Dest-EGFP		Am	attR1-attR2	na	Promoter analysis (GFP)	(119)
p2CGW7		Am	attR1-attR2, RfA	na	N-terminal fusion to CFP tag (35S pro)	(57)
p2FGW7		Am	attR1-attR2, RfA	na	N-terminal fusion to GFP tag (35S pro)	(57)
p2RGW7		Am	attR1-attR2, RfA	na	N-terminal fusion to RFP tag (35S pro)	(57)
p2YGW7		Am	attR1-attR2, RfA	na	N-terminal fusion to YFP tag (35S pro)	(57)
YFP attR		Am	attR1-attR2	na	N-terminal fusion to YFP tag (35S pro)	[g]

RLUC attR	Am	attR1-attR2	na	N-terminal fusion to LUC tag (35S pro)	g
hRLUC-attR	Am	attR1-attR2	na	N-terminal fusion to LUC tag (35S pro)	g
pSAT6-Dest-EGFP-C1	Am	attR1-attR2	na	N-terminal fusion to GFP tag (35S pro)	(119)
p2GWC7	Am	attR1-attR2, RfA	na	C-terminal fusion to CFP tag(35S pro)	(57)
p2GWF7	Am	attR1-attR2, RfA	na	C-terminal fusion to GFP tag(35S pro)	(57)
p2GWR7	Am	attR1-attR2, RfA	na	C-terminal fusion to RFP tag(35S pro)	(57)
p2GWY7	Am	attR1-attR2, RfA	na	C-terminal fusion to YFP tag(35S pro)	(57)
attR-RLUC	Am	attR1-attR2	na	C-terminal fusion to LUC tag(35S pro)	g
attR-hRLUC[a]	Am	attR1-attR2	na	C-terminal fusion to LUC tag(35S pro)	g
attR-YFP	Am	attR1-attR2	na	C-terminal fusion to YFP tag(35S pro)	g
pSAT6-Dest-EGFP-N1	Am	attR1-attR2	na	C-terminal fusion to GFP tag (35S pro)	(119)
pUC-SPYNE[G]	Am	attR1-attR2 RfB	na	Bimolecular fluorescence complementation (35S pro)	(101)
pUC-SPYCE[G]	Am	attR1-attR2 RfB	na	Bimolecular fluorescence complementation (35S pro)	(101)

(continued)

Table 8.1 (continued)

Name[a]	Vector backbone	BSM[c]	Gateway Cassette Rf[d]	T-DNA marker[e]	Applications	References
p35S-GAD-GW		Am	attR1-attR2	na	Plant two-hybrid (35S pro)	(120)
p35S-GBD-GW		Am	attR1-attR2	na	Plant two-hybrid (35S pro)	(120)
p35S-HA-GW		Am	attR1-attR2	na	Plant two-hybrid (35S pro)	(120)
pCLCVA-GW,007		Am	attR1-attR2	na	Virus-induced gene silencing (CaLCuV)	i
Pm42GW7,3		Am	attR4-attR2	na	MultiSite (two-fragment recombination with terminator)	(62)

[a] The asterisk refers to the DNA selectable markers as indicated in the T-DNA marker column.
[b] RLUC, Renilla luciferase; hRLUC, humanized luciferase.
[c] BSM, bacterial selectable markers: Am, ampicillin; Cm, chloramphenicol; Km, kanamycin; Sm, streptomycin; Sp, spectinomycin.
[d] RfA, RfB, or RfC, Gateway reading frames.
[e] T-DNA selectable markers: B, Basta; H, hygromycin; K, kanamycin; NM, no marker.
[f] http://www.pi.csiro.au/rnai/vectors.htm
[g] http://www.bio.utk.edu/vonarnim/vectors.shtml
[h] http://www.psb.ugent.be/gateway/
[i] M. Karimi, S. Bernacki, P.Hilson, and D. Robertson, unpublished results.
Other useful web resources are
http://www.biology.wustl.edu/pikaard/Vectors%20homepage.html
http://www.cambia.org/daisy/cambia/home.html
http://www.unizh.ch/botinst/Devo_Website/curtisvector/

many different destination vectors, each intended for a distinct functional assay. For an ORF, these assays could be Y2H protein interaction mapping, protein production in vitro, in bacteria, fungi, and insect cells, phenotypic complementation in yeast or plants, and subcellular localization of proteins fused with a fluorescent tag.

Furthermore, the availability of Gateway recombination site variants has led to a notable technological improvement, dubbed the MultiSite Gateway system that enables the simultaneous assembly of multiple DNA fragments in one single LR clonase reaction through three or more distinct and incompatible *att* site series (54, 55) (**Fig. 8.2**). MultiSite cloning schemes are attractive for gene function studies requiring the repeated modular arrangement of multiple elements, such as promoters, coding sequences, and terminators. MultiSite plant binary T-DNA destination vectors have been designed to take advantage of this technological improvement (62, 63). Building blocks that fit into these recipient vectors have been formatted as entry clones that carry sequences commonly used in plant molecular biology, including regulatory sequences (promoters and terminators), enzymatic reporters, and tags coding for epitopes, fluorescent proteins, and affinity purification peptides (61, 64).

In the primary implementation of the MultiSite cloning system, the basic collection of modular entry clone cassettes can be assembled in two-fragment or three-fragment LR reactions in the order *att*4-*att*1/*att*1-*att*2 or *att*4-*att*1/*att*1-*att*2/*att*2-*att*3, from the 5′ to 3′ orientation relative to transcription (**Fig. 8.3A** and **B**). In this framework, the promoters or enhancer motifs are flanked by the *att*L4 and *att*R1 recombination sites, ORFs by the *att*L1 and *att*L2 sites, and terminators by the *att*R2 and *att*L3 sites. These entry clones are recombined into plant binary T-DNA destination vectors, carrying recipient *ccdB* cassettes flanked by the *att*R4-*att*R2 or *att*R4-*att*R3 sites (62).

However, the MultiSite Gateway technology can also be used in configurations in which *att* sites are positioned so that several DNA fragments are recombined into separate cassettes. In this instance, genes captured in distinct entry clones, for example *att*L1-gene1-*att*L2, *att*L4-gene2-*att*L3, and *att*L6-gene3-*att*L5, can be placed in a single LR reaction in the same T-DNA binary vector but each under the control of different promoters and terminators (61) (**Fig. 8.2C**). This flexible scaffold facilitates the stacking and simultaneous testing of multiple transgenes in transformed plants to explore multigene traits in model plant species or crops.

Finally, it is worth emphasizing that the expression units resulting from LR reactions are no dead ends. Indeed, cassettes built in expression clones and flanked by specific *att*B sites can serve as a template for the reverse BP recombination with the corresponding donor vectors, resulting in the replacement of a

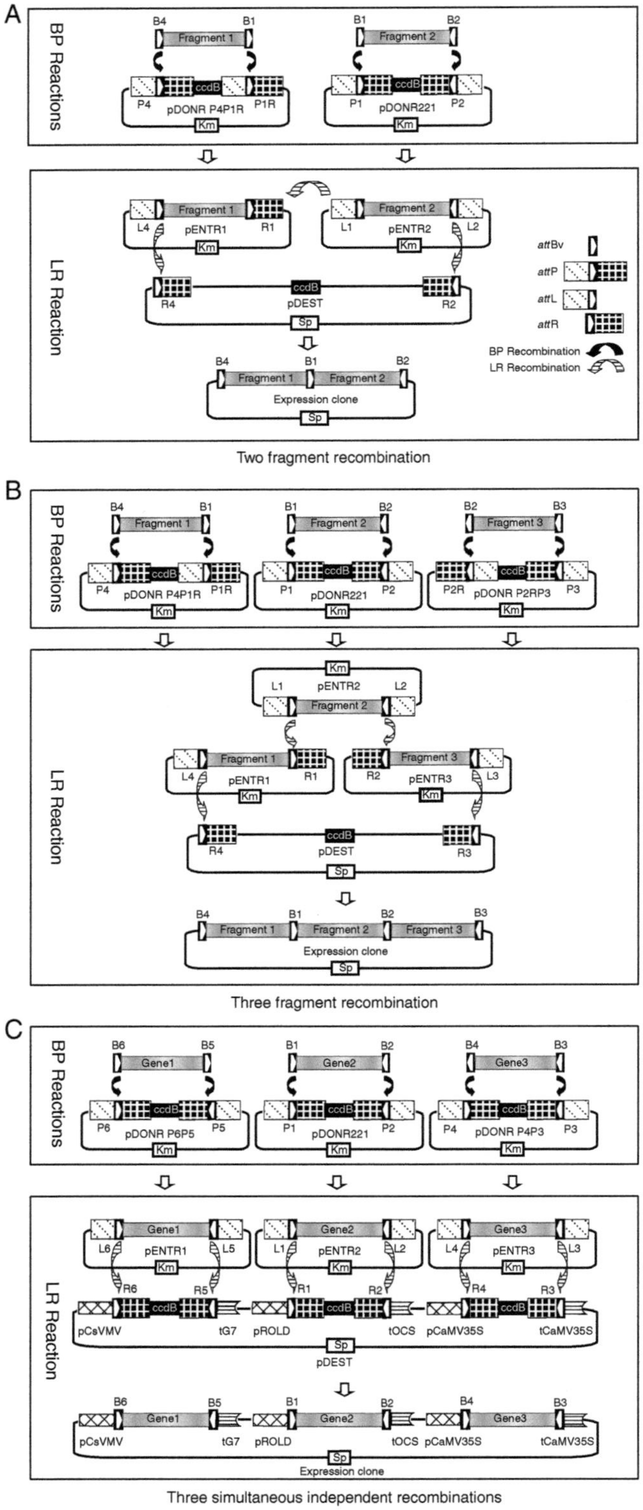

Fig. 8.2. MultiSite Gateway recombinational cloning strategy. In all three schemes, each of the entry clones is produced in vitro in a BP clonase reaction that transfers a PCR amplicon or plasmid segment flanked by the appropriate *att*B sites

DNA fragment formatted in the Gateway format by its corresponding *ccdB* cassette (61) (**Fig. 8.3C**). Modular destination vectors generated by reverse BP cloning are particularly advantageous when series of constructs need to be created in which only one of the elements varies. Once the variable segment is replaced by a *ccdB* cassette flanked by *att*L and/or *att*R sites, the resulting destination vector can be used to generate additional expression clones, bypassing the need to perform MultiSite Gateway LR clonase reactions involving three or more plasmids and requiring complex assembly validation. Of course, similar vectors can be built via classical restriction/ligation strategies, but such cloning may be cumbersome with large plasmids and does not allow the downstream modular manipulation of elements in the resulting expression clones by alternating reverse BP and LR reactions.

6. Cloned Sequence Repertoires in Plants

6.1. Protein-Coding Sequences

The most practical manner to handle a protein for a variety of research purposes is to create a recombinant DNA plasmid containing the corresponding complementary DNA (cDNA) or ORF. With this basic intermediate material, the protein can be synthesized in vitro or in any species and cell type, provided the appropriate expression cassettes and transformation procedures are available. If the ORF is trimmed, without 5′ or 3′ untranslated regions, the encoded protein can be produced as a translational fusion with any chosen peptide moiety, such as purification or marker tags, by positioning the ORF in an expression unit in frame with the tag sequence. Because genome annotation and downstream functional studies require the characterization of protein-coding sequences, large-scale initiatives focusing on model plant species isolated and sequenced full-length cDNA and ORF clones (65–69). As listed in **Table 8.2**, most of the available *Arabidopsis* clone repertoires have been captured in Gateway *att*L1-gene-*att*L2 entry clones. In the case of ORFeomes, the original stop codon may be either maintained (closed

Fig. 8.2. (continued) into one of three donor vectors. Subsequently, two (**A**) or three (**B**) fragments are assembled contiguously in vitro in a single MultiSite LR clonase reaction, by transfer from the two or three entry plasmids into a destination vector to form an expression clone, respectively. (**C**) Alternatively, three genes of interest can be transferred simultaneously in three independent transcription units. In all cases, the products of the BP or LR reactions are introduced into *E. coli* cells and the entry or expression vectors are selected in bacteria grown on kanamycin (Km) or spectinomycin (Sp) medium, respectively. Symbols as indicated in panel A. Adapted from refs. 61 and 62, with permission from Elsevier © and from the American Society of Plant Biologists ©.

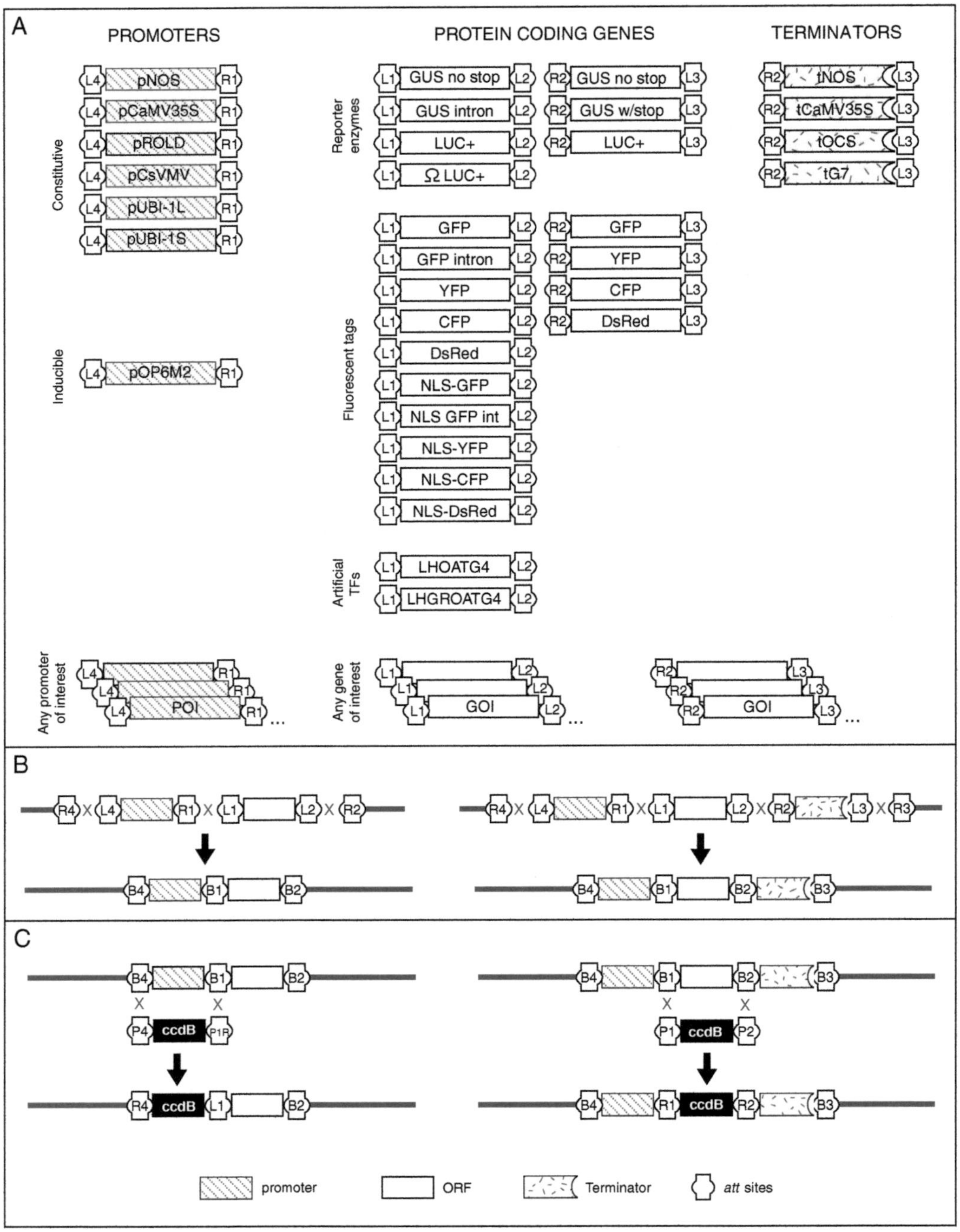

Fig. 8.3. Building blocks, DNA assembly, and replacement strategies for MultiSite Gateway cloning. (**A**) Entry clones either available as shared resources or generated to characterize specific sequences of interest. (**B**) Assembly of expression clones by recombination of two (left) or three (right) entry clones with a destination vector in LR clonase reactions. (**C**) Examples for the replacement of genetic elements by the counterselectable *ccdB* cassette within expression clones for the creation of modular destination vectors in reverse BP reactions. Box shape annotation is as indicated at the bottom and name of *att* sites as labeled within the corresponding box symbols. Ref. 61; This figure was originally published in Plant Physiology, 145, Karimi, M., Bleys, A., Vanderhaeghen, R. and Hilson, P., 2007, page 1185, and reproduced with permission from the American Society of Plant Biologists ©.

Table 8.2
Cloned sequence repertoires of *Arabidopsis*

Creator	Format	Focus	Validation	Scale	URL	Stock center	References
ORF clones							
SSP consortium and Salk Institute	Univector pUNI51		Full sequence	14,000	signal.salk.edu/cdnastatus.html	ABRC	(67)
Ecker/ Invitrogen	Gateway entry		Full sequence	12,000	signal.salk.edu/cdnastatus.html		
TIGR	Gateway entry	Hypothetical genes	Full sequence	3,000	www.tigr.org/tdb/hypos/	ABRC	(70, 121, 122)
Peking-Yale Joint Center	Gateway entry	Transcription factors	5′ and 3′ end sequence	1,150		ABRC	(76)
Dinesh-Kumar et al.	Gateway expression (from Peking-Yale JC)	TAP-tagged transcription factor		1,100		ABRC	
REGIA	Gateway entry	Transcription factors	5′ and 3′ end sequence	1,000	gabi.rzpd.de/materials/	GABI/ RZPD	(75)
Dinesh-Kumar et al.	Gateway entry, no stop	Plant protein chips	5′ and 3′ end sequence	5,100	plants.gersteinlab.org/	ABRC	(111)
ATOME 1	Gateway entry		5′ and 3′ end sequence	2,000	urgv.evry.inra.fr/orfeome/	CNRGV	
ATOME 2	Gateway entry, no stop	Originates from SSP	5′ and 3′ end sequence	3,500	same	CNRGV	

(continued)

Table 8.2(continued)

Creator	Format	Focus	Validation	Scale	URL	Stock center	References
Doonan et al.	Gateway Expression (from SSP)	GFP fusion for subcellular location	5′ and 3′ end sequence	100_s	data.jic.bbsrc.ac.uk/cgi-bin/gfp/	ABRC	(88)
Callis et al.	Gateway entry	Protein ubiquitination	Full sequence	100_s	plantsubq.genomics.purdue.edu	ABRC	(123)
Sheen et al.	Expression	Epitope-tagged MAPK	Full sequence	100	genetics.mgh.harvard.edu/sheenweb/category_genes.html	ABRC	
cDNA clones							
RIKEN/SSP/ Salk Institute	λ ZAP or λ PS		Full sequence	17,000	www.brc.riken.go.jp/lab/epd/Eng/order/order.shtml	BRC	(65)
MPI-MG	Gateway expression		5′ end sequence	4,500	gabi.rzpd.de/materials/	GABI/ RZPD	(108)
Génoscope/LTI	Gateway entry		Full single pass sequence	29,000	www.genoscope.cns.fr/Arabidopsis	CNRGV	(124)
RNAi clones							
AGRIKOLA	Gateway entry		PCR-sized insert	28,000	http://www.agrikola.org	NASC	(79, 125, 126)
AGRIKOLA	Constitutive hp RNA expression		PCR-sized insert	26,000	http://www.agrikola.org	NASC	(79)
AGRIKOLA	Gateway entry		Pure, sequence validated	1,000	bccm.belspo.be/db/lmbp_gst_clones/	BCCM/ LMBP	

AGRIKOLA	Constitutive hp RNA expression		Pure, sequence validated	800	bccm.belspo.be/db/ lmbp_gst_clones/	BCCM/ LMBP	
CFGC	ds RNA expression	Chromatin remodeling	Single pass sequence	200	http://www.chromdb.org	ABRC	(127, 128)
amiRNA Central	Artificial miRNA		Full sequence	10,000	2010.cshl.edu	Open Bio-systems Inc.	

Stock centers distributing *Arabidopsis* clone repertoires:

- Arabidopsis Biological Resource Center (ABRC, USA), http://www.biosci.ohio-state.edu/pcmb/Facilities/abrc/abrchome.htm
- RIKEN BioResource Center (BRC, Japan), http://www.brc.riken.jp/lab/epd/Eng/catalog/pDNA.shtml
- GABI Primary Database (GABI/RZPD, Germany), http://gabi.rzpd.de/
- National Resources Centre for Plant Genomics (CNRGV, France), http://cnrgv.toulouse.inra.fr/ENG/index.html
- European Arabidopsis Stock Centre (NASC, United Kingdom), http://arabidopsis.info/
- BCCM/LMBP Plasmid and DNA library collection (BCCM/LMBP, Belgium), http://bccm.belspo.be/db/lmbp_gst_clones/
- Open Biosystems Inc., www.openbiosystems.com/

Clone collections from plant species other than *Arabidopsis* are also distributed via the RIKEN BioResource Center, the Rice Genome Resource Center (http://www.rgrc.dna.affrc.go.jp/), and the GABI primary database.

Updated version from ref. 60 with permission from Elsevier ©.

configuration), resulting in the production of the native protein, or removed, enabling the addition of C-terminal peptides (open configuration). Some cloning protocols have been adapted so that open and closed ORFs are isolated simultaneously (70).

6.2. Non-coding Sequences

Cis-acting regulatory sequences may be isolated from the genome either as entire intergenic or promoter regions or as discrete binding sites recognized by chromatin-associated factors. A library of approximately 20,000 *Arabidopsis* promoter amplicons has been created (www.psb.ugent.be/SAP/) that can be captured readily as *att*L4-promoter-*att*R1 entry clones compatible with MultiSite Gateway LR reactions (71). Accessions from this versatile resource have already been used for the detailed mapping of transcript patterns in planta (unpublished results). This *Arabidopsis* promoterome may also be exploited for the study of transcriptional networks via one-hybrid screens either in yeast (72) or in plant cells (73, 74), in combination with TF libraries (75, 76).

6.3. Tags for RNAi

Recently two main methods have emerged as most practical for post-transcriptional gene silencing in higher plants. First, hairpin RNAs (hpRNAs) that harbor a ds stem of a few hundred base pairs and a loop with an intronic sequence have been shown to efficiently knock down expression (77). Gateway recombinational schemes have helped streamline the production of the corresponding silencing constructs. In particular, a recombinant DNA plasmid that carries the inverted repeat coding for the two complementary strands of the hpRNA can be created in vitro in a single LR clonase reaction, provided the gene region targeted for RNAi is available as a Gateway entry clone (58, 78, 79). Second, amiRNAs also induce potent RNAi in various plant species (17, 80, 81). This approach entails the expression of an endogenous plant microRNA precursor engineered to yield a 21-nucleotide ds amiRNA selected to target the degradation of one or several transcripts of interest according to experimentally defined target selection parameters (wmd2.weigelworld.org).

Because amiRNA target recognition relies on a short 21-nucleotide sequence, amiRNAs yield more specific silencing than the numerous siRNAs derived from the long (hundreds of base pairs) ds hpRNA molecules. The knockdown of off-target genes that share (even small) stretches of sequence homology with the silencing RNA sequences complicates the analysis of the resulting phenotypes and the identification of the causal gene(s). Unfortunately, specific amiRNAs or hpRNAs do not always result in significant downregulation and current algorithms lack predictive power of silencing success. Large-scale Gateway clone collections are available for knocking down *Arabidopsis* genes with either of the two methods (**Table 8.2**). Although the silencing tags are primarily transcribed from transgenes constructed for constitutive silencing, they can also serve for virus-mediated, inducible or tissue-specific RNAi (82, 83).

7. Functional Screens

7.1. Genetic Perturbations

In vivo homologous recombination is still not routinely achievable in higher plants. Mutant alleles are mainly generated either via mutagenesis with chemical or physical agents that alter the DNA integrity or after the insertion of engineered T-DNAs or transposable elements at random positions into the chromosomes. These untargeted approaches have been extremely valuable to study genome elements without any prior knowledge. Insertional lines will remain key assets for plant geneticists because they provide stably tagged mutations in genes of interest. However, the limitations are that thousands of mutant lines must be generated to reach a satisfactory genome coverage, the location of the mutated sites must be determined, and phenotypes need to be genetically linked to each mutation. Reverse genetics screens based on available sequence repertoires offer complementary advantages. Gain-of-function and loss-of-function phenotypes can be produced by overexpression and RNAi, respectively, and only a few transgenic individuals need to be produced for phenotypes to be attributed to the construct encoding a particular genetic perturbation.

The full-length cDNA over-expressing (FOX) gene-hunting system is one implementation of this strategy. It consists in the bulk transformation into *Arabidopsis* of thousands of randomly selected and normalized RIKEN Arabidopsis full-length (RAFL) cDNA sequences driven by the cauliflower mosaic virus 35S promoter (84). The initial large-scale phenotypic analysis of first-generation (T1) transformed lines indicated that a high fraction showed altered morphology (9 %) and that the mutant phenotypes were in most cases transmitted as a dominant or semi-dominant trait through the next generations. In another attempt to develop a batch procedure for reverse genetics in planta, a library of 32 ethylene response transcription factor (ERF) ORFs was transferred as a pool in an overexpression vector, then in *Agrobacterium* and *Arabidopsis* plants via the floral dip procedure. The original complexity of the ERF collection was maintained throughout the successive steps and the resulting transgenic plants were screened for enhanced abiotic stress tolerance phenotypes, leading to the identification of several leads (85) (www.ubpb.gwdg.de/wdllab/index.html). When compared to transforming one construct at a time, the batch approach reduces costs, labor, and greenhouse space for the build-up of a library of overexpressing plants. Note that the constitutive expression of a target gene might sometime leads to its co-suppression and mutant phenotypes can be wrongly attributed to enhanced rather than decreased activity.

As the major bottleneck in reverse genetics screens is the phenotypical analysis of large mutant plant populations, it might be advantageous to investigate particular segments of signaling or biochemical pathways in cultured cell lines, even though such simplified systems are not appropriate to address the complexity of entire processes that take place in whole organisms. Typically, cell-based screens yield numerous candidates suspected to act in the biological process under scrutiny and have to be followed by in-depth studies to determine the precise role of the identified hits. A high-throughput procedure for the introduction of RAFL cDNA fragments into binary vectors was developed for the production of gain-of-function *Arabidopsis* T87 suspension cell lines (86). Since RAFL cDNA fragments are cloned into a cDNA vector carrying rare oriented restriction sites (65, 87), a new entry vector, pRAFLENTR, was designed enabling transfer into the same restriction sites and between the Gateway *att*L1 and *att*L2 sequences. Once pRAFLENTR derivatives were generated by restriction/ligation cloning, the RAFL cDNA sequence could easily be transferred through Gateway cloning into various destination vectors (86). Note that the same strategy was applied to transfer 12,000 *Arabidopsis* ORF sequences originally captured as pUNI vectors (67) into Gateway entry clones and then into Y2H vectors for production of proteins as translational fusions either with the GAL4 transcriptional activation domain or with the DNA-binding domain (P. Braun, D. Hill, M. Vidal, and J. Ecker, personal communication).

When a genetic screen is chosen, it is important to keep in mind that the characterization of stable mutants, such as T-DNA insertional lines, only provides information on the final state resulting from the mutations affecting all cells at all times in the mutated individuals. By contrast, transgenes derived from master clones can be designed to drive the production of proteins or silencing RNAs in specific cell types or in response to exogenous inducers, so that the direct consequences of the triggered genetic perturbations can be studied in the course of development or in small time windows.

7.2. Protein Localization

A key characteristic of a protein function is its subcellular localization. To investigate this property in a systematic fashion, hundreds of GFP–ORF fusions were expressed in *Arabidopsis* cell cultures after transient transformation with a hypervirulent *Agrobacterium* strain (88). To facilitate localization studies, a series of fluorescent *Arabidopsis* organelle markers have been developed, highlighting the endoplasmic reticulum, the Golgi apparatus, the tonoplast, peroxisomes, mitochondria, plastids, and the plasma membrane (89). All markers were generated with four different fluorescent proteins to allow flexible combinations in co-localization experiments.

7.3. Protein–Protein Interactions

Plant PPIs have been studied with various techniques, each with specific advantages and limitations (90, 91). Of all, the Y2H system is by far the easiest to scale up. Although Y2H has not been utilized yet to analyze plant interactomes at the global scale, small interaction maps focusing on a limited set of potential interactors have already been assembled to tackle specific biological questions. For example, a comprehensive study investigated the interactions involving over 100 *Arabidopsis* MADS box TFs based on a matrix-mating approach. These TFs form dimers with different regulatory properties and the systematic pairwise PPI screen was helpful to identify factors involved in the same developmental programs because of their similar interaction profiles (92).

But membrane protein associations are not accurately reported in classical Y2H screens in which the reconstituted TF must reach the nucleus. The split-ubiquitin system was specifically developed to circumvent this pitfall (93). In this alternative yeast heterologous configuration, PPI occurs at the cytosolic side of yeast membranes. Upon protein interaction, two ubiquitin fragments are brought together, forming a functional ubiquitin molecule and triggering the action of endogenous ubiquitin-specific proteases. Cleavage of the reconstituted ubiquitin from the fused membrane proteins releases a TF that activates transcription of marker genes. This system was improved for high-throughput application by using a mating approach to bring bait and prey together in one yeast cell. It has been applied to characterize *Arabidopsis* potassium channels (93) and is now implemented for the systematic study of associations between *Arabidopsis* membrane proteins (www.associomics.org).

Recently, a promising split-protein technology has been developed to identify PPIs directly in plant cells: the firefly luciferase (LUC) complementation imaging (LCI) assay (94). The assembly of the non-functional LUC N- and C-terminal fragments is detected by a luminometer or by a low-light imaging device after addition of the substrate luciferin. LCI is particularly attractive for plant studies because the luminescence is measured in the dark and is not affected by autofluorescence. LCI constructs were tested in protoplasts and intact leaves with protein pairs known to interact with different affinities in plant cells (94).

Several methods exist to localize microscopically PPIs in living cells, although their use in high-throughput mode still needs to be demonstrated. With Förster or fluorescence resonance energy transfer (FRET), two candidate proteins are labeled with compatible donor and acceptor chromophores (fluorophores) (95). When the fused donor and acceptor fluorophores are brought together via the association of their carrier proteins, intermolecular FRET results predominantly in emission from the acceptor chromophore. The FRET pair most commonly used in biological assays consists of the cyan fluorescent protein (CFP) and yellow fluorescent protein (YFP). A limitation of FRET is the need to excite the donor fluorophore by light, which can lead to photobleaching, autofluorescence, and direct excitation of the acceptor fluorophore. Furthermore, some tissues can be damaged by the excitation light or might be directly photoresponsive, as is the case for many plant tissues. The bioluminescence resonance energy transfer (BRET) technique avoids these drawbacks. It is based on a bioluminescent LUC that produces blue light in the presence of a substrate, excitating the YFP when the two hybrid proteins interact and resulting in an easily detected yellow shift in the luminescence spectrum (96). A series of recombinational cloning vectors were generated to accelerate the production of proteins tagged with LUC or YFP in plants (97).

Bimolecular fluorescence complementation (BiFC) is another powerful technique to determine the subcellular localization of interacting partners in vivo and in real time (98–100). In this configuration, a signal is detectable only when the two fragments of a split fluorescent protein are brought together by association of the fused partners. Complementary sets of expression vectors were designed for BiFC analyses in transiently or stably transformed plant cells (101). Similarly, *Agrobacterium* multigene expression binary vectors have been constructed carrying the BiFC expression cassettes for co-production of the two protein fusions with either the N- or C-terminal fragment of YFP, together with an additional fluorescent protein that serves as an internal transformation control or as a marker for specific subcellular compartments (102). The BiFC technique is not restricted to the use of YFP fragments (103). For example, multicolor BiFC enables the simultaneous visualization of multiple protein interactions in the same cell and the comparison of the efficiencies of complex formation with alternative interaction partners (99, 104).

Finally, TAP and MS protocols, originally developed for the characterization of yeast protein complexes, have been adapted to *Arabidopsis* cell suspension cultures (64). This platform necessitated the construction of Gateway vectors for the expression of TAP-tagged proteins, a streamlined procedure for the fast generation (8 weeks) of transgenic suspension cultures, and TAP modified for plant cells. Although the throughput for the study of *Arabidopsis* complexes cannot rival yeast-based procedures, already more than 150 baits have been processed so far with multiple biological replicates (G. De Jaeger, personal communication).

7.4. Protein Arrays

Novel techniques based on ORFeome resources have been successfully applied to enhance the scale at which biochemical assays can be conducted. Enzymatic reactions have been miniaturized and carried out on microarrays printed either with proteins purified from microorganisms and in vitro expression cocktails (105), or synthesized directly on the chip (106). Protein microarrays can also be used to detect interactions between proteins, with nucleic acids, lipids, and other compounds (107). They have already been applied to identify potential targets of protein kinases among *Arabidopsis* proteins purified from *E. coli* (108, 109), to characterize specificity and cross-reactivity of monoclonal antibodies or polyclonal sera (110), and to investigate the calmodulin/calmodulin-like interactome with an *Arabidopsis* protein microarray containing 1,133 proteins (111).

8. Conclusion

The creation and quality control of genome-scale sequence repertoires is an expensive and laborious task. Nevertheless, a lot can be gained in building reference clone collections based on versatile configurations suitable for as many downstream applications as possible. Such resources can be exploited by scientists interested in the functional characterization of only a few genes. They are also the necessary starting point for systematic approaches focusing on the analysis of large gene or protein sets. Although sometimes lagging in comparison to achievements in other eukaryotic model species, large-scale clone-based plant functional genomics projects are now under way, including binary PPIs mapping, protein complex association mapping, and screens based on protein arrays. Information resulting from these projects, together with data sets originating from structural genomics (genome sequence, epigenome modifications) and profiling (mRNA, siRNA, miRNA, proteins, metabolites) surveys, will provide the necessary foundation for quantitative modeling and systems biology.

Acknowledgments

This work was supported by the 6th European Integrated Projects AGRON-OMICS (grant no. LSHG-CT-2006-037704). We thank Martine De Cock for help in preparing the manuscript.

References

1. Hieter, P. and Boguski, M. (1997) Functional genomics: it's all how you read it. *Science.* **278**, 601–602.
2. Fleischmann, R.D., Adams, M.D., White, O., et al. (1995) Whole-genome random sequencing and assembly of *Haemophilus influenzae* Rd. *Science.* **269**, 496–512.
3. Goffeau, A., Barrell, B.G., Bussey, H., et al. (1996) Life with 6,000 genes. *Science.* **274**, 546, 563–567.
4. C. elegans Sequencing Consortium (1998) Genome sequence of the nematode *C. elegans*: a platform for investigating biology. *Science.* **282**, 2012–2018.
5. Adams, M.D., Celniker, S.E., Holt, R.A., et al. (2000) The genome sequence of *Drosophila melanogaster*. *Science.* **287**, 2185–2195.
6. Venter, J.C., Adams, M.D., Myers, E.W., et al. (2001) The sequence of the human genome. *Science.* **291**, 1304–1351.
7. The Arabidopsis Genome Initiative (2000) Analysis of the genome sequence of the flowering plant *Arabidopsis thaliana*. *Nature.* **408**, 796–815.
8. International Rice Genome Sequencing Project (2005) The map-based sequence of the rice genome. *Nature.* **436**, 793–800.
9. Tuskan, G.A., DiFazio, S., Jansson, S., et al. (2006) The genome of black cottonwood, *Populus trichocarpa* (Torr. & Gray). *Science.* **313**, 1596–1604.
10. Jaillon, O., Aury, J.M., Noel, B., et al. (2007) The grapevine genome sequence suggests ancestral hexaploidization in major angiosperm phyla. *Nature.* **449**, 463–467.
11. Smith, V., Botstein, D., and Brown, P.O. (1995) Genetic footprinting: a genomic strategy for determining a gene's function given its sequence. *Proc. Natl. Acad. Sci. USA.* **92**, 6479–6483.
12. Smith, V., Chou, K.N., Lashkari, D., Botstein, D., and Brown, P.O. (1996) Functional analysis of the genes of yeast chromosome V by genetic footprinting. *Science.* **274**, 2069–2074.
13. Giaever, G., Chu, A.M., Ni, L., et al. (2002) Functional profiling of the *Saccharomyces cerevisiae* genome. *Nature.* **418**, 387–391.
14. Yuan, D.S., Pan, X., Ooi, S.L., et al. (2005) Improved microarray methods for profiling the yeast knockout strain collection. *Nucleic Acids Res.* **33**, e103.
15. Tong, A.H., Evangelista, M., Parsons, A.B., et al. (2001) Systematic genetic analysis with ordered arrays of yeast deletion mutants. *Science.* **294**, 2364–2368.
16. Tong, A.H., Lesage, G., Bader, G.D., et al. (2004) Global mapping of the yeast genetic interaction network. *Science.* **303**, 808–813.
17. Ossowski, S., Schwab, R., and Weigel, D. (2008) Gene silencing in plants using artificial microRNAs and other small RNAs. *Plant J.* **53**, 674–690.
18. Perrimon, N. and Mathey-Prevot, B. (2007) Applications of high-throughput RNA interference screens to problems in cell and developmental biology. *Genetics.* **175**, 7–16.
19. Scherr, M. and Eder, M. (2007) Gene silencing by small regulatory RNAs in mammalian cells. *Cell Cycle.* **6**, 444–449.
20. Boutros, M., Kiger, A.A., Armknecht, S., et al. (2004) Genome-wide RNAi analysis of growth and viability in *Drosophila* cells. *Science.* **303**, 832–835.
21. Ramadan, N., Flockhart, I., Booker, M., Perrimon, N., and Mathey-Prevot, B. (2007) Design and implementation of high-throughput RNAi screens in cultured *Drosophila* cells. *Nat. Protoc.* **2**, 2245–2264.
22. Berns, K., Hijmans, E.M., Mullenders, J., et al. (2004) A large-scale RNAi screen in human cells identifies new components of the p53 pathway. *Nature.* **428**, 431–437.
23. Du, G., Yonekubo, J., Zeng, Y., Osisami, M., and Frohman, M.A. (2006) Design of expression vectors for RNA interference based on miRNAs and RNA splicing. *FEBS J.* **273**, 5421–5427.
24. Paddison, P.J., Silvam, J.M., Conklin, D.S., et al. (2004) A resource for large-scale RNA-interference-based screens in mammals. *Nature.* **428**, 427–431.
25. Fields, S. and Song, O. (1989) A novel genetic system to detect protein–protein interactions. *Nature.* **340**, 245–246.
26. Ito, T., Chiba, T., Ozawa, R., Yoshida, M., Hattori, M., and Sakaki, Y. (2001) A comprehensive two-hybrid analysis to explore the yeast protein interactome. *Proc. Natl. Acad. Sci. USA.* **98**, 4569–4574.
27. Uetz, P., Giot, L., Cagney, G., et al. (2000) A comprehensive analysis of protein–protein interactions in *Saccharomyces cerevisiae*. *Nature.* **403**, 623–627.

27a. Yu, H., Braun, P., Yildirim, M.A., et al. (2008) High-quality binary protein interaction map of the yeast interactome network. *Science* **322**, 104–110.

28. Giot, L., Bader, J.S., Brouwer, C., et al. (2003) A protein interaction map of *Drosophila melanogaster*. *Science.* **302**, 1727–1736.

29. Li, S., Armstrong, C.M., Bertin, N., et al. (2004) A map of the interactome network of the metazoan *C. elegans*. *Science.* **303**, 540–543.

30. Gandhi, T.K., Zhong, J., Mathivanan, S., et al. (2006) Analysis of the human protein interactome and comparison with yeast, worm and fly interaction datasets. *Nat. Genet.* **38**, 285–293.

31. Rual, J.F., Venkatesan, K., Hao, T., et al. (2005) Towards a proteome-scale map of the human protein–protein interaction network. *Nature.* **437**, 1173–1178.

32. Stelzl, U., Worm, U., Lalowski, M., et al. (2005) A human protein–protein interaction network: a resource for annotating the proteome. *Cell.* **122**, 957–968.

33. Rossi, F., Charlton, C.A., and Blau, H.M. (1997) Monitoring protein–protein interactions in intact eukaryotic cells by β-galactosidase complementation. *Proc. Natl. Acad. Sci. USA.* **94**, 8405–8410.

34. Olson, K.R. and Eglen, R.M. (2007) β galactosidase complementation: a cell-based luminescent assay platform for drug discovery. *Assay Drug Dev. Technol.* **5**, 137–144.

35. Galarneau, A., Primeau, M., Trudeau, L.E., and Michnick, S.W. (2002) β-lactamase protein fragment complementation assays as *in vivo* and *in vitro* sensors of protein protein interactions. *Nat. Biotechnol.* **20**, 619–622.

36. Cabantous, S., Terwilliger, T.C., and Waldo, G.S. (2005) Protein tagging and detection with engineered self-assembling fragments of green fluorescent protein. *Nat. Biotechnol.* **23**, 102–107.

37. Remy, I. and Michnick, S.W. (1999) Clonal selection and *in vivo* quantitation of protein interactions with protein-fragment complementation assays. *Proc. Natl. Acad. Sci. USA.* **96**, 5394–5399.

38. Remy, I., Campbell-Valois, F.X., and Michnick, S.W. (2007) Detection of protein–protein interactions using a simple survival protein-fragment complementation assay based on the enzyme dihydrofolate reductase. *Nat. Protoc.* **2**, 2120–2125.

39. Fetchko, M. and Stagljar, I. (2004) Application of the split-ubiquitin membrane yeast two-hybrid system to investigate membrane protein interactions. *Methods.* **32**, 349–362.

40. Thaminy, S., Miller, J., and Stagljar, I. (2004) The split-ubiquitin membrane-based yeast two-hybrid system. *Methods Mol. Biol.* **261**, 297–312.

41. Gavin, A.C., Bosche, M., and Krause, R., et al. (2002) Functional organization of the yeast proteome by systematic analysis of protein complexes. *Nature.* **415**, 141–147.

42. Gavin, A.C., Aloy, P., Grandi, P., et al. (2006) Proteome survey reveals modularity of the yeast cell machinery. *Nature.* **440**, 631–636.

43. Yashiroda, Y., Matsuyama, A., and Yoshida, M. (2008) New insights into chemical biology from ORFeome libraries. *Curr. Opin. Chem. Biol.* **12**, 55–59.

44. Aslanidis, C. and de Jong, P.J. (1990) Ligation-independent cloning of PCR products (LIC-PCR). *Nucleic Acids Res.* **18**, 6069–6074.

45. Dong, Y., Burch-Smith, T.M., Liu, Y., Mamillapalli, P., and Dinesh-Kumar, S.P. (2007) A ligation-independent cloning tobacco rattle virus vector for high-throughput virus-induced gene silencing identifies roles for NbMADS4-1 and -2 in floral development. *Plant Physiol.* **145**, 1161–1170.

46. Nour-Eldin, H.H., Hansen, B.G., Norholm, M.H., Jensen, J.K., and Halkier, B.A. (2006) Advancing uracil-excision based cloning towards an ideal technique for cloning PCR fragments. *Nucleic Acids Res.* **34**, e122.

47. Geu-Flores, F., Nour-Eldin, H.H., Nielsen, M.T., and Halkier, B.A. (2007) USER fusion: a rapid and efficient method for simultaneous fusion and cloning of multiple PCR products. *Nucleic Acids Res.* **35**, e55.

48. Bitinaite, J., Rubino, M., Varma, K.H., Schildkraut, I., Vaisvila, R., and Vaiskunaite, R. (2007) USER friendly DNA engineering and cloning method by uracil excision. *Nucleic Acids Res.* **35**, 1992–2002.

49. Hamann, T. and Møller, B.L. (2007) Improved cloning and expression of cytochrome P450s and cytochrome P450 reductase in yeast. *Protein Expr. Purif.* **56**, 121–127.

50. Nagano, Y., Takao, S., Kudo, T., Iizasa, E., and Anai, T. (2007) Yeast-based recombineering of DNA fragments into plant transformation vectors by one-step transformation. *Plant Cell Rep.* **26**, 2111–2117.

51. Liu, Q.H., Li, M.Z., Leibham, D., Cortez, D., and Elledge, S.J. (1998) The univector plasmid-fusion system, a method for rapid construction of recombinant DNA without restriction enzymes. *Curr. Biol.* **8**, 1300–1309.
52. Benoit, R.M., Wilhelm, R.N., Scherer-Becker, D., and Ostermeier, C. (2006) An improved method for fast, robust, and seamless integration of DNA fragments into multiple plasmids. *Protein Expr. Purif.* **45**, 66–71.
53. Hartley, J.L., Temple, G.F., and Brasch, M.A. (2000) DNA cloning using *in vitro* site-specific recombination. *Genome Res.* **10**, 1788–1795.
54. Cheo, D.L., Titus, S.A., Byrd, D.R., Hartley, J.L., Temple, G.F., and Brasch, M.A. (2004) Concerted assembly and cloning of multiple DNA segments using *in vitro* site-specific recombination: functional analysis of multi-segment expression clones. *Genome Res.* **14**, 2111–2120.
55. Sasaki, Y., Sone, T., Yoshida, S., et al. (2004) Evidence for high specificity and efficiency of multiple recombination signals in mixed DNA cloning by the Multisite Gateway system. *J. Biotechnol.* **107**, 233–243.
56. Curtis, M.D. and Grossniklaus, U. (2003) A Gateway cloning vector set for high-throughput functional analysis of genes *in planta*. *Plant Physiol.* **133**, 462–469.
57. Karimi, M., Inzé, D., and Depicker, A. (2002) GATEWAY^{TM} vectors for *Agrobacterium*-mediated plant transformation. *Trends Plant Sci.* 7, 193–195.
58. Wesley, S.V., Helliwell, C.A., Smith, N.A., et al. (2001) Construct design for efficient, effective and high-throughput gene silencing in plants. *Plant J.* **27**, 581–590.
59. Earley, K.W., Haag, J.R., Pontes, O., et al. (2006) Gateway-compatible vectors for plant functional genomics and proteomics. *Plant J.* **45**, 616–629.
60. Hilson, P. (2006) Cloned sequence repertoires for small- and large-scale biology. *Trends Plant Sci.* **11**, 133–141.
61. Karimi, M., Bleys, A., Vanderhaeghen, R., and Hilson, P. (2007) Building blocks for plant gene assembly. *Plant Physiol.* **145**, 1183–1191.
62. Karimi, M., De Meyer, B., and Hilson, P. (2005) Modular cloning in plant cells. *Trends Plant Sci.* **10**, 103–105.
63. Wakasa, Y., Yasuda, H., and Takaiwa, F. (2006) High accumulation of bioactive peptide in transgenic rice seeds by expression of introduced multiple genes. *Plant Biotechnol. J.* **4**, 499–510.
64. Van Leene, J., Stals, H., Eeckhout, D., et al. (2007) A tandem affinity purification-based technology platform to study the cell cycle interactome in *Arabidopsis thaliana*. *Mol. Cell Proteomics.* **6**, 1226–1138.
65. Seki, M., Narusaka, M., Kamiya, A., et al. (2002) Functional annotation of a full-length *Arabidopsis* cDNA collection. *Science.* **296**, 141–145.
66. Kikuchi, S., Satoh, K., Nagata, T., et al. (2003) Collection, mapping, and annotation of over 28,000 cDNA clones from *japonica* rice. *Science.* **301**, 376–379.
67. Yamada, K., Lim, J., Dale, J.M., et al. (2003) Empirical analysis of transcriptional activity in the *Arabidopsis* genome. *Science.* **302**, 842–846.
68. Nanjo, T., Futamura, N., Nishiguchi, M., Igasaki, T., Shinozaki, K., and Shinohara, K. (2004) Characterization of full-length enriched expressed sequence tags of stress-treated poplar leaves. *Plant Cell Physiol.* **45**, 1738–1748.
69. Thao, S., Zhao, Q., Kimball, T., et al. (2004) Results from high-throughput DNA cloning of *Arabidopsis thaliana* target genes using site-specific recombination. *J. Struct. Funct. Genomics.* **5**, 267–276.
70. Underwood, B.A. Vanderhaeghen, R., Whitford, R., Town, C.D., and Hilson, P. (2006) Simultaneous high-throughput recombinational cloning of open reading frames in closed and open configurations. *Plant Biotechnol. J.* **4**, 317–324.
71. Benhamed, M., Martin-Magniette, M.L., Taconnat, L., et al. Genome scale Arabidopsis promoter array identifies targets of the histone acetyltransferase GCN5. Submitted.
72. Deplancke, B., Mukhopadhyay, A., Ao, W., et al. (2006) A gene-centered *C. elegans* protein-DNA interaction network. *Cell.* **125**, 1193–1205.
73. De Sutter, V., Vanderhaeghen, R., Tilleman, S., et al. (2005) Exploration of jasmonate signalling via automated and standardized transient expression assays in tobacco cells. *Plant J.* **44**, 1065–1076.
74. Berger, B., Stracke, R., Yatusevich, R., Weisshaar, B., Flugge, U.I., and Gigolashvili, T. (2007) A simplified method for the analysis of transcription factor-promoter interactions that allows high-throughput data generation. *Plant J.* **50**, 911–916.
75. Paz-Ares, J. and the REGIA Consortium (2002) REGIA, an EU project on functional

genomics of transcription factors from *Arabidopsis thaliana*. *Comp. Funct. Genom.* **3**, 102–108.

76. Gong, W., Shen, Y.P., Ma, L.G., et al. (2004) Genome-wide ORFeome cloning and analysis of *Arabidopsis* transcription factor genes. *Plant Physiol.* **135**, 773–782.
77. Smith, N.A. Singh, S.P., Wang, M.B., Stoutjesdijk, P.A., Green, A.G., and Waterhouse, P.M. (2000) Total silencing by intron-spliced hairpin RNAs. *Nature.* **407**, 319–320.
78. Helliwell, C. and Waterhouse, P. (2003) Constructs and methods for high-throughput gene silencing in plants. *Methods.* **30**, 289–295.
79. Hilson, P., Allemeersch, J., Altmann, T., et al. (2004) Versatile gene-specific sequence tags for Arabidopsis functional genomics: transcript profiling and reverse genetics applications. *Genome Res.* **14**, 2176–2189.
80. Alvarez, J.P., Pekker, I., Goldshmidt, A., Blum, E., Amsellem, Z., and Eshed, Y. (2006) Endogenous and synthetic microRNAs stimulate simultaneous, efficient, and localized regulation of multiple targets in diverse species. *Plant Cell.* **18**, 1134–1151.
81. Schwab, R., Ossowski, S., Riester, M., Warthmann, N., and Weigel, D. (2006) Highly specific gene silencing by artificial microRNAs in *Arabidopsis*. *Plant Cell.* **18**, 1121–1133.
82. Robertson, D. (2004) VIGS vectors for gene silencing: many targets, many tools. *Annu. Rev. Plant Biol.* **55**, 495–519.
83. Wielopolska, A., Townley, H., Moore, I., Waterhouse, P., and Helliwell, C. (2005) A high-throughput inducible RNAi vector for plants. *Plant Biotechnol. J.* **3**, 583–590.
84. Ichikawa, T., Nakazawa, M., Kawashima, M., et al. (2006) The FOX hunting system: an alternative gain-of-function gene hunting technique. *Plant J.* **48**, 974–985.
85. Weiste, C., Iven, T., Fischer, U., Onate-Sanchez, L., and Droge-Laser, W. (2007) *In planta* ORFeome analysis by large-scale over-expression of GATEWAY®-compatible cDNA clones: screening of ERF transcription factors involved in abiotic stress defense. *Plant J.* **52**, 382–390.
86. Ogawa, Y., Dansako, T., Yano, K., et al. (2008) Efficient and high-throughput vector construction and *Agrobacterium*-mediated transformation of *Arabidopsis thaliana* suspension-cultured cells for functional genomics. *Plant Cell Physiol.* **49**, 242–250.
87. Seki, M., Carninci, P., Nishiyama, Y., Hayashizaki, Y., and Shinozaki, K. (1998) High-efficiency cloning of *Arabidopsis* full-length cDNA by biotinylated CAP trapper. *Plant J.* **15**, 707–720.
88. Koroleva, O.A., Tomlinson, M.L., Leader, D., Shaw, P., and Doonan, J.H. (2005) High-throughput protein localization in Arabidopsis using *Agrobacterium*-mediated transient expression of GFP-ORF fusions. *Plant J.* **41**, 162–174.
89. Nelson, B.K., Cai, X., and Nebenfuhr, A. (2007) A multicolored set of *in vivo* organelle markers for co-localization studies in Arabidopsis and other plants. *Plant J.* **51**, 1126–1136.
90. Lalonde, S., Ehrhardt, D.W., Loque, D., Chen, J., Rhee, S.Y., and Frommer, W.B. (2008) Molecular and cellular approaches for the detection of protein–protein interactions: latest techniques and current limitations. *Plant J.* **53**, 610–635.
91. Miernyk, J.A. and Thelen, J.J. (2008) Biochemical approaches for discovering protein–protein interactions. *Plant J.* **53**, 597–609.
92. de Folter, S., Immink, R.G., Kieffer, M., et al. (2005) Comprehensive interaction map of the Arabidopsis MADS box transcription factors. *Plant Cell.* **17**, 1424–1433.
93. Obrdlik, P., El-Bakkoury, M., Hamacher, T., et al. (2004) K+ channel interactions detected by a genetic system optimized for systematic studies of membrane protein interactions. *Proc. Natl. Acad. Sci. USA.* **101**, 12242–12247.
94. Chen, H., Zou, Y., Shang, Y., et al. (2008) Firefly luciferase complementation imaging assay for protein–protein interactions in plants. *Plant Physiol.* **146**, 368–376.
95. Hink, M.A., Bisselin, T., and Visser, A.J. (2002) Imaging protein–protein interactions in living cells. *Plant Mol. Biol.* **50**, 871–883.
96. Subramanian, C., Xu, Y., Johnson, C.H., and von Arnim, A.G. (2004) *In vivo* detection of protein–protein interaction in plant cells using BRET. *Methods Mol. Biol.* **284**, 271–286.
97. Subramanian, C., Woo, J., Cai, X., et al. (2006) A suite of tools and application notes for *in vivo* protein interaction assays using bioluminescence resonance energy transfer (BRET). *Plant J.* **48**, 138–152.

98. Hu, C.D., Chinenov, Y., and Kerppola, T.K. (2002) Visualization of interactions among bZIP and Rel family proteins in living cells using bimolecular fluorescence complementation. *Mol. Cell.* **9**, 789–798.
99. Bhat, R.A., Lahaye, T., and Panstruga, R. (2006) The visible touch: *in planta* visualization of protein–protein interactions by fluorophore-based methods. *Plant Methods.* **2**, 12.
100. Hu, C.D., Grinberg, A.V., and Kerppola, T.K. (2006) Visualization of protein interactions in living cells using bimolecular fluorescence complementation (BiFC) analysis. *Curr. Protoc. Cell Biol.* Chapter 21: Unit 21.3.
101. Walter, M., Chaban, C., Schutze, K., et al. (2004) Visualization of protein interactions in living plant cells using bimolecular fluorescence complementation. *Plant J.* **40**, 428–438.
102. Citovsky, V., Lee, L.Y., Vyas, S., et al. (2006) Subcellular localization of interacting proteins by bimolecular fluorescence complementation *in planta*. *J. Mol. Biol.* **362**, 1120–1131.
103. Beauchemin, C., Boutet, N., and Laliberte, J.F. (2007) Visualization of the interaction between the precursors of VPg, the viral protein linked to the genome of *Turnip Mosaic Virus*, and the translation eukaryotic initiation factor iso 4E in planta. *J. Virol.* **81**, 775–782.
104. Guo, H.-S., Fei, J.-F., Xie, Q., and Chua, N.-H. (2003) A chemical-regulated inducible RNAi system in plants. *Plant J.* **34**, 383–392.
105. Zhu, H., Bilgin, M., Bangham, R., et al. (2001) Global analysis of protein activities using proteome chips. *Science.* **293**, 2101–2105.
106. Ramachandran, N., Hainsworth, E., Bhullar, B., et al. (2004) Self-assembling protein microarrays. *Science.* **305**, 86–90.
107. LaBaer, J. and Ramachandran, N. (2005) Protein microarrays as tools for functional proteomics. *Curr. Opin. Chem. Biol.* **9**, 14–19.
108. Feilner, T., Hultschig, C., Lee, J., et al. (2005) High-throughput identification of potential *Arabidopsis* MAP kinases substrates. *Mol. Cell Proteomics.* **4**, 1558–1168.
109. Feilner, T. and Kersten, B. (2007) Phosphorylation studies using plant protein microarrays. *Methods Mol. Biol.* **355**, 379–390.
110. Kersten, B. and Feilner, T. (2007) Generation of plant protein microarrays and investigation of antigen–antibody interactions. *Methods Mol. Biol.* **355**, 365–378.
111. Popescu, S.C., Popescu, G.V., Bachan, S., et al. (2007) Differential binding of calmodulin-related proteins to their targets revealed through high-density *Arabidopsis* protein microarrays. *Proc. Natl. Acad. Sci. USA.* **104**, 4730–4755.
112. Himmelbach, A., Zierold, U., Hensel, G., et al. (2007) A set of modular binary vectors for transformation of cereals. *Plant Physiol.* **145**, 1192–1200.
113. Joubès, J., De Schutter, K., Verkest, A, Inzé, D., and De Veylder, L. (2004) Conditional, recombinase-mediated expression of genes in plant cell cultures. *Plant J.* **37**, 889–896.
114. Brand, L., Horler, M., Nuesch, E., et al. (2006) A versatile and reliable two-component system for tissue-specific gene induction in Arabidopsis. *Plant Physiol.* **141**, 1194–1204.
115. Brown, A.P., Affleck, V., Fawcett, T., and Slabas, A.R. (2006) Tandem affinity purification tagging of fatty acid biosynthetic enzymes in *Synechocystis* sp PCC6803 and *Arabidopsis thaliana*. *J. Exp. Bot.* **57**, 1563–1571.
116. Rohila, J.S., Chen, M., Chen, S., et al. (2006) Protein–protein interactions of tandem affinity purification-tagged protein kinases in rice. *Plant J.* **46**, 1–13.
117. Liu, Y.L., Schiff, M., and Dinesh-Kumar, S.P. (2002) Virus-induced gene silencing in tomato. *Plant J.* **31**, 777–786.
118. Marjanac, G., De Paepe, A., Peck, I., Jacobs, A., De Buck, S., and Depicker, A. (2008) Evaluation of CRE-mediated excision approaches in *Arabidopsis thaliana*. *Transgenic Res.* **17**, 239–250.
119. Tzfira, T., Tian, G.W., Lacroix, B., et al. (2005) pSAT vectors: a modular series of plasmids for autofluorescent protein tagging and expression of multiple genes in plants. *Plant Mol. Biol.* **57**, 503–516.
120. Ehlert, A., Weltmeier, F., Wang, X., et al. (2006) Two-hybrid protein–protein interaction analysis in Arabidopsis protoplasts: establishment of a heterodimerization map of group C and group S bZIP transcription factors. *Plant J.* **46**, 890–900.
121. Xiao, Y.L., Malik, M., Whitelaw, C.A., and Town, C.D. (2002) Cloning and sequencing of cDNAs for hypothetical genes from chromosome 2 of Arabidopsis. *Plant Physiol.* **130**, 2118–2128.
122. Xiao, Y.L., Smith, S.R., Ishmael, N., et al. (2005) Analysis of the cDNAs of hypothetical genes on Arabidopsis chromosome 2

reveals numerous transcript variants. *Plant Physiol.* **139**, 1323–1337.

123. Stone, S.L., Hauksdottir, H., Troy, A., Herschleb, J., Kraft, E., and Callis, J. (2005) Functional analysis of the RING-type ubiquitin ligase family of Arabidopsis. *Plant Physiol.* **137**, 13–30.
124. Castelli, V., Aury, J.-M., Jaillon, O., et al. (2004) Whole genome sequence comparisons and "full-length" cDNA sequences: a combined approach to evaluate and improve *Arabidopsis* genome annotation. *Genome Res.* **14**, 406–413.
125. Thareau, V., Déhais, P., Serizet, C., Hilson, P., Rouzé, P., and Aubourg, S. (2003) Automatic design of gene-specific sequence tags for genome-wide functional studies. *Bioinformatics.* **19**, 2191–2198.
126. Sclep, G., Allemeersch, J., Liechti, R., et al. (2007) CATMA, a comprehensive genome-scale resource for silencing and transcript profiling of Arabidopsis genes. *BMC Bioinformatics.* **8**, 400.
127. Kerschen, A., Napoli, C.A., Jorgensen, R.A., and Muller, A.E. (2004) Effectiveness of RNA interference in transgenic plants. *FEBS Lett.* **566**, 223–228.
128. McGinnis, K., Chandler, V., Cone, K., et al. (2005) Transgene-induced RNA interference as a tool for plant functional genomics. *Methods Enzymol.* **392**, 1–24.

Section II

Networks, Pathways, Statistical Issues, and Novel Computational Tools

Chapter 9

Challenges and Approaches to Statistical Design and Inference in High-Dimensional Investigations

Gary L. Gadbury, Karen A. Garrett, and David B. Allison

Abstract

Advances in modern technologies have facilitated high-dimensional experiments (HDEs) that generate tremendous amounts of genomic, proteomic, and other "omic" data. HDEs involving whole-genome sequences and polymorphisms, expression levels of genes, protein abundance measurements, and combinations thereof have become a vanguard for new analytic approaches to the analysis of HDE data. Such situations demand creative approaches to the processes of statistical inference, estimation, prediction, classification, and study design. The novel and challenging biological questions asked from HDE data have resulted in many specialized analytic techniques being developed. This chapter discusses some of the unique statistical challenges facing investigators studying high-dimensional biology and describes some approaches being developed by statistical scientists. We have included some focus on the increasing interest in questions involving testing multiple propositions simultaneously, appropriate inferential indicators for the types of questions biologists are interested in, and the need for replication of results across independent studies, investigators, and settings. A key consideration inherent throughout is the challenge in providing methods that a statistician judges to be sound and a biologist finds informative.

Key words: FDR, genomics, high-dimensional, microarray, multiple testing, statistics.

1. Introduction

The present genomic era (1) has ushered in new challenges in high-dimensional study design and analysis. Draft sequences of several genomes coupled with new technologies allow study of entire genomes rather than isolated single genes. Questions from such high-dimensional investigations involve multiplicity at unprecedented scales. These questions may involve thousands of

Dmitry A. Belostotsky (ed.), *Plant Systems Biology, vol. 553*

DOI 10.1007/978-1-60327-563-7_9 Springerprotocols.com

genetic polymorphisms, gene expression levels, protein measurements, genetic sequences, or any combination of these and their interactions.

This chapter is targeted to statisticians, biologists, and those whose expertise bridges the interface. Questions that biologists want to ask from high-dimensional experiment (HDE) data require novel analytic approaches. It is important that the statistical methods applied to HDE data are aimed at drawing inferences biologists are interested in and also that these analytic methods have sound statistical foundations. There is now a relatively large and quickly growing body of statistical literature on the design of HDEs and on the analysis of resulting data from HDEs. Here we summarize some key methodological developments from statisticians as they pertain to HDEs. Included is some review of statistical foundations related to the interpretation of statistical evidence (e.g., a *P*-value), sampling variability, and aspects of a study design.

In the next section, we discuss design, analysis, and inference in the context of a single gene (i.e., a single hypothesis test). An example of a microarray experiment is used to illustrate the ideas. In **Section 3** we extend the discussion to high-dimensional studies where many hypotheses are simultaneously investigated. **Section 4** focuses more on the false discovery rate (FDR) (2) and related quantities that have garnered increased interest when analyzing high-dimensional data. **Section 5** discusses some other topics related to HDEs.

2. Statistical Inference for a Single Gene

A variable, Y, will be used to denote the information of interest in an HDE. In a microarray experiment, Y will be a measure of genetic expression after perhaps background correction, normalization, or transformation. These latter pre-processing choices are generally determined by the technology used in the experiment, potential biases induced in the measurement due to factors involved in the experiment, and characteristics regarding the statistical distribution of Y. In the discussion that follows, Y will be the genetic expression after pre-processing and, in this section, we will consider the analysis for differential expression for a single gene. Later, the issues encountered in high-dimensional settings will be discussed. For some references on pre-processing of gene expression data, see (3–6).

2.1. Discussion of Designs

Consider an experiment where there is one treatment factor with T levels. The goal is to determine if a gene is differentially expressed across the levels of the treatment. In many earlier studies

$T = 2$ and the two levels were a test treatment versus a control treatment. Travers et al. (7), for example, compared gene expression in plants experiencing ambient precipitation patterns and plants that experienced precipitation altered following a pattern predicted by models of climate change in the United States Great Plains. More generally, T can be greater than 2, such as a set of T different precipitation patterns, a set of T plant genotypes, an infection by T plant pathogens or combinations of plant pathogens, or planting in a set of T different soil types.

In a completely randomized design comparing T levels of a treatment, a total of N samples are randomly divided into groups of size $n_1, n_2, \ldots, n_T$ with each group receiving one of the levels of the treatment. In such a design the number of possible treatment assignments is

$$C = \frac{N!}{n_1!n_2!\ldots n_T!}, \qquad [1]$$

where $N = \sum_i n_i$. Observed data are represented by Y_{ij}, i.e., the expression of a gene for the jth sample in the ith treatment group where $i = 1, \ldots, T, j = 1, \ldots, n_i$, and n_i is the number of samples assigned to the ith treatment group. The one-way analysis of variance (ANOVA) model that is often used to model the data is of the form, $Y_{ij} = \mu_i + \varepsilon_{ij}$, where μ_i is the population mean response of genetic expression for samples exposed to the ith treatment level and ε_{ij} is a random error term. In a hypothesis test to determine if there is any mean differential expression due to the treatment, the ANOVA model above is compared to a reduced null model $Y_{ij} = \mu + \varepsilon_{ij}$. The null hypothesis that all means are equal, Ho: $\mu_1 = \mu_2 = \cdots = \mu_T$, is tested against an alternative that there is at least one difference between means, stated as Ho: $\mu_i \neq \mu_{i'}$ for some $i \neq i'$. Tests of linear combinations of means may also be of interest. These are of the form, Ho: $\sum_{i=1}^{T} c_i\mu_i = 0$ versus an alternative Ha: $\sum_{i=1}^{T} c_i\mu_i \neq 0$ where $c_1, \ldots, c_T$ are constants chosen by the investigator, e.g., $c_1 = 1$, $c_2 = -1$ and all other constants equal to 0 would be a test of $\mu_1 - \mu_2$.

More complex designs may be needed in contexts such as ecological studies, where the design may be influenced by conditions in the field where samples are obtained. In the Travers et al. (7) example, the emphasis of the analysis was on comparing gene expression for plants in replicate plots experiencing ambient precipitation patterns and plants in replicate plots experiencing precipitation patterns altered to follow a climate change prediction. But this study was performed in a pre-existing field experiment that had its own experimental structure. The precipitation treatments were applied in the field in a randomized complete block design, where plants within a block were likely to be somewhat more similar genetically and to have somewhat more similar interactions with other organisms. Time of day is also very important in determining levels of expression for some genes. Since collecting plant samples and preserving them for later processing is time-consuming, this model also included time of day as a predictor in a strip-plot design. This field study also included other treatment structures, such as a temperature treatment with two levels, ambient and increased, applied to subplots. Milliken et al. (8) address the issues involved in choosing among designs for pairing samples (in 2-dye experiments) collected from pre-existing split-plot experiments such as this one, where it may not be feasible to include comparisons of all treatment combinations and all dye combinations on the same microarrays. More discussion of microarray designs is given in (9).

Missing data can arise in microarray experiments and some designs are more robust to missing data than others. Field samples may be prone to missing data because of potentially more degraded tissue. Cross-species hybridization may also result in more missing data because of lower homology between species for genes that are less highly conserved. Consideration of potential sources of missing data in an HDE may aid in the choice of a design that minimizes some of the negative consequences of missing data while maintaining adequate statistical efficiency to detect effects that are of interest.

2.2. Statistical Tests

A statistical test involves a metric, say δ, that can be computed from observed data. This metric quantifies a departure from the null hypothesis and compares the size of this metric to what could have been observed by chance if the null hypothesis, *Ho*, were true. This assessment of chance is quantified by the *P*-value that is computed as a probability of observing a value of the metric as extreme (i.e., favoring the alternative *Ha*) or more extreme if *Ho* were true. Small *P*-values represent evidence in favor of *Ha*. *P*-values can be computed in two ways: under a random sampling framework or a random treatment assignment framework.

A random treatment assignment compares the observed metric with what would have been observed under different treatment assignments if *Ho* were true. Consider a two-sample completely randomized design ($T=2$) that is testing *Ho*: $\mu_1 - \mu_2 = 0$ versus a two-tailed alternative. One metric would be the usual estimate of $\mu_1 - \mu_2$, $\delta = \bar{y}_1. - \bar{y}_2$, where $\bar{y}_i. = \frac{1}{n_i}\sum_{j=1}^{n_i} y_{ij}$, $i = 1, 2$. The statistical test is the usual Fisher randomization test (10). Under *Ho*, values of Y_{ij} are permuted across the assigned treatments resulting in C (equation [1], for $T=2$) values of $\bar{y}_1. - \bar{y}_2$. The proportion of these $\bar{y}_1. - \bar{y}_2$ that are further away from zero than the observed value of $\bar{y}_1. - \bar{y}_2$ is the randomization-based *P*-value. More details of this test, and the required assumptions, are in Mehta et al. (11).

The metric δ need not be a mean difference. It could be a standardized mean difference, i.e., the usual t-statistic or a modified t-statistic as described in (12). Pepe et al. (13) proposed a metric derived from receiver operating characteristic (ROC) curves. The Wilcoxon rank-sum test is a randomization test based on the ranks of gene expression. A limitation of randomization-based P-values is their discreteness in small samples (14, 15). For example, if $N = 6$ and $n_1 = n_2 = 3$, there are only 10 possible two-tailed P-values resulting from the randomization test. This discreteness limits follow-up work that may involve ranking the most promising results in genetic expression studies and/or estimation of FDR.

Randomization tests can be extended for T treatment groups in a completely randomized design. The metric must quantify how different the group means are from an overall grand mean. One possibility is $\delta = \sum_{i=1}^{T} n_i(\bar{y}_{i\cdot} - \bar{y}_{\cdot\cdot})^2$ where $\bar{y}_{\cdot\cdot}$ is the mean of all N observations. The value of δ computed from observed data is then compared with the C possible values obtained by permuting the observed data across treatment groups. Other metrics are possible and the computation for a randomization test becomes more complex. Mielke and Berry (16) have details regarding permutation-type tests for more involved designs. When sample sizes are small, the discreteness issue that was present for two treatment groups is still present. When samples are larger, the number of possible randomizations becomes extremely large and computation of a P-value may require Monte Carlo approximation. Parametric tests can also be used to approximate a randomization-based P-value. The common parametric tests are the two-sample t-test for two treatment groups or the ANOVA-based F-test for multiple treatment groups, or two-way and higher order ANOVA designs when multiple treatments or blocking variables are used.

In a random sampling framework for comparing T levels of a treatment, the data for the expression of a particular gene are assumed to be a random sample from a larger population. For example, in an ecological study involving big bluestem plants, some samples might be drawn from a population of diseased big bluestem plants while other samples may be drawn from a population of non-diseased big bluestem plants. While big bluestem plants may be distributed through a large part of the United States, it may only be realistic to sample from a single state or even from a single prairie and assume that the sample is roughly representative of a larger target population that is of interest in the study. The resulting data are then assumed to be obtained from a statistical population model, that is, for the ith treatment group, $Y_{i1}, Y_{i2}, \ldots, Y_{in_i} \sim F_Y(y; \mu_i, \sigma_i)$ where F_Y is some population distribution that is used as a model for the response variable. If F_Y is a normal distribution or if sample sizes are large enough, then the appropriate statistical tests are the usual *F*-tests for ANOVA models and, in the case of just two treatments, the usual two-sample *t*-test.

For more complicated designs, mixed effects models for gene expression data (17) can include random effects. Blocks may be extremely important in biological studies, in the greenhouse, growth chamber, or in the field, and are often reasonably included as random effects. Many factors influencing gene expression are not yet understood but may be dealt with to some extent by blocking. Blocking may be done across space, across time, and across individual scientists working with samples. Other treatments of more direct interest may also be included as random effects. For example, if expression in multiple genotypes is compared and the genotypes are randomly selected, it would be reasonable to treat genotype as a random effect.

Regardless of the design and model used to analyze resulting data, ultimately some hypothesis will be of interest such as determining if a gene is differentially expressed across two or more treatment conditions, or testing a contrast in a more complex model. If required assumptions are met regarding the distribution of data, the appropriate test results in a "valid" *P*-value.

2.3. Discussion of P-values

Exploiting the properties of a *P*-value, as a random variable (18), has recently become popular in methods that analyze high-dimensional data (19–21), though the idea goes back further (22). A key feature of a valid *P*-value is that its distribution, when the null hypothesis is true, is uniform on the interval from 0 to 1. This leads to a convenient interpretation of a *P*-value as a measure of the making of a type I error when rejecting a null hypothesis (i.e., rejecting a true null hypothesis). What a *P*-value is and what it is not are best illustrated with simple probability statements. Let $\{Ho\}$ be the event that the null hypothesis is true and $\{\overline{Ho}\}$ the event that it is false, and suppose that a null hypothesis will be rejected if a *P*-value is less than some threshold, τ. Denote a *P*-value, as a random variable, as *P*. Then $\Pr[P \leq \tau | \{Ho\}] = \tau$, that is, the probability of rejecting a true null hypothesis is equal to the threshold at which it was rejected. This is a probability of committing a type I error. Replacing the threshold with the actual observed *P*-value from a test then allows the *P*-value to be interpreted as the chance of a type I error.

A small *P*-value is often interpreted as evidence against the null hypothesis and is, thus, often misinterpreted. Berger and Sellke (23) discuss this and show some examples where a *P*-value is equal to 0.05, but the probability that the null hypothesis is false, given the data, is closer to 0.50. However, it is this latter probability that is more intuitive to investigators (11). Stated as a probability this is $\Pr[\{Ho\}|P \leq \tau]$, the probability the null hypothesis is true given that a P-value falls below a given threshold. Interpretation in a high-dimensional setting is as follows: if null hypotheses are to be rejected when the corresponding P-values are less than or equal to τ, $\Pr[\{Ho\}|P \leq \tau]$ is an expected proportion of those "discoveries" that are false. An issue in computing this probability is that a prior probability, $\Pr[\{Ho\}]$, is needed. Computing this prior probability from P-values obtained from multiple hypothesis tests has been the focus of many methods that analyze high-dimensional data (cf., 20, 24, 25).

3. Statistical Inference in High-Dimensional Experiments

3.1. Multiple Test Statistics and Multiple P-Values

In a high-dimensional experiment there are, say K, observations per sampled unit and data from a completely randomized design comparing T levels of a treatment are of the form $Y_{ij} = (Y_{1ij}, \ldots, Y_{Kij})'$ for the jth sample in the ith treatment group. Randomization-based inference follows as discussed in **Section 2** except the entire vector of observations for the jth unit is permuted across treatment conditions, $i = 1, \ldots, T$. In the random sampling framework the sample for the ith treatment group is $Y_{i1}, \ldots, Y_{in_i} \sim F_Y(y; \mu_i, \Sigma_i)$, where μ_i is a K-dimensional vector of the mean expression for each gene represented on the arrays, and Σ_i is a $K \times K$ variance–covariance matrix. The earlier discussions for testing a single gene for differential expression across the T treatment conditions would, with this multivariate structure, now be a test on a marginal distribution for a single gene; there are K marginal distributions, one for each gene. The result from gene-specific tests is a distribution of K test statistics or P-values. The most "interesting" genes are determined by a ranking procedure or assignment of a posterior probability of being differentially expressed, i.e., using the notation from **Section 2** this would be $\Pr[\{\overline{Ho}\}|P \leq \tau]$ as defined in (20).

Figure 9.1 illustrates two distributions of P-values obtained from an experiment that evaluated the effect of two treatments, drought stress and infection by a rust fungus, in a factorial design (26). The drought treatment levels consisted of the presence or absence of drought stress. The pathogen treatment levels consisted of presence or absence of rust infection. The distribution of P-values for a drought effect shows a stronger "signal" than that for a rust effect because more P-values seem to be clustering toward zero than would be expected under a global null hypothesis, i.e., a "global null hypothesis" that there were no genes

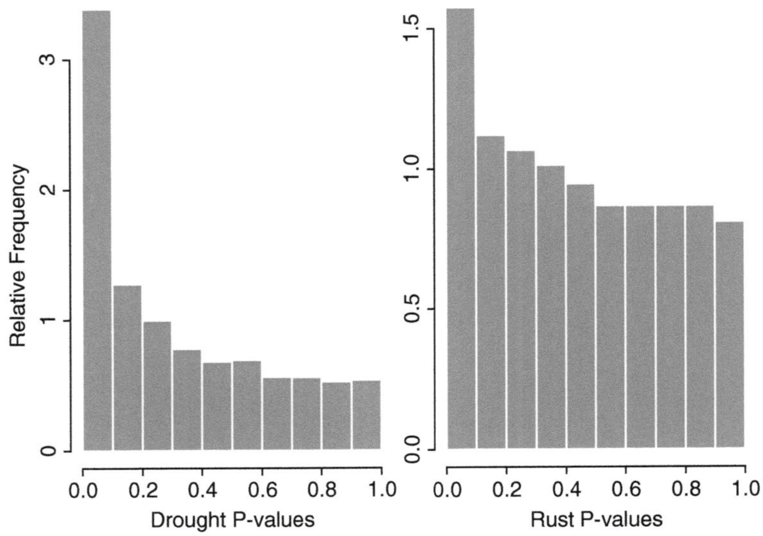

Fig. 9.1. Distribution of *P*-values from tests for differential expression due to a drought effect *(left panel)* and a rust effect *(right panel)*.

differentially expressed. If a global null hypothesis were true, one would expect the histogram of *P*-values to be relatively flat on the interval from 0 to 1. If genes were to be declared "statistically significant" if a *P*-value is less than the threshold $\tau = 0.01$, then an estimate of the "true-positive" probability $\Pr\left[\{\overline{Ho}\}|P \leq \tau\right]$ is 0.952 for a drought effect and 0.673 for a rust effect. These probabilities were called true-positive probabilities in (20), and the method reported there was used in the computations here. Subtracting this probability from 1 is related to the false discovery rate to be discussed in a later section. So of these genes that are declared significant for a drought effect, most are expected to be true discoveries, but only a little over two-thirds are expected to be true discoveries when looking at a rust effect. Lowering the threshold will increase this true-positive probability but at a cost of a smaller set of genes with *P*-values below the threshold.

3.2. Sampling Variability and Replication

Sampling variability in HDEs can arise from multiple sources (27, 28). A figure in Gadbury et al. (29, p. 81) and accompanying discussion illustrated sources of variability affecting a distribution of *P*-values. Technical variability of pixel effects on a spot was discussed by Brody et al. (30), and other design issues affecting technical variability have been considered by others, for example (31). Whether to spot one gene multiple times on a single microarray or to have repeated microarrays for a single tissue sample are aspects of assessing technical variability within and across arrays (32). Ultimately, statistical inference generally is targeted to some defined population of organisms and it is biological variability that is of primary interest and is, in fact, essential for drawing valid inferences to a larger population of organisms (33). If the cost of obtaining replicate biological samples is not large versus that of obtaining a measurement (i.e., running a microarray), then there are design advantages of obtaining biological replicates versus expending resources on repeated measurements (9). Moreover, increasing sample size (number of biological samples or replicates) can increase the true-positive probability discussed above and increase the chances of discovering true results in an HDE, i.e., a higher expected discovery rate (34).

Hereafter, replication in an HDE will refer to biological replication, i.e., distinct tissue samples that are appropriately considered replicates in the context of the experiment being performed (35). In some microarray experiments, a tissue sample will correspond to a microarray or to a dye on a microarray in the case of dye swap experiments. In randomization tests or resampling procedures such as the bootstrap (36), the biological tissue or the microarray represented by the data vector $Y_{ij} = (Y_{1ij}, \ldots, Y_{Kij})'$ is the unit of randomization or resampling. As mentioned earlier, randomization tests can produce coarse distributions of test statistics (and, hence, *P*-values), making it impossible to identify a list of the most promising candidate genes. It is tempting, therefore, to take advantage of the large number of genes and permute or resample genes themselves. However, genes are not exchangeable and variance of gene expression values is not homogeneous across genes. Methods involving mixed effects models and/or empirical Bayesian methods involving variance shrinkage have been proposed to address inferential issues associated with unequal variances across genes (17, 37, 38).

Gene-specific hypothesis tests are often carried out for each gene and variance estimates are computed for each gene. Correlation structure among measurements on a tissue sample (e.g., for co-regulation of certain sets of genes in a microarray experiment) leads to correlated *P*-values from multiple hypothesis tests, and this correlation structure cannot be estimated from observed data due to the high dimensionality. Yet this correlation can increase sampling variability leading to increased variance of estimates obtained from an HDE such as the true-positive probability defined in **Section 3.1**.

3.3. An Illustration of Correlated Tests

In an HDE, there are quantities of interest to the investigator that summarize results over thousands of tests. One quantity is the number of *P*-values below a threshold given that the global null hypothesis is true. Another is the proportion of all hypotheses tests for which the null hypothesis is true. Yet another is the true-positive probability discussed above or the analogous quantity, the false discovery rate. Many methods that estimate these quantities may perform well, on average, but some estimates that are produced can have high variance when there is correlation of gene expression values leading to correlated *P*-values (39–41).

Figure 9.2 shows the effect of correlation on the sampling variability in a distribution of 10,000 *P*-values when the global null hypothesis is true. The data that would have produced the distributions of *P*-values would correspond to a situation where there was no mean difference in expression across two treatment groups for any genes. The test statistics that produced the *P*-values were standard normal in all four plots. However, all test statistics were independent in **Fig. 9.2(A)** but were correlated in **Fig. 9.2(B)–(D)** by a correlation matrix, Σ. In **Fig. 9.2(A)**, Σ was the identity matrix meaning that all tests were independent and that resulting *P*-values were uncorrelated. The histogram of *P*-values is nearly uniform, as would be expected. Repeated sampling from this model and computing a distribution of *P*-values will result in plots resembling that in **Fig. 9.2(A)**. In the other parts of **Fig. 9.2**, Σ was block diagonal where 20 blocks of size 500 were used. Correlation between all pairs of genes within a block was set to 0.5 and correlation of genes in different blocks was 0.

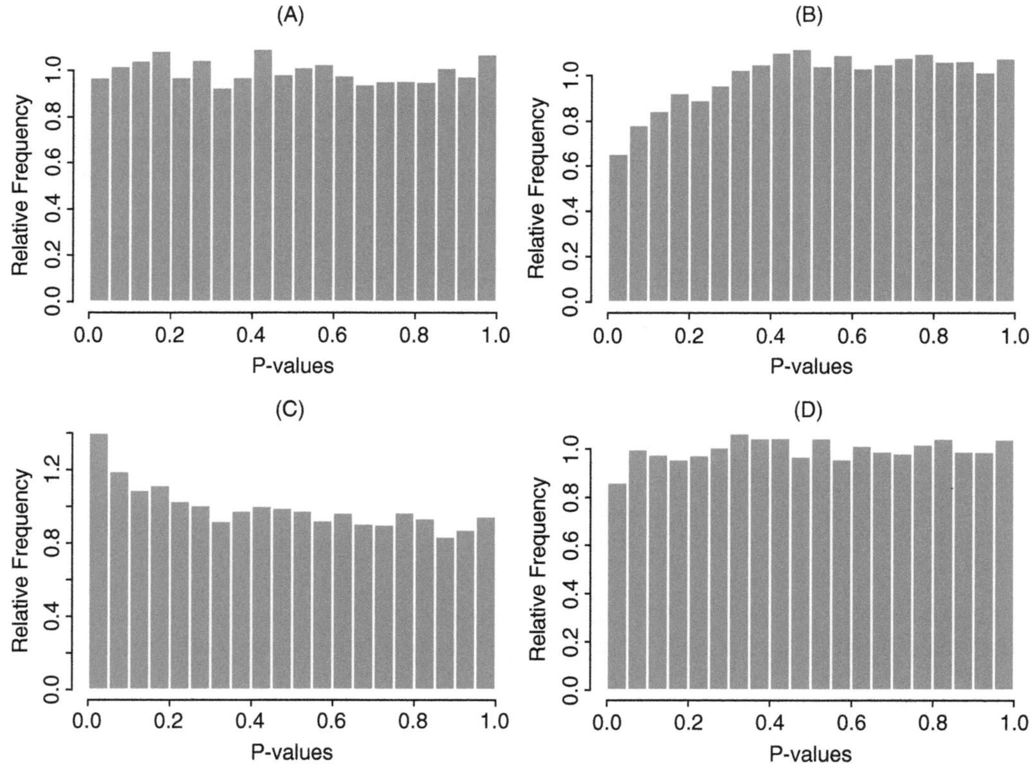

Fig. 9.2. 10,000 *P*-values under the global null hypothesis. *P*-values are uncorrelated in **(A)** but correlated in **(B)–(D)** using 20 blocks of size 500 equicorrelation matrices where the common correlation is 0.5.

Figure 9.2(B)–(D) shows three different distributions of P-values computed from three simulations of data from this model. **Figure 9.2(D)** looks similar to **Fig. 9.2(A)** where genes are independent. However, **Fig. 9.2(B)** and **(C)** shows how patterns in the distribution can arise due to sampling variability, even though the global null hypothesis is true.

A simple statistic based on a distribution of P-values is the number of P-values below a given threshold, say 0.01. Let $N(p=0.01)$ represent this number. Since there are 10,000 P-values and the global null hypothesis is true, we expect $N(p=0.01)$ to be, on average, 100 regardless of the correlation structure. In the individual samples in **Fig. 9.2**, it is equal to 90, 36, 145, and 86 in parts **(A)–(D)**, respectively. $N(p=0.01)$ is expected to be 100 in all of the plots but the standard deviation is 9.95 in part (A) and 67.48 in the other plots. A technique for deriving the standard deviation for this statistic was outlined in (22) with further details given in (42). The standard deviation of $N(p=0.01)$, as a statistic, increases by a factor of 6 due to the correlation structure. What correlation structure is reasonable in a genetic expression study may depend on the organism and application. The one illustrated here might be rather extreme. The standard deviation of $N(p=0.01)$ will decrease as the block size for correlated data decreases and/or as the value of the correlation decreases toward zero.

The performance of statistical methods for estimating quantities of interest in HDEs has typically been evaluating using simulations that include simulating data with various correlation structures (14, 20, 43). The key point here is that a weak signal in a distribution of P-values may be due to some genes differentially expressed or that the effect of correlation on sampling variability is producing the observed signal. It is unlikely that any correlation could produce a strong signal like that in **Fig. 9.1** for a drought effect.

Recent papers have appeared that deal in more detail on correlation structure among genes and the effect that it might have on conclusions from a study (40, 44). A topic discussed later and one of active research interest (33) is gene class testing where certain classes of genes, rather than individual genes, are tested for differential expression. The methodological issues that arise with these tests and their ability to overcome some of the issues associated with testing single genes are beginning to be investigated (45).

3.4. Multiple Testing in High-Dimensional Experiments

One of the topics given the most attention has been the issue of multiple testing in HDE settings. The multiple comparisons problem becomes especially acute when thousands of tests are being conducted simultaneously and one wants to guard against type I errors, i.e., rejecting a true null hypothesis. For example, as discussed in the illustration above using **Fig. 9.2** where there are 10,000 tests for which the null hypothesis is true, one would expect to find 100 "statistically significant" results at a threshold of 0.01 and 500 at the threshold of 0.05. These numbers are the number of type I errors that would be committed if declaring a test significant at the two respective thresholds. One obvious technique to control the number of type I errors is to lower the threshold at which significance is declared. Development of statistical methods to control for the number or proportion of type I errors in multiple testing situations is its own area of research with entire texts devoted to the topic, for example (46).

A common method that controls for the probability of a single type I error is the Bonferroni adjustment. Suppose that only one test were to be conducted and statistical significance is set to be at a level 0.05, so that a *P*-value below this number is significant. When there are K tests being conducted simultaneously, a single test is declared significant, using a Bonferroni adjustment, at a *P*-value below $0.05/K$. The probability of one or more type I errors among all K tests is then less than or equal to 0.05. In each of the *P*-value distributions in **Fig. 9.2** where the global null hypothesis was true, K =10,000 and in all four cases there were no *P*-values below $0.05/K$, so no type I errors would be committed using a Bonferroni adjustment. Thus this adjustment did what it was supposed to do.

Now consider the two distributions of *P*-values shown in **Fig. 9.1** where there appears to be a signal in each. There are $K = 7550$ *P*-values representing a drought effect and 7471 representing a rust effect. For the drought effect, there are only 14 *P*-values below $0.05/K$ and zero below this for the rust effect. With a Bonferroni adjustment one would find 14 statistically significant results for a drought effect and none for a rust effect. The adjustment is extremely conservative and there are very likely many true findings that are being missed, i.e., many type II errors. In fact, the method in (20) estimates that around 46% of the null hypotheses are false for a drought effect and around 16% for a rust effect. Many of the modern methods for HDE data seek a balance between controlling for a certain proportion of type I errors and detecting truly significant results out of thousands of possible tests. Many of these methods are focused on the false discovery rate (FDR) first discussed by Benjamini and Hochberg (2).

4. The False Discovery Rate and Related Quantities in High-Dimensional Experiments

As stated earlier, FDR is similar to one minus the true-positive probability discussed earlier. Much work in statistical methods development has focused on a mathematical definition of FDR and methods to either bound it or estimate it. Many of these methods work on the distribution of *P*-values from multiple tests, so herein we discuss FDR in this context. Stated in words, FDR is an expected proportion of hypothesis tests that are declared statistically significant, but that are false discoveries, i.e., the null hypothesis is actually true. **Table 9.1** shows quantities of interest in an HDE where there are a total of *K* hypothesis tests.

Table 9.1
Quantities of interest in an HDE. The total number of tests is equal to *K*. The row totals are known but column totals are not, nor are the individual values *A, B, C, D*

	Ho true	Ho false	Total
Tests that are *not* declared significant	A	B	$K-R$
Test that are declared significant	C	D	R
Total	M	$K-M$	K

4.1. Definitions of the False Discovery Rate

There are two approaches to using FDR in an HDE. One is to specify a desired FDR (or an upper bound for it) and select a threshold for statistical significance based on this desired upper limit. Another is to specify a threshold (i.e., significance level for a *P*-value) at which a hypothesis test will be declared significant, and then estimate the FDR at that threshold. We discuss the latter and show some ways that FDR can be estimated at a given threshold for significance.

In **Table 9.1**, the row totals are known. Once a threshold, τ, is set by the investigator, R is the number of *P*-values below that threshold. The number C is unknown and this is the number of false discoveries out of the total R rejected null hypotheses. The quantity C/R is the proportion of false discoveries. We can also note other quantities such as $B/(K–R)$ which is the proportion of null hypotheses that are false but that were not detected in the test (i.e., the *P*-value was above τ).

The proportion C/R is an unknown quantity from an HDE. FDR is defined with respect to this proportion as a parameter in an HDE for which estimates can be derived. Benjamini and Hochberg (2) defined *FDR* as follows:

$$FDR = E\left[C/R \;\; I_{\{R>0\}\}}\right] = E\left[C/R \middle| R>0\right]P(R>0), \quad [2]$$

where $I_{\{R>0\}}$ is an indicator function equal to 1 if $R>0$ and zero otherwise, and where $E()$ is an expectation operator representing a population average. Storey (19) defined the positive *FDR* as

$$pFDR = E\left[C/R \middle| R>0\right]. \quad [3]$$

Since $P(R>0) \geq 1-(1-\tau)^K$, and since K is usually very large, $FDR \approx pFDR$. For example, for $K=10{,}000$, $P(R>0) \geq 0.99995$ when $\tau = 0.001$ so we do not distinguish between FDR and $pFDR$ as the parameter being estimated and simply refer to it as FDR with estimates denoted by $\widehat{FDR}$. In fact there are other versions of FDR that have been defined that differ in the way the expectation is taken on the ratio C/R. Other examples are the marginal FDR, the empirical FDR, and the conditional FDR, but in many cases these different versions of FDR are numerically close with some being equivalent under certain conditions (47).

4.2. Estimating the False Discovery Rate and Related Quantities

The proportion M/K is the proportion of true null hypotheses among all K tests, a quantity that is unknown in an HDE and must be estimated. An estimate of this proportion (or an upper bound for it) is needed in order to produce an estimate of FDR, and many methods have produced estimates for this proportion (e.g., (20, 48, 49)). Let $\pi_0 = M/K$ and an estimate of this as $\hat{\pi}_0$, and define $P_R = R/K$, the proportion of rejected null hypotheses at a threshold τ, and note that P_R is a known quantity in an HDE. There are two (at least) basic techniques that are used to estimate FDR. One set of techniques produce an estimate of π_0 and then estimate FDR at a selected threshold τ using,

$$\widehat{FDR} = \frac{\tau\hat{\pi}_0}{P_R} \qquad [4]$$

These methods differ in how $\hat{\pi}_0$ is obtained with many methods focused on producing a conservative estimate. Clearly, $\hat{\pi}_0 = 1$ would be the most conservative and, if the distribution of P-values from multiple tests looks like that shown in **Fig. 9.2(A)**, then perhaps π_0 is close to 1. However, if distributions look like those in **Fig. 9.1**, then π_0 should be less than 1. Many methods that estimate π_0 use algorithms that assess how much the distribution of P-values deviates from a uniform distribution like in **Fig. 9.2(A)**.

Another set of techniques uses a mixture model framework to produce estimates of FDR. The mixture model (usually a two-component mixture) approach on a distribution of P-values uses a model of the form

$$F(p;\pi_0,\theta) = \pi_0 F_0(p) + (1-\pi_0)F_1(p), \qquad [5]$$

where F is a cumulative distribution function (CDF), p a P-value, F_0 a distribution of a P-value under the null hypothesis, F_1 a distribution of a P-value under the alternative hypothesis, π_0 is interpreted as before, and θ a (possibly vector) parameter of the distribution. Since valid P-values are assumed, F_0 is a uniform distribution. Estimating the components of the model in [5] yields estimates of FDR. The equation for FDR in a mixture model framework is

$$FDR = \frac{\pi_0 \tau}{\pi_0 \tau + (1 - \pi_0) F_1(\tau)} \quad [6]$$

Equation [6] has been defined as the positive FDR (19) but, as stated earlier, the different versions of FDR are close when K is large. Methods based on the mixture model framework differ in how the components of equation [6] are computed. Note that the only difference between equations [4] and [6] is the denominator. In [6], the denominator is the distribution function of a P-value and some have used a parametric form. Allison et al. (20) used a mixture of a uniform distribution and a beta distribution. The denominator in equation [4] is a version of the empirical distribution function which is a step function with increments of $1/K$ at each observed P-value. Another quantity called the local FDR (LFDR, (50)) can be directly defined from the mixture model in [5]. The definition is similar to [6] except that CDFs are replaced by the corresponding probability density function (pdf):

$$LFDR = \frac{\pi_0}{\pi_0 + (1 - \pi_0) f_1(\tau)}. \quad [7]$$

The interpretation of LFDR is the posterior probability that a test with a P-value equal to the threshold τ is a test for which the null hypothesis is true. As with FDR, LFDR will be smaller at smaller values of τ. Also, FDR can be thought of as a type of averaging of LFDR over all tests with a P-value $\leq \tau$ so, as a result, values of LFDR will be greater than FDR at a given τ. Estimates of FDR and LFDR are obtained in statistical methods by estimating the components in equations [6] and [7], respectively. Computing an estimate at thresholds equal to each observed P-value gives an FDR (LFDR) curve that is seen to be an increasing function of the P-values. **Figure 9.3** shows FDR and LFDR curves for the distribution of P-values shown for the drought effect in **Fig. 9.1**. The estimates were obtained using the mixture model method of Allison et al. (20). One can see that LFDR values are greater than FDR values. From the plot one can see (roughly) that for tests with a P-value smaller than 0.05, one would expect a proportion of around 0.10 false discoveries. For a test with a P-value equal to 0.05, one might estimate the posterior probability (LFDR) that the null hypothesis is actually true to be around 0.20. User-friendly software for fitting the mixture model of Allison et al. (20) and computing quantities based on the model was reported in Trivedi et al. (51) and is available at http://www.ssg.uab.edu/hdbstat.

In **Fig. 9.3** one can see that the FDR curve is a monotonically increasing function of the P-values. That is, FDR is smaller at smaller P-values. This does not necessarily happen when FDR is computed using equation [4] because the denominator is not a continuous function of the observed P-values. The FDR "curve" for the 100 smallest P-values obtained from the same data set is

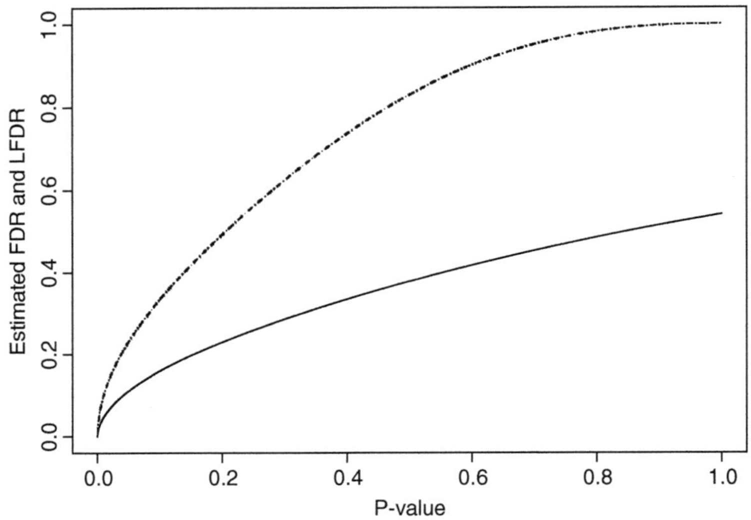

Fig. 9.3. FDR *(solid line)* and LFDR *(dashed line)* for the distribution of *P*-values in **Fig. 9.1** for the drought effect. Estimated quantities for the plots were obtained using the method in Allison et al. (20).

shown in the left panel of **Fig. 9.4**. There are some cases where a smaller *P*-value yields an increased estimate of FDR. Storey (52) defined the *Q*-value and interpreted it as a Bayesian posterior *P*-value, that is, it is a measure of the strength of an observed statistic (or *P*-value) with respect to the positive FDR. Estimates

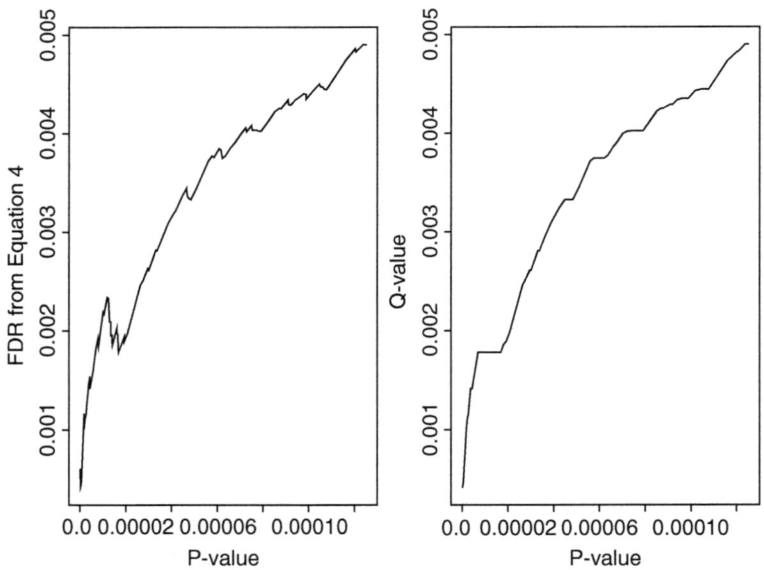

Fig. 9.4. The FDR for the smallest 100 *P*-values *(left panel)* using equation [4] and the *Q*-value *(right panel)* for the same *P*-values representing a drought effect (**Fig. 9.1**). The values for the plots were obtained using the smoothing method in Storey and Tibshirani (53).

of the Q-value from observed data should be monotonically increasing with the P-value. Storey (19) showed the algorithm to compute a Q-value from observed data and a plot of this Q-value is shown in the right panel of **Fig. 9.4**. A Q-value in **Fig. 9.4** computed at a given P-value is never larger at a smaller P-value. When there is no "signal" in a distribution of P-values (as seen in those in **Fig. 9.2**), the Q-value may remain large for all P-values, that is, for any list of tests that are rejected at a particular threshold, the proportion of false discoveries may be high. The software for computing Q-values is available as an R library called qvalue, available at www.r-project.org.

4.3. Sample Size Considerations for the False Discovery Rate and Related Quantities

Sometimes computed values of FDR can be large at even very small thresholds, and these large values may be due to the small sample sizes that are often common in HDEs. Gadbury et al. (34) presented a method to evaluate the role of sample size in bringing quantities like FDR to desired levels when the design is a comparison of two treatments. They also defined the expected discovery rate (EDR) which was the expected proportion of true alternative hypotheses that will be discovered in an HDE, i.e., the expected proportion $D/(K - M)$. Larger sample sizes yielded smaller values of FDR and larger values of EDR.

Recall from **Fig. 9.1** that the signal for a rust effect was not strong. Suppose that a two-treatment comparison study for differential expression due to a rust effect was being planned, and the signal present in **Fig. 9.1** for a rust effect was to be used as a pilot data set for planning sample size requirements for the new study. The P-values in **Fig. 9.1** were actually obtained from a two-factorial design structure, but for convenience and for purposes of illustration, we use this distribution as if it was obtained from a simple two treatment comparison study. **Figure 9.5** shows the technique reported in (34) that uses the distribution of P-values for a rust effect as a template but extrapolates EDR for various sample sizes and reports it for three different thresholds at which a null hypothesis is rejected. A smaller threshold yields a smaller EDR since fewer null hypotheses will be rejected; however, a smaller threshold yields a lower FDR because one is more certain that those null hypotheses that are rejected are true discoveries. One might notice that the EDR values in **Fig. 9.5** do not rise to the level of traditional power analyses in planning experiments. In HDEs there are many thousands of hypotheses being tested and an investigator might be content of discovering a smaller fraction of truly expressed genes for the purpose of follow-up research. A tool for implementing the method in (34) was reported in Page et al. (54) and is available online at www.poweratlas.org.

There are other results in the literature regarding sample size requirements in HDEs. Lee and Whitmore (55) investigated sample size requirements on type I and type II error probabilities.

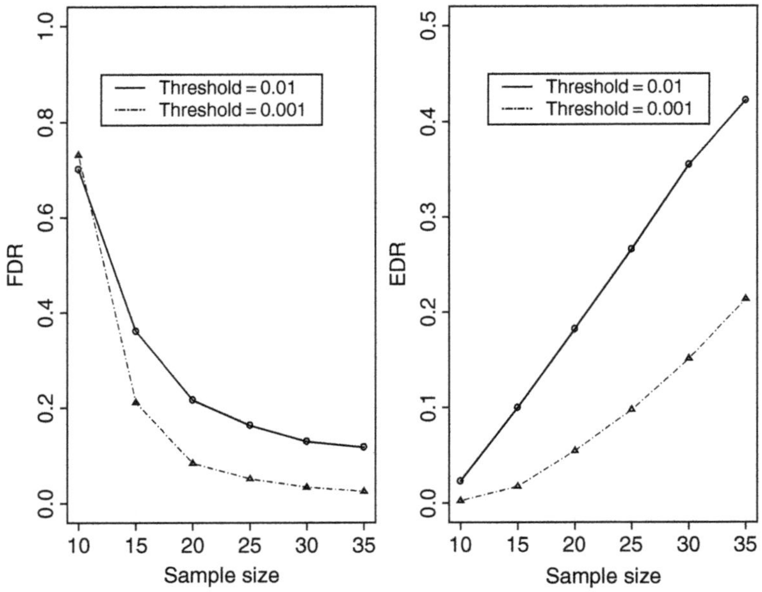

Fig. 9.5. Sample size analysis illustrated using *P*-values in **Fig. 9.1** for a rust effect as a pilot data set. Sample sizes reflect the number of microarrays in each of the two treatment groups in a two-treatment comparison study.

"Power" was equal to 1 − *P*(type II error) which is analogous to the *EDR* in Gadbury et al. (34). Lee and Whitmore (55) also extended their results to situations where there may be more than two treatment groups where interest is in determining differential expression among several treatment groups. Pan et al. (56) used a *t*-type statistic to quantify differential expression and presented a model that, when fitted to a "pilot" data set, could be used to assess the number of replicates required to achieve desired power at a given threshold. The fitted model was considered fixed, a type I error was specified, and power computed for any specified effect size, e.g., standardized difference in mean expression levels between two groups.

Zein et al. (57) considered sample size effects on pairwise comparisons of different groups and discussed the role of both technical and biological variabilities. Actual data sets were used to develop parameter specifications for simulated data sets. They used the term sensitivity as analogous to *EDR*, and specificity that is analogous to 1 − *FDR*. They evaluated the effect of varying sample sizes on these two quantities for various simulated data sets and using different types of statistical tests for differential expression, e.g., *t*-tests and a rank-based test. More recently Shao and Tseng (58) presented a sample size calculation with an adjustment for dependence among tests in microarray studies. More discussion of power and sample size in HDEs is in (29).

5. Classification and Validation Strategies and Some Remarks on Future Developments

Thus far we have discussed design and inference issues in HDEs and focused on the multiple testing issues when many thousands of tests are being conducted simultaneously. Here we conclude this chapter by discussing some other topics and techniques.

5.1. Clustering High-Dimensional Data

Clustering is one of the earlier techniques and has been and is still popular (33). Cluster analysis attempts to group data into classes based on some similarity metric. An early illustration of cluster analysis on data from an HDE is Eisen et al. (59). They used clustering on two time course gene expression data sets and showed that some genes of similar functions would cluster together. They also applied a type of randomization procedure to assess whether the clusters were real or whether they were an artifact of the clustering procedure. Even in HDE data that are completely random (i.e., data are generated so that there are not real clusters), a clustering routine will find clusters, so one cannot always be sure that a cluster is real without some technique to assess its repeatability in similar experiments. Some have referred to this as stability of clustering and have compared the stability of different cluster routines under different conditions (60).

Attempts to evaluate the stability of clustering techniques have generally used resampling techniques such as the bootstrap. Kerr and Churchill (61) assessed the reliability of conclusions obtained using clustering on data from microarray experiments. They used a clustering technique on a data set and assessed the stability of clusters on simulated data sets. The simulated data sets were created by fitting an ANOVA model to data and resampling residuals from the model in a bootstrap routine. Another resampling approach was used to evaluate the number of clusters present in an HDE data set where mixture models were used as a basis for clustering (62). Kapp and Tibshirani (63) assessed the reproducibility of clusters by defining a "cluster quality measure" that is related to prediction accuracy, that is, the ability of a new datum to be classified to a previously defined cluster.

To reflect the variability in an experiment due to biological samples, the resampling unit should be at the level of the biological specimen, i.e., a microarray in a microarray experiment (11). However, sometimes sample sizes are too small to use resampling at this level to evaluate the reliability of clusters. Garge et al. (60) conducted a simulation study of four clustering techniques and found that all four techniques produced low stability scores when evaluated on microarray types of data sets. Although obtaining reproducible clusters in an HDE with relatively small samples may have challenges, clustering methods can still be useful as an exploratory method for obtaining a general description of how genes covary with respect to their gene expression levels (33). One key advantage to a reliable clustering of data in a gene expression experiment is a reduction of dimension from one of many thousands of genes to a dimension of a smaller number of clusters, providing the clusters contain meaningful information about the function or classification of certain sets of genes.

5.2. Gene Class Testing

One challenge when analyzing data from an HDE such as a gene expression experiment is finding ways to successfully interpret the enormous number of results that are obtained (64). A type of analysis has emerged that appears designed to help address this challenge. Such analyses recognize that genes can be and have been placed into a priori categories and they use this categorical information in analytic strategies that can reduce the number of results about individual genes to a smaller number of more interpretable findings concerning classes or families. Gene class testing is a relatively recent technique, with some methods for implementing it still in development (45).

Many methods use Gene Ontology (GO) terms for assignment of genes to classes, though other knowledge bases are available (*see* (65, 66) for discussion and illustration). The idea of gene class testing in an HDE is to identify classes or sets of genes that are differentially expressed across one or more treatment conditions, or that are associated with some phenotype. Pavlidis et al. (67) compared two computational methods for associating gene expression changes with age for selected sets of genes using GO classes. The two methods used different techniques to evaluate what GO classes are most associated with aging. Mootha et al. (68) presented a gene set enrichment analysis (GSEA) and illustrated its use on a gene expression study using human diabetic muscle. The technique used an enrichment score to quantify association of a gene set to a phenotypic class. The method was also illustrated on some cancer-related data sets that included leukemia and lung cancer (69). Goeman et al. (70) proposed their global test procedure and illustrated it on two examples. Their test produces one *P*-value for a group that is being tested. These are just a few of the methods that have been proposed for testing classes of genes for association with a phenotype or phenotypic class, e.g., "treatment" condition. Pan (71) proposed fitting mixture models to classes of genes and using these sub-mixture models within classes to determine differential expression, somewhat similar in concept to other approaches using mixture models that were fit to all genes, for example (20). A limitation of mixture models is that many measurements are usually needed to obtain a good fit (72) and, in this case, gene classes would need to be relatively large.

The fact that much recent development of statistical methods has occurred in gene class testing suggest its potential usefulness and promise as a tool for the analysis of HDE data. Many of the current methods have prompted some concerns (73) and others suffer from at least one flaw (33). Goeman and Buhlmann (45) review assumptions and limitations of some of the recent methods for gene class testing. Undoubtedly this area of research will continue as one of active interest.

5.3. Validating Methods Using Simulations

A general discussion of validity of findings in HDEs was given in Mehta et al. (74) and in Allison et al. (33). Here we discuss validity in the context of statistical methods and the results that they are designed to produce as was also done in Mehta et al. (11). Validity of results from an HDE data analysis depends on many of the same assumptions that are required for a valid analysis of data from a traditional experiment. Examples are assumptions about distributions of data (or residuals), the choice of the model used, and assumptions about random sampling and/or treatment assignment.

Many statistical methods that analyze data from HDEs produce conservative estimates of π_0 and FDR (i.e., estimates tend to be biased high). The properties of certain methods and the estimates that they produce can sometimes be evaluated using mathematical derivations and proofs. One example is Genovese and Wasserman (75) who looked at the FDR controlling procedure of Benjamini and Hochberg (2). The performance of many methods and comparisons of methods have been evaluated using computer simulation experiments.

One technique for simulating microarray data considered sources of variability in such data and created simulated data sets based on some knowledge of this variability gleaned from real data sets (27). Many methods simulate data sets using statistical distributions, often normal distributions (e.g., (49, 76)). Correlation structure, if considered at all, has been implemented in simulated data using a block-diagonal correlation matrix as was done, for example, in (20, 43).

Concern about how well-simulated data correspond to reality has generated interest in simulated data that are derived from actual data sets. Cattell and Jaspars (77) used the term plasmode to describe data that are constructed to reflect some aspect of reality. Mehta et al. (11) described a plasmode as a real (i.e., not computer-simulated but from actual biological specimens) data set for which some aspect of the truth is known. Plasmodes can be used to learn about the validity and lack of validity of certain statistical methods for microarray analysis. The great advantage of plasmodes is that, unlike with computer simulations, one need not question whether the particular distributions or correlations are realistic because they are taken directly from real data. Plasmodes are beginning to show promise as a valuable resource for the scientific community.

One example of a plasmode is a real microarray data set with specific mRNAs spiked-in (cf., (37, 78)). Evaluating whether a particular method can correctly detect the spiked mRNAs gives information about the method's ability to detect gene expression. Affycomp (37) is a set of tools and plasmode (spike-in) data sets on an integrated web site that allows investigators to analyze the same benchmark data sets using a new method.

Plasmodes could also be derived from a real data set in a manner for which some truth is known. Gadbury et al. (79) have explored techniques to do this in the context of a microarray experiment for a two-treatment comparison study. A distribution of realistic "effect sizes" in an HDE can be obtained by analyzing a real data set. A data set for which the global null hypothesis is true (the null hypothesis is true for all tests) may be obtained by dividing the data for one treatment group into two pseudo-treatment groups. Differentially expressed genes can be created by sampling effect sizes from the experiment and incorporating them into the data for one of the pseudo-treatment groups for a proportion $1 - \pi_0$ of genes. In the resulting plasmode data set a true value of π_0 and a true value of FDR at a particular threshold can be known. Methods can then be evaluated on their ability to estimate these quantities.

5.4. Future Developments Related to High-Dimensional Experiments

Experiments investigating genome-wide gene expression may shift to greater use of sequencing in place of microarrays as sequencing becomes less expensive. In this case, rather than evaluating expression for the set of genes represented on a microarray, any genes expressed may be analyzed by determining the frequency of occurrence of corresponding RNA in samples. This may make it easier to discover new genes that are differentially expressed, but it may also make it more difficult to study genes with low levels of expression. One difference for statistical analyses will be that only a certain (large) number of sequences can be evaluated from each sample, so a higher frequency of sequences corresponding to one gene will be associated with a lower frequency of sequences corresponding to other genes.

Proteomics, lipidomics, and metabolomics are becoming more approachable for more plant systems. The data sets generated in these new "Omics" fields may often be modeled using approaches similar to those for studies of genome-wide gene expression (transcriptomics). Ultimately a new challenge for statistics will be the development of good comparisons of responses to treatments across these different types of data sets. Biological questions may be answered using different sets of findings, possibly from different Omics experiments. As noted in Allison et al. (33), how best to examine intersections between sets of findings is a needed area of research as is how to evaluate complex multi-component hypotheses. Bayesian approaches might be helpful in these areas.

Acknowledgments

D. Allison and G. Gadbury acknowledge the support from NIH Grant U54 CA100949. K. Garrett acknowledges the support from NSF Grants DBI-0421427, DEB-0516046, and EF-0525712, and DOE Grant DE-FG02-04ER63892. Gadbury and Garrett are grateful to programs supporting research from the Ecological Genomics Institute, Kansas State University, and the NSF Long Term Ecological Research Program at Konza Prairie, Kansas, and the Kansas Agricultural Experiment Station (Contribution No. 09-118-B).

References

1. Wolfsberg, T.G., Wetterstrand, K.A., Guyer, M.S., Collins, F.S., and Baxevanis, A.D. (2002) A user's guide to the human genome. *Nature Genetics Supplement* **32**, 1–79.
2. Benjamini, Y. and Hochberg, Y. (1995) Controlling the false discovery rate: a practical and powerful approach to multiple testing. *Journal of the Royal Statistical Society, Series B* **57**, 289–300.
3. Schadt, E.E., Li, C., Ellis, B., and Wong, W.H. (2001) Feature extraction and normalization algorithms for high-density oligonucleotide gene expression array data. *Journal of Cellular Biochemistry, Supplement* **37**, 120–125.
4. Quackenbush, J. (2002) Microarray data normalization and transformation. *Nature Genetics* **32**, 496–501.
5. Smyth, G.K. and Speed, T. (2003) Normalization of cDNA microarray data. *Methods* **31**, 265–273.
6. Ekstrom, C.T., Bak, S., Kristensen, C., and Rudemo, M. (2004) Spot shape modelling and data transformations for microarrays. *Bioinformatics* **20**, 2270–2278.
7. Travers, S.E., Smith, M.D., Bai, J.F., Hulbert, S.H., Leach, J.E., Schnable, P.S., Knapp, A.K., Milliken, G.A., Fay, P.A., Saleh, A., and Garrett, K.A. (2007) Ecological genomics: making the leap from model systems in the lab to native populations in the field. *Frontiers in Ecology and the Environment* **5**, 19–24.
8. Milliken, G.A., Garrett, K.A., and Travers, S.E. (2007) Experimental design for two-color microarrays applied in a pre-existing split-plot experiment. *Statistical Applications in Genetics and Molecular Biology* **6**, Article 20.
9. Kerr, M.K. (2003) Design considerations for efficient and effective microarray studies. *Biometrics* **59**, 822–828.
10. Fisher, R.A. (1966) The Design of Experiments, 8th edition. Hafner Publishing Company: New York.
11. Mehta, T.S., Zakharkin, S.O., Gadbury, G.L., and Allison, D.B. (2006) Epistemological issues in omics and high-dimensional biology: give the people what they want. *Physiological Genomics* **28**, 24–32.
12. Cui, X. and Churchill, G.A. (2003) Statistical tests for differential expression in cDNA microarray experiments. *Genome Biology* **4**, 21.
13. Pepe, M.S., Longton, G., Anderson, G.L., and Schummer, M. (2003) Selecting differentially expressed genes from microarray experiments. *Biometrics* **59**, 133–142.
14. Gadbury, G.L., Page, G.P., Heo, M., Mountz, J.D., and Allison, D.B. (2003) Randomization tests for small samples: an application for genetic expression data. *Journal of the Royal Statistical Society, Series C (Applied Statistics)* **52**, 365–76.
15. Xu, R. and Li, X. (2003) A comparison of parametric versus permutation methods with applications to general and temporal microarray gene expression data. *Bioinformatics* **19**, 1284–1289.
16. Mielke, P.W. and Berry, K.J. (2007) Permutation Methods: A Distance Function Approach. Springer: New York.
17. Wolfinger, R.D., Gibson, G., Wolfinger, E.D., Bennett, L., Hamadeh, H., Bushel, P., Afshari, C., and Paules, R.S. (2001) Assessing gene significance from cDNA microarray expression data via mixed models. *Journal of Computational Biology* **8**, 625–663.

18. Sackrowitz, H. and Samuel-Cahn, E.P. (1999) P values as random variables—expected P values. *The American Statistician* **53**, 326–331.
19. Story, J.D. (2002) A direct approach to false discovery rates, *Journal of the Royal Statistical Society, Series B* **64**, 479–498.
20. Allison, D.B., Gadbury, G.L., Heo, M., Fernandez, J.R., Lee, C., Prolla, T.A., and Weindruch, R.A. (2002) Mixture model approach for the analysis of microarray gene expression data. *Computational Statistics and Data Analysis* **39**, 1–20.
21. Ruppert, D., Nettleton, D., and Hwang, J.T.G. (2007) Exploring the information in P-values for the analysis and planning of multiple-test experiments. *Biometrics* **63**, 487–495.
22. Schweder, T. and Spjøtvoll, E. (1982) Plots of P-values to evaluate many tests simultaneously. *Biometrika* **69**, 493–502.
23. Berger, J.O. and Sellke, T. (1987) Testing a point null hypothesis: The irreconcilability of P values and evidence. *Journal of the American Statistical Association* **82**, 112–122.
24. Broberg, P. (2004) A new estimate of the proportion unchanged genes in a microarray experiment. *Genome Biology* **5**, P10.
25. Langaas,M., Lindqvist, B.H., and Ferkingstad, E. (2005) Estimating the proportion of true null hypotheses, with application to DNA microarray data. *Journal of the Royal Statistical Society, Series B* **67**, 555–572.
26. Frank, E.E. (2007) The effects of drought and pathogen stress on gene expression and phytohormone concentrations in *Andropogon gerardii*. M.S. Thesis; Kansas State University: Manhattan, KS.
27. Singhal, S., Kyvernitis, C.G., Johnson, S.W., Kaisera, L.R., Leibman, M.N., and Albelda, S.M. (2003) Microarray data simulator for improved selection of differentially expressed genes. *Cancer Biology and Therapy* **2**, 383–391.
28. Zakharkin, S.O., Kim, K., Mehta, T., Chen, L., Barnes, S., Scheirer, K.E., Parrish, R.S., Allison, D.B., and Page, G.P. (2005) Sources of variation in Affymetrix microarray experiments. *BMC Bioinformatics* **29**, 214.
29. Gadbury, G.L., Xiang, Q., Edwards, J.W., Page, G.P., and Allison, D.B. (2006) The role of sample size on measures of uncertainty and power. In: Allison, D.B., Page, G.P., Beasley, T.M., Edwards, J.W., ed. DNA Microarrays and Related Genomics Techniques. Boca Raton: Chapman & Hall/CRC: 77–94.
30. Brody, J.P., Williams, B.A., Wold, B.J., and Quake, S.R. (2002) Significance and statistical errors in the analysis of DNA microarray data. *Proceedings of the National Academy of Sciences of the United States of America* **99**(20), 12975–12978.
31. Nguyen, D.V., Arpat, A.B., Wang, N., and Caroll, R.G. (2002) DNA microarray experiments: biological and technical aspects. *Biometrics* **58**, 701–717.
32. Rosa Guilherme, J.M., Steibel, J.P., and Tempelman, R.J. (2005) Reassessing design and analysis of two-colour microarray experiments using mixed effects models. *Comparative and Functional Genomics* **6**(3), 123–131.
33. Allison, D.B., Cui, X., Page, G.P., and Sabripour, M.(2006) Microarray data analysis: From disarray to consolidation and consensus. *Nature Review Genetics* **7**, 55–65.
34. Gadbury, G.L., Page, G.P., Edwards, J.W., Kayo, T., Prolla, T.A., Weindruch, R., Permana, P.A., Mountz, J., and Allison, D.B. (2004) Power analysis and sample size estimation in the age of high dimensional biology: a parametric bootstrap approach illustrated via microarray research. *Statistical Methods in Medical Research* **13**, 325–38.
35. Hurlbert, S.H. (1984) Pseudoreplication and the design of ecological field experiments. *Ecological Monographs* **54**, 187–211.
36. Efron, B. and Tibshirani, R.J. (1993) An Introduction to the Bootstrap. Boca Raton, FL: CRC Press.
37. Irizarry, R.A., Wu, Z., and Jaffee, H.A. (2006) Comparison of Affymetrix GeneChip expression measures. *Bioinformatics* **22**, 789–794.
38. Ishwaran, H., Rao, J.S., and Kogalur, U.B. (2006) BAMarray: Java software for Bayesian analysis of variance for microarray data. *BMC Bioinformatics* **7**(1), 59.
39. Qiu, X., Klebanov, L., and Yakovlev, A. (2005) Correlation between gene expression levels and limitations of the empirical Bayes methodology for finding differentially expressed genes. *Statistical Applications in Genetics and Molecular Biology* **4**, Article 34.
40. Qiu, X., Xiao, Y., Gordon, A., and Yakovlev, A. (2006) Assessing stability of gene selection in microarray data analysis. *BMC Bioinformatics* **7**, 50.

41. Owen, A. (2005) Variance in the number of false discoveries. *Journal of the Royal Statistical Society, Series B* **67**, 411–426.
42. Hu, X. (2007) Distributional aspects of P-value and their use in multiple testing situations. Ph.D. Dissertation. University of Missouri – Rolla: Rolla, Missouri.
43. Nettleton, D., Hwang, G.J.T., Caldro, R.A., and Wise, R.P. (2006) Estimating the number of true null hypotheses from a histogram of p-values. *Journal of Agricultural, Biological, and Environmental Statistics* **11**, 337–356.
44. Efron, B. (2007) Correlation and large-scale simultaneous significance testing. *Journal of the American Statistical Association* **102**, 93–103.
45. Goeman, J.J. and Buhlmann, P. (2007) Analyzing gene expression data in terms of gene sets: methodological issues. *Bioinformatics* **23**, 980–987.
46. Hochberg, Y., and Tamhane, A.C. (1987) Multiple Comparisons Procedures. New York: John Wiley & Sons, Inc.
47. Tsai, C., Hsueh, H., and Chen, J.J. (2003) Estimation of false discovery rates in multiple testing: application to gene microarray data. *Biometrics* **59**, 1071–1081.
48. Pounds, S. and Morris, S.W. (2003) Estimating the occurrence of false positive and false negative in microarray studies by approximating and partitioning the empirical distribution of p-values. *Bioinformatics* **19**(10), 1236–1242.
49. Nguyen, D. (2004) On estimating the proportion of true null hypotheses for false discovery rate controlling procedures in exploratory DNA microarray studies. *Computational Statistics & Data Analysis* **47**, 611–637.
50. Efron, B. (2004) Large-scale simultaneous hypothesis testing: The choice of a null hypothesis. *Journal of the American Statistical Association* **99**, 96–104.
51. Trivedi, P., Edwards, J.W., Wang, J., Gadbury, G.L., Srinivasasainagendra, V., Zakharkin, S.O., Kim, K., Mehta, T., Brand, J.P.L., Patki, A., Page, G.P., and Allison, D.B. (2005) HDBStat!: A platform-independent software suite for statistical analysis of high dimensional biology data. *BMC Bioinformatics* **6**, 86.
52. Storey, J.D. (2003) The positive false discovery rate: A Bayesian interpretation and the q-value. *The Annals of Statistics* **31**, 2013–2035.
53. Storey, J.D. and Tibshirani, R. (2003) Statistical significance for genomewide studies. *Proceedings of the National Academy of Sciences* **100**, 9440–9445.
54. Page, G.P., Edwards, J.W., Gadbury, G.L., Yelisetti, P., Wang, J., Trivedi, P., Allison, D.B. (2006) The PowerAtlas: a power and sample size atlas for microarray experimental design and research. *BMC Bioinformatics* **7**, 84.
55. Lee, M.L.T. and Whitmore, G.A. (2002) Power and sample size for DNA microarray studies. *Statistics in Medicine* **21**, 3543–3570.
56. Pan, W., Lin, J., and Le, C.T. (2002) How many replicates of arrays are required to detect gene expression changes in microarray experiments? A mixture model approach. *Genome Biology* **3**(5), 1–10.
57. Zien, A., Fluck, J., Zimmer, R., and Lengauer, T. (2003) Microarrays: how many do you need? *Journal of Computational Biology* **10**, 653–667.
58. Shao, Y. and Tseng, C.-H. (2007) Sample size calculation with dependent adjustment for FDR-control in microarray studies. *Statistics in Medicine* **26**, 4219–4237.
59. Eisen, M.B., Spellman, P.T., Brown, P.O., and Botstein, D. (1998) Cluster analysis and display of genome-wide expression patterns. *Proceedings of the National Academy of Science* **95**, 14863–14868.
60. Garge, N.R., Page, G.P., Sprague, A.P., Gorman, B.S., and Allison, D.B. (2005) Reproducible clusters from microarray research: Wither? *BMC Bioinformatics* **6**(Suppl 2), S10.
61. Kerr, M.K. and Churchill, G.A. (2001) Bootstrapping cluster analysis: assessing the reliability of conclusions from microarray experiments. *Proceedings of the National Academy of Science* **98**, 8961–8965.
62. McLachlan, G.J. and Khan, N. (2004) On a resampling approach for tests on the number of clusters with mixture model-based clustering of tissue samples. *Journal of Multivariate Analysis* **90**, 90–105.
63. Kapp, A.V. and Tibshirani, R. (2007) Are clusters found in one dataset present in another dataset? *Biostatistics* **8**, 9–31.
64. Breitling, R., Amtmann, A., and Herzyk, P. (2004) Iterative Group Analysis (iGA): a simple tool to enhance sensitivity and facilitate interpretation of micro array experiments. *BMC Bioinformatics* **5**(1), 34.

65. Osier, M.V. (2006) Postanalysis interpretation: "What do I do with this gene list?" In: Allison DB, Page GP, Beasley TM, Edwards JW, ed. DNA Microarrays and Related Genomics Techniques. Chapman & Hall. CRC: Boca Raton, FL, 321–333.
66. Osier, M.V., Zhao, H., and Cheung, K.-H. (2004) Handling multiple testing while interpreting microarrays with the gene ontology database. *BMC Bioinformatics* **5**, 124.
67. Pavlidis, P., Qin, J., Arango, V., Mann, J.J., and Sibille, E. (2004) Using the gene ontology for microarray data mining: a comparison of methods and application to age effects in human prefrontal cortex. *Neurochemical Research* **29**(6), 1213–1222.
68. Mootha, V.K., Lindgren, C.M., Eriksson, K.F., Subramanian, A., Sihag, S., Lehar, J., Puigserver, P., Carlsson, E., Ridderstrale, M., Laurila, E., Houstis, N., Daly, M.J., Patterson, N., Mesirov, J.P., Golub, T.R., Tamayo, P., Spiegelman, B., Lander, E.S., Hirschhorn, J.N., Altshuler, D., and Groop, L.C. (2003) PGC-1alpha-responsive genes involved in oxidative phosphorylation are coordinately down-regulated in human diabetes. *Nature Genetics* **34**(3), 267–273.
69. Subramanian, A., Tamayo, P., Mootha, V.K., Mukherjee, S., Ebert, B.L., Gillette, M.A., Paulovich, A., Pomeroy, S.L., Golub, T.R., Lander, E.S., and Mesirov, J.P. (2005) Gene set enrichment analysis: A knowledge-based approach for interpreting genome-wide expression profiles. *Proceedings of the National Academy of Science* **43**, 15545–15550.
70. Goeman, J.J., van de Geer, S.A., de Kort, F., and van Houwelingen, H.C. (2004) A global test for groups of genes: testing association with a clinical outcome. *Bioinformatics* **20**(1), 93–99.
71. Pan, W. (2005) Incorporating gene functional annotations in detecting differential gene expression. *Journal of the Royal Statistical Society*, Series C-Applied Statistics **55**, 301–316.
72. Xiang, Q., Edwards, J.W., and Gadbury, G.L. (2006) Interval estimation in a finite mixture model: Modeling P-values in multiple testing applications. *Computational Statistics and Data Analysis* **51**, 570–586.
73. Damian, D. and Gorfine, M. (2004) Statistical concerns about the GSEA procedure. *Nature Genetics* **36**, 663.
74. Mehta, T., Tanik, M., and Allison, D.B. (2004) Towards sound epistemological foundation of statistical methods for high-dimensional biology. *Nature Genetics* **36**, 943–947.
75. Genovese, C. and Wasserman, L. (2002) Operating characteristics and extensions of the false discovery rate procedure. *Journal of the Royal Statistical Society*,Series B **64**, 499–517.
76. Hsueh, H., Chen, J.J., and Kodell, R.L. (2003) Comparison of methods for estimating the number of true null hypotheses in multiplicity testing. *Journal of Biopharmaceutical Statistics* **13**(94), 675–689.
77. Cattell ,R.B. and Jaspars, J. (1967) A general plasmode (No. 30-10-5-2) for factor analytic exercises and research. *Multivariate Behavioral Research Monographs* **67**, 1–212.
78. Choe, S.E., Boutros, M., Michelson, A.M., Church, G.M., and Halfon, M.S. (2005) Preferred analysis methods for Affymetrix GeneChips revealed by a wholly defined control dataset. *Genome Biology* **6**(2), R16.
79. Gadbury, G.L., Xiang, Q., Yang, L., Barnes, S., Page, G.P., Allison, D.B. (2007) Evaluating Statistical Methods Using Plasmode Data Sets in the Age of Massive Public Databases: An Illustration using False Discovery Rates. *Plos Genetics* **4(6)**, e1000098.

Chapter 10

Discrete Dynamic Modeling with Asynchronous Update, or How to Model Complex Systems in the Absence of Quantitative Information

Sarah M. Assmann and Réka Albert

Abstract

A major aim of systems biology is the study of the inter-relationships found within and between large biological data sets. Here we describe one systems biology method, in which the tools of network analysis and discrete dynamic (Boolean) modeling are used to develop predictive models of cellular signaling in cases where detailed temporal and kinetic information regarding the propagation of the signal through the system is lacking. This approach is also applicable to data sets derived from some other types of biological systems, such as transcription factor-mediated regulation of gene expression during the control of developmental fate, or host defense responses following pathogen attack, and is equally applicable to plant and non-plant systems. The method also allows prediction of how elimination of one or more individual signaling components will affect the ultimate outcome, thus allowing the researcher to model the effects of genetic knockout or pharmacological block. The method also serves as a starting point from which more quantitative models can be developed as additional information becomes available.

Key words: Boolean model, computational biology, dynamic modeling, discrete model, network analysis, signal transduction, systems biology.

1. Introduction

In recent years, technical advances have allowed the generation of large data sets that describe the genomes, transcriptomes, proteomes, and metabolomes of organisms, particularly those of model prokaryotic, plant, and animal species. Methods are also rapidly advancing for the large-scale analysis of how the components of these data sets interact both with each other and across

Dmitry A. Belostotsky (ed.), *Plant Systems Biology, vol. 553*

DOI 10.1007/978-1-60327-563-7_10 Springerprotocols.com

biological levels. For example, methods such as mass spectrometric identification of proteins following co-immunoprecipitation (1, 2), and yeast two-hybrid analysis (3–5) and its variants (6), have allowed researchers to build networks that portray the protein interactomes of model species (3, 7–9). Chromatin immunoprecipitation followed by query of whole-genome microarrays, the "ChIP-chip method" (10, 11) (also see the chapters by Morohashi, Xie, and Grotewold and by Barkan in this volume), as well as protein microarrays (12) has allowed discovery of the global targets of key transcription factors, allowing connections to be drawn between the proteome and the transcriptome (13–15).

Rapid cell signaling pathways comprise one of the most challenging types of networks to identify and model. These pathways usually involve at least two biological levels, typically the metabolome and the proteome. In many situations, information on the behavior of the signaling molecules in the intact living cell is not experimentally accessible and cannot be deduced post hoc because of the dynamic and often reversible nature of cell signaling events. In addition, such networks typically include post-translational events which alter the activity of, but not necessarily the level of, key signaling proteins. Global methods for identification of post-translational modifications such as protein phosphorylation are still improving (16–18) and, more importantly, assessment of the impact of such modifications on protein behavior (enzyme kinetics, subcellular localization, protein turnover, etc.) is often not available.

Here we describe a method by which cellular signaling networks can be assembled and modeled in predictive fashion, despite the above limitations. Such networks consist of an input (the initial signal or triggering event), an output (the ultimate target or outcome of the signaling cascade), and a variable, often large, number of internal nodes, consisting of all the known secondary messengers, enzymes, and metabolites involved in conveying the signal. The approach compensates for lack of detailed kinetic and temporal information on signal propagation by considering all biologically relevant starting combinations of the states of all internal nodes, with each node allowed only two possible states, either "active" (ON) or "not active" (OFF). For each one of these starting combinations, the input is turned on, and the status of each internal node is then changed or "updated" based on the statuses of the nodes feeding into it. The internal nodes are updated in a random order. Once every internal node has been updated in this fashion, the outcome (i.e., the status of the output node) is assessed.

This type of analysis is termed "discrete dynamic modeling with asynchronous update." The model is called dynamic because it follows the change in the status of the nodes in time. It is discrete because of the assumption of two discrete levels of activity instead of a continuum of levels. Asynchronous update refers to the fact that the status of the internal nodes is updated one by one, in a non-synchronous manner.

Below is a summary of the standard steps necessary for carrying out this analysis:

1. Thoroughly read and assimilate the relevant literature concerning the signal transduction pathway of interest.
2. Construct a table that formalizes the components ("nodes") of the system and the relationships ("edges") between them.
3. Based on this table, construct the simplest possible network that incorporates all of the information.
4. For each node, develop an equation that describes the necessary condition for the node to be ON, using the Boolean operators AND, OR, and NOT.
5. Select a status for the input node and a starting condition for the internal nodes.
6. Update the status of the internal nodes for an increasing number of steps to find the long-term behavior of the output node.
7. Do replicate simulations and summarize the observed outcomes.
8. Assess whether the model accurately predicts known experimental results. If not, revise the network and/or the Boolean rules.
9. Assess the robustness of the model to changes in interactions or in Boolean rules.
10. Use the model as desired to predict the outcome when specific nodes are deleted (always OFF) or overexpressed (always ON), and use these outcomes in the planning of new wet bench experiments. Use the results from new wet bench experiments to revise and extend the model.

2. Materials

1. *Information:* Collate all available information on the biological system of interest. This information can be obtained by keyword searches using tools such as PubMed and Google Scholar, as well as targeted keyword searches of the databases of journals known to publish on the topic of interest. As described in more detail under Methods, the information obtained should be compiled in a standardized fashion, using formal rules.
2. *Software:* Networks can be generated manually. For example, one of the more complex networks to be modeled using the methods of this chapter, a network describing induction of

stomatal closure by the plant hormone abscisic acid (ABA), was manually generated (19). Since the publication of that work a custom software package NET-SYNTHESIS has been developed (available at http://www.cs.uic.edu/~dasgupta/network-synthesis/) for finding the simplest representation of the signal transduction network. The theoretical underpinnings of the algorithms are explained in detail in (20). The input to NET-SYNTHESIS is a list of relationships among biological components and its output is a network diagram and a text file with the edges of the signal transduction network.

The SmartDraw charting software (http://www.smartdraw.com/) can be used to draw the network. Another good alternative for generation and automatic layout of a range of different diagrams and networks is yED (http://www.yworks.com/en/products_yed_about.htm).

Among several software packages available for graph analysis of signal transduction networks, we recommend the Python library NetworkX (https://networkx.lanl.gov/wiki).

Implementation of the model can be done with code generated in-house, and two of the programming languages most often used for this purpose are Python (http://www.python.org/) and C (http://www.cprogramming.com/). Alternatively, we recommend using the recently developed software application BooleanNet (http://code.google.com/p/booleannet/) for simulation of Boolean models. The input to the software is a set of Boolean rules in a simple text format and thus this software requires minimal programming expertise. The software can be run via a web interface or as a Python library to be used through an application programming interface.

3. Methods

3.1. Thoroughly Read the Relevant Literature Concerning the Signal Transduction Pathway of Interest

After reading all available literature on the topic, assess whether sufficient information is on hand such that modeling would be informative. If detailed qualitative information is available, but quantitative information is lacking, proceed with the method of this chapter. If sufficient quantitative information is available, the method describe in this chapter can be superseded by other methods, such as continuous models based on ordinary differential equations (21, 22) which can incorporate this quantitative information. A third possibility is that there may be so little information available that, while modeling could be done, it would not provide information beyond that which could be readily deduced without a formal evaluation. For example, if all that is known about a system is that process X activates process Y which in turn activates process Z, one can draw a simple linear network and deduce that knockout of Y will eliminate signaling, but a formal analysis is hardly required.

In assessing the literature, the modeler should especially focus on experiments that provide information of the type relevant to network construction. Experiments that identify nodes belonging to a signaling pathway are of several main types, including (1) in vivo or in vitro experiments which show that the properties (e.g., activity or subcellular localization) of a protein change upon application of the input signal or upon modulation of components already definitively known to be associated with the input signal; (2) experiments that directly assay a small molecule or metabolite (e.g., imaging of cytosolic Ca^{2+} concentrations) and show that the concentration of that metabolite changes upon application of the input signal or modulation of its associated elements; (3) pharmacological experiments which demonstrate that the output of the pathway of interest is altered in the presence of an inhibitory agent that blocks signaling from the candidate intermediary node (e.g., a pharmacological inhibitor of an enzyme or strong buffering of an ionic species); (4) experiments which show that artificial addition of the candidate intermediary node (e.g., exogenous provision of a metabolite) alters the output of the signaling pathway; (5) experiments in which genetic knockout or overexpression of a candidate node is shown to affect the output of the signaling pathway.

Some of the major types of experiments that identify edges (i.e., relationships between nodes) are (1) experiments that demonstrate physical interaction between two nodes, such as data on protein–protein interaction obtained from yeast two-hybrid assays or in vitro or in vivo co-immunoprecipitation and (2) experiments that demonstrate genetic epistasis between two genes/gene products.

3.2. Construct a Table That Formalizes the Components ("Nodes") of the System and the Relationships ("Edges") Between Them

To standardize and formalize the information available from the literature, it is valuable to next construct a table that summarizes the elements that contribute to the signaling pathway, and the relationships between them. In a signal transduction pathway, there is typically an input, perceived by a receptor or "input node," followed by a series of elements or internal nodes through which the signal percolates to the output node, which represents the final outcome of the signal transduction process. For a cellular signal transduction pathway not involving alterations in gene expression, elements or "nodes" often consist of proteinaceous receptors, intermediary signaling proteins and metabolites, effector proteins, and a final output node which represents the ultimate combined effect of the effector proteins. However, other types of biological macromolecules participate in other types of signal transduction pathways. For example, in a qualitative model describing regulation of the transcript level of a particular gene, the gene itself and the transcription factors that regulate it, as well as any small RNAs that regulate the transcript' abundance, would all be intermediate signaling elements, with the final output being presence or absence of transcript.

For the purposes of illustration, we will demonstrate the method of dynamic modeling with asynchronous update using a model system consisting of just four nodes: an input node "A," two intermediate nodes, "B" and "C," and an output node "D." While such a system might in reality fall into the category of "formal modeling not required," we have deliberately chosen a simple system in order to allow us to clearly illustrate the fundamental principles of the method. For our simple example, let us say that the relationships between our four nodes are described by the statements found in **Table 10.1**.

Table 10.1
Compile a table that uses simple terms to describe the relationships between nodes. In this table, positive interactions are indicated by arrows and negative interactions are indicated by blunted lines. When a node provides input into the relationship of two other notes, that relationship is described by the words "promotes" or "inhibits"

	A → B	
	A ---\ C	
	B → D	
	C --\ D	
B	A → D	promotes
C	A → D	inhibits

Note that some choices may have to be made in constructing a summary table, especially in the case where there are two conflicting reports in the literature. For example, imagine that in one report it is stated that proteins X and Y do not physically interact based on yeast two-hybrid analysis, while in a second report, it is described that proteins X and Y do interact, based on co-immunoprecipitation from the native (e.g., plant) tissue. The modeler will need to decide which information is more reliable and proceed accordingly. Such aspects dictate that human intervention will inevitably be an important component of the literature curation process, even as automated text search engines such as GENIES (23–25) grow in sophistication.

3.3. Based on This Table, Construct the Simplest Possible Network (i.e., the Network with the Fewest Edges) That Incorporates All of the Information

According to **Table 10.1**, the simplest possible network is the one depicted in **Fig. 10.1**. While it is relatively easy to see how the network of **Fig. 10.1** is constructed from the data of **Table 10.1**, modeling of a multi-component biological network is considerably more complicated. Simplifying assumptions that should be used are illustrated in **Fig. 10.2**; these assumptions can also be tailored to specific networks. These simplifying assumptions are formalized and incorporated in the software package NET-SYNTHESIS (*see* Materials).

Additional information can also be added to the network. For example, if two proteins have been shown to interact genetically, but physical interaction has not been demonstrated, an intermediate node, consisting of a small black filled circle (cf. **Fig. 10.3**), can be added to the network. The presence of this intermediate node leaves open the possibility that additional

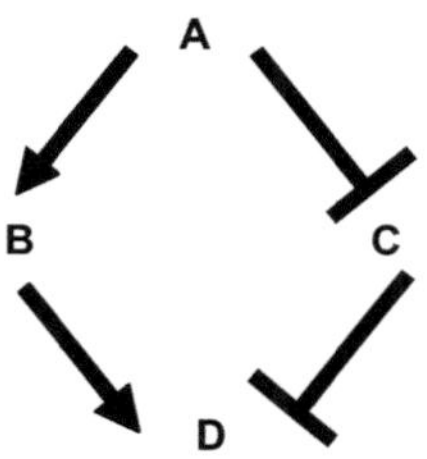

Fig. 10.1. An example of a simple four-node network. In this four-node network example, node A is the input and node D is the output. Nodes B and C are intermediate or internal nodes. Interactions between the nodes are represented by edges (*lines*). Positive interactions are indicated by *arrows* and negative interactions are indicated by *blunted lines*.

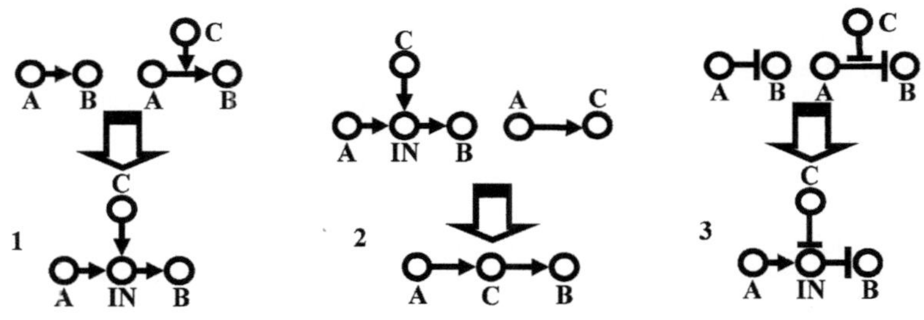

Fig. 10.2. Simplifying interference rules for network reconstruction.

1. If A →B and C → process(A →B), where A →B is not a biochemical reaction such as an enzyme catalyzed reaction or protein–protein/small molecule interaction, we assume that C is acting on an intermediary node (IN) of the A–B pathway.
2. If A →B, A →C, and C →process(A →B), where A →B is not a direct interaction, the most parsimonious explanation is that C is a member of the A–B pathway, i.e., A →C →B.
3. If A --| B and C --| process(A --|B), where A –|B is not a direct interaction, we assume that C is inhibiting an intermediary node (IN) of the A–B pathway. Note that A→IN--| B is the only logically consistent representation of the A–B pathway.

This figure and figure legend are reproduced from (19).

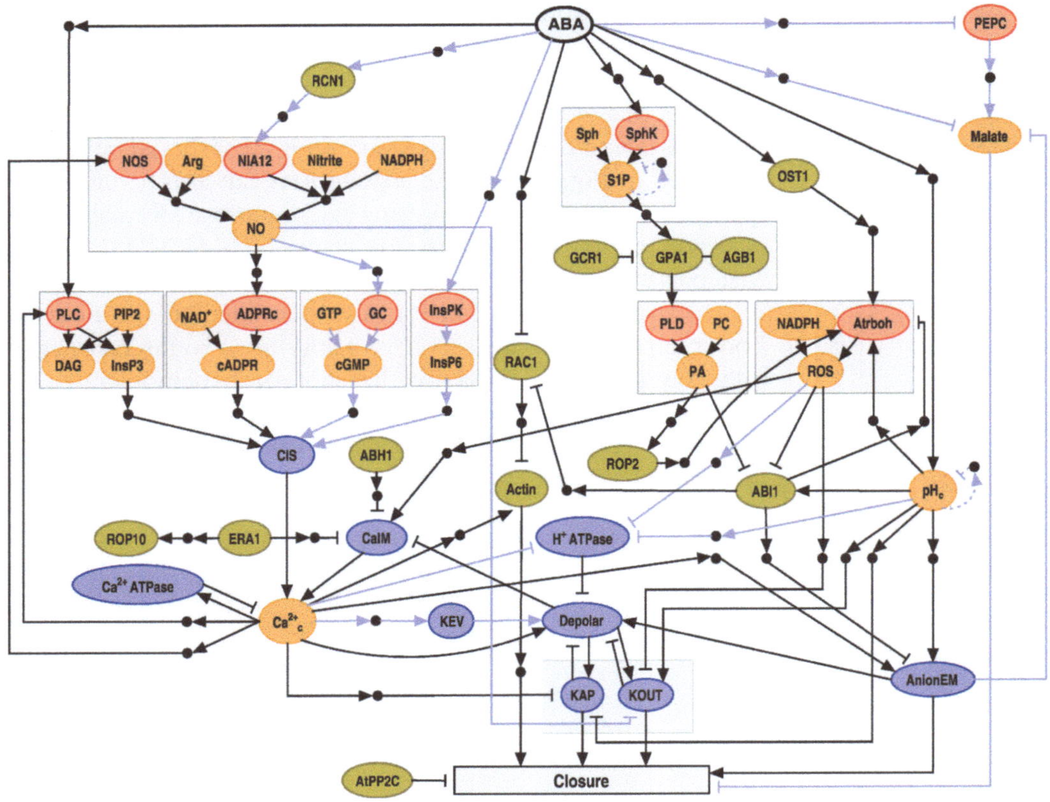

Fig. 10.3. An example of a complex signaling network. The network for induction of stomatal closure by the plant hormone ABA based on data available as of 2006.
Nodes in this graph include
Input = signal hormone ABA
Enzymes (red): ADP ribose cyclase (ADPRc), guanyl cyclase (GC), nitric oxide synthase (NOS), nitrate reductase (NIA12), NADPH oxidase (Atrboh), phospholipase C (PLC), phospholipase D (PLD), sphingosine kinase (SphK), phosphoenolpyruvate carboxylase (PEPC), inositol polyphosphate kinase (InsPK);
Signal transduction proteins (green): farnesyl transferase (ERA1), heterotrimeric G protein α (GPA1) and β component (AGB1), protein kinase (OST1), protein phosphatase 2A (RCN1), protein phosphatase 2C (ABI1/2), protein phosphatase 2C (AtPP2C), putative GPCR (GCR1), small GTPases (ROP2/ROP10/RAC1), actin cytoskeleton disruption (actin), mRNA cap binding protein (ABH1);
Membrane transport (blue): anion efflux at the plasma membrane (AnionEM), Ca^{2+} influx to the cytosol from intracellular stores (CIS), Ca^{2+} influx across the plasma membrane (CalM), potassium efflux through rapidly activating K^+ channels (AP channels) at the plasma membrane (KAP), K^+ efflux through slowly activating outwardly rectifying K^+ channels at the plasma membrane (KOUT), K^+ efflux from the vacuole to the cytosol (KEV), H^+ ATPase at the plasma membrane (HATPase), Ca^{2+} efflux from the cytosol (Ca^{2+} ATPase);
Secondary messengers and small molecules (orange): cytosolic Ca^{2+} increase ($Ca^{2+}{}_i$), Arg (arginine), cADPR, cGMP, DAG, GTP, InsP3, InsP6, nitrite, NO, PIP2, PA (phosphatidic acid), S1P (sphingosine-1-phosphate), Sph (sphingosine), PC (phosphatidyl choline), malate.
Output = stomatal closure
Small black filled circles represent putative intermediary nodes mediating indirect regulatory interactions. *Arrowheads* represent activation; *short perpendicular bars* indicate inhibition. *Purple lines* denote interactions derived from species other than Arabidopsis or inferences made during the network synthesis process. Nodes involved in the same metabolic pathway or protein complex are bordered by a *gray box*; only those arrows that point into or out of the box signify information flow.
This figure and figure legend are reproduced from (19).

intervening elements can be specified as information becomes available. If information has been compiled from different biological species, edges (connecting lines) can be color-coded to indicate the biological species from which the information was obtained (cf. **Fig.** 10.3).

In some instances it is beneficial to simplify an overly complex network by reducing the number of nodes that are of lesser interest. Simply deleting "uninteresting" nodes would eliminate the indirect connections among interesting nodes mediated by them and is thus not an option. We can however collapse pairs of uninteresting nodes if they mediate the exact same relationships among interesting nodes and this can be executed using the software NET-SYNTHESIS.

At this point, the network has been constructed, and some useful information can already be derived. For example, one can assess whether the network includes hubs, which are nodes with a much greater-than-average number of inputs and/or outputs, which may then prove more essential to the functioning of the network. One can also determine the minimum path length between input and output and assess the presence or absence of distinct regulatory motifs, such as negative and positive feedback loops. One can also evaluate the extent to which nodes are interconnected; if there is little interconnectivity (crosstalk), such that there are a number of completely independent (parallel) paths between the input and output nodes, this implies that the system has greater redundancy and thus greater resilience to perturbation, with the caveat that such path analysis cannot assess possible synergistic effects on the output of two independent paths.

3.4. For Each Node, Develop an Equation That Describes the Necessary Condition for That Node to Be on, Using the Boolean Operators AND, OR, and NOT

To proceed beyond a static network description of the system to actual modeling, the next step is to formally describe the state of each node (active or inactive) based on the states of the nodes that supply inputs to it. The state of each regulated node can change if the state of any of the input nodes changes; thus one needs to specify the change in state, i.e., the next state of the regulated node, as a function of the current state of its regulators. In the absence of quantitative information, this formal description is achieved using the Boolean operators AND, OR, and NOT. In those cases where there is only one input to a node, it is straightforward to describe the Boolean rule for that node. For example, in the network of **Fig.** 10.1, node B will be ON if node A is ON, and node C will be OFF if node A is ON. We can therefore write

$$\mathrm{B}* = \mathrm{A} \qquad [1]$$

$$C* = NOT\ A \qquad [2]$$

where the asterisks signify the next states of nodes B and C, respectively.

However, when nodes have multiple inputs, describing the Boolean rule may be more complicated. For example, in our network of **Fig.** 10.1, two possible Boolean rules could be written to describe the activation of node D:

$$D* = B\ AND\ (NOT\ C) \qquad [3]$$

$$D* = B\ OR\ (NOT\ C) \qquad [4]$$

The scientist must rely on information available from the literature, combined with her or his expert knowledge, to devise the Boolean rule that best describes the actual biological situation. For the purposes of illustration we will assume that the AND rule (Eq. [**3**]) reflects the true biological situation: for node D to be turned on, both activation from B and loss of inhibition by C are required. However, if there were actual experimental data demonstrating that node D still could be turned on when A was activated in a knockout mutant of node B, then, out of the two rules given above, the OR rule (Eq. [**4**]) would be more likely to reflect biological reality.

In addition to compiling information in equations such as those given above, the outputs from a Boolean rule can also be summarized in a truth table (**Table** 10.2), and constructing such a truth table (which can be done manually or with a simple computer code) is indeed needed for the actual modeling process. The truth table of a Boolean rule indicates the next state of the regulated node (the node on the left hand side of the equation) for every combination of the states of its inputs (i.e., the nodes that appear on the right hand side of the equation). Note that such a table has 2^n entries where n equals the number of inputs. For simplicity, and in accordance to Boolean algebra, the ON state is usually represented as 1 and the OFF state is represented as 0 in truth tables.

Table 10.2
Truth tables for the three regulated nodes of the network illustrated in Fig. 10.1, based on the three Boolean rules: (1) B* = A; (2) C* = NOT A; (3) D* = B AND (NOT C). A 0 signifies that the node is OFF; a 1 signifies that the node is ON

Entry	A	B*
1	0	0
2	1	1

Entry	A	C*
3	0	1
4	1	0

Entry	B	C	D*
5	0	0	0
6	0	1	0
7	1	0	1
8	1	1	0

3.5. Select a Status for the Input Node and a Starting Condition for the Internal Nodes

The input node represents the signal to be transduced by the network. Usually setting its status to be continuously on once the simulation has commenced is the best representation of the biological process to be modeled. One could also explore scenarios where the signal changes in time (following a prescribed pattern or stochastically). It is also beneficial to consider the case where the input is off to verify whether the signal is required to observe the correct output in the model. A comparison of the model's behavior with the experimentally observed behavior will indicate whether changes are needed to the model's assumptions (see next steps).

It usually is most realistic to assume that the output node is initially off. In addition to setting the status of the input and output nodes, one needs to set a starting condition (state) for the internal nodes. If information is available on the resting state of these nodes, it should be incorporated in this initial status, for example all internal nodes could be set to OFF if it is known that prior to receiving the signal they have low abundances and/or activities. If an all-OFF initial status is not realistic but there is no specific information on the initial condition of internal nodes, the modeler should set the initial condition of each internal node randomly and sample over a large number of initial conditions to find an overall behavior that does not depend on the details of the initial condition (e.g., (19)).

3.6. Update the Status of the Internal Nodes for an Increasing Number of Steps to Find the Long-Term Behavior of the Output Node

The interactions and regulatory relationships described by the Boolean rule of each node will usually cause a change in the state of the node from the state specified in the initial condition. The new (updated) state of each node can be looked up from the truth table for that node using as inputs the current state of the node's regulators. For example, if in the network of **Fig. 10.1** the status of the input node A is 1 (ON), the updated state of node B will also be 1 (ON), regardless of the initial state of B.

The order in which each node's status is updated can have a considerable effect on the dynamics of the system; thus the modeler needs to select the update method best fitting the information available about the system. The simplest and traditionally used formalism assumes that the processes represented as edges in the network have similar durations, and correspondingly node states are updated simultaneously at multiples of a fixed timestep. This update method is called synchronous update. For example, if in our four-node network the status of the input node is $A = 1$ and the initial condition for the internal and output nodes is $B_0 = C_0 = D_0 = 0$, at the first update the states specified in the initial condition are used to determine the first-timestep state of the nodes. Substituting the input and initial states into the truth table (**Table 10.2**) we find $B_1 = 1$ (from entry 2), $C_1 = 0$ (from entry 4), and $D_1 = 0$ (from entry 5). The second update uses the status of the nodes after the first update as inputs and thus leads to $B_2 = 1$ (from entry 2), $C_2 = 0$ (from entry 4), and $D_2 = 1$ (from entry 7). Note that the states of nodes B and C have not changed after the update. Update is not synonymous with change in this context.

Synchronous update cannot properly account for the different time scales over which various events take place in a biological system. Most often these time scales are not known at all; nonetheless imposing the equality of all time scales, as the synchronous model does, introduces an artificial constraint. We can extend the basic model to account for different timescales by instead performing the updates in a non-synchronous order within each iteration. This update method is called asynchronous update. If there is insufficient timing information available, the update order can be selected randomly from all possible permutations of the internal/output nodes. In our four-node example there are $3! = 3*2*1 = 6$ permutations of the three updatable nodes (B, C, D). Let us say that the first update order is B, then D, then C. Starting from the same initial condition $A = 1, B_0 = C_0 = D_0 = 0$ as before, update of B uses $A = 1$ and yields $B_1 = 1$ (from entry 2), update of D uses $B_1 = 1$ (because node B has already been updated) and $C_0 = 0$ and yields $D_1 = 1$ (from entry 7), update of C uses $A = 1$ and yields $C_1 = 0$ (from entry 4). Let us now choose C, then B, then D as the next update order. Update of C uses $A = 1$ and yields $C_2 = 0$ (from entry 4), update of B uses $A = 1$ and yields $B_2 = 1$ (from entry 2), update of D uses $B_2 = 1$ and $C_2 = 0$ and yields $D_2 = 1$ (from entry 7). Note that the sequence of states for nodes B and C was the same in both synchronous and asynchronous update, because they only depend on the input node A, but the sequence of the output node's states was different in the two cases. Other update orders, for example B then C then D in the first update round, would give the same sequence as synchronous update. Thus, random choice of update orders introduces stochasticity into the evolution of the system and allows it to sample all timescales.

In many cases after several rounds of update the status of some or all nodes stabilizes and does not change anymore. If the whole system's state stabilizes, the resulting state is called a steady state. The steady states of a Boolean model depend only on the Boolean rules (truth tables) describing the regulation of nodes. Importantly, the steady states do not depend on the update order of the nodes, because the updates do not lead to any state changes in the steady state. The steady states allowed by a Boolean model can be determined analytically by noting that in the steady state the "next" state of each node equals its current state; thus the update rules become a system of equations that can be solved. In the case of our example network this system of equations is

$$B = A$$

$$C = NOT\ A$$

$$D = B\ AND\ (NOT\ C)$$

and admits two solutions, one for each state of the input node A (*see* **Table 10.3**).

Table 10.3
Steady states for the network illustrated in Fig. 10.1, based on the three Boolean rules: (1) B* = A; (2) C* = NOT A; (3) D* = B AND (NOT C). A 0 signifies that the node is OFF; a 1 signifies that the node is ON

Entry	A	B	C	Output (D)
1	0	0	1	0
2	1	1	0	1

There is no inherent property of Boolean models that requires them to achieve a steady state. Boolean models may also reach a repeating sequence of states called a cycle. This cycle can be very long, but ultimately every sequence of states must be cyclical because the total number of states in the network is finite and equals 2^N, where N is the number of nodes in the network. Fixed states or repeating state sequences are both denoted attractors. The same network may reach a steady state or a cycle depending on the initial conditions. Thus each attractor is associated with a set of states (called its domain of attraction) that, if used as an initial condition, converge into that attractor. Asynchronously updated Boolean models have the same steady states as their synchronously updated versions, but they usually have more and longer cycles (26). In signal transduction networks we are sometimes less interested in the state of internal nodes and more interested in the state of the output node(s). In such instances, it is sufficient to observe the long-term behavior (attractors) of the output node(s).

3.7. Do Replicate Simulations and Summarize the Observed Outcomes

Synchronous models starting with a given initial condition will always reach the same state after the same number of steps. This is due to the fact that the system is entirely deterministic (reproducible). The domains of attraction of each attractor can be determined by doing repeated simulations starting from every relevant initial condition. In synchronous Boolean models the domains of attraction of each attractor are non-overlapping; thus in principle one can completely map the state space of these models. In practice this is very time consuming for networks with more than ten nodes.

In contrast to the synchronous model, in an asynchronous model, because of the added complexity due to different update orders, the domains of attraction can be overlapping. In other words, the same initial condition can lead to different attractors for different update orders. Fortunately, this is less frequent in models of signal transduction networks because they tend to be highly (although not completely) directional and the value of the signal strongly channels the dynamics of the system. For example, in our simple network of **Fig.** 10.1, all initial conditions coupled with the input A = 1 lead to the steady state under entry 2 in **Table** 10.3, and an output state D = 1, regardless of the update order of the nodes.

The dynamic behavior and attractors of a Boolean model are outcomes that cannot be directly specified when formulating the model. The inputs to the model are the nodes and update rules (truth tables). If there is a known steady state or cycle in the system that is being modeled and the model does not reflect it, the modeler needs to reassess the network and the update rules and then rerun the simulation to check whether the desired dynamic behavior has been obtained (see next step).

In the network for ABA-induction of stomatal closure which we modeled (**Fig.** 10.3; (19)), there are 38 internal nodes, but only one attractor, a steady state, for the output node (closure), for each state of the input node (ABA). Specifically, closure = 1 for ABA = 1 and closure = 0 for ABA = 0. This agrees with the biological reality formulated qualitatively as "ABA signaling causes closure of previously open stomata." Different initial conditions for the internal nodes and different update orders only affect the time when the output node reaches the correct steady state, but all replicate simulations stabilize within eight rounds of update. The number of rounds of update required for stabilization of all replicate simulations is a rough indication of the "worst-case" timing in the real system (i.e., the time necessary to reach the correct outcome from the less advantageous initial conditions). The unit of time, i.e., the duration of a round of update, corresponds to the duration of the longest process that is represented by an edge in the network. In a manner of speaking, this number represents the maximum number of dynamical steps needed for the signal to propagate to the output. Eight rounds of update, as compared to the number of internal nodes, indicate that the signal has more efficient routes than propagating step by step through every internal node.

3.8. Assess Whether the Model Accurately Predicts Known Experimental Results. If Not, Revise the Network and/or the Boolean Rules

The validity of the model can be assessed by comparing the attractors in the model as well as their domains of attraction, with relevant experimental information. For example, both the four-node model and the ABA-induced stomatal closure model of Li et al. (19) indicate that the steady state of the output node only depends on the value of the input node. If there is a discrepancy between the actual biological outcomes and the "in silico" outcomes predicted by the model, then the model needs to be revised.

For example, if for the real biological system an attractor exists in which the output node is on even though the input node is off, or in which the output node is off even though the input node is on, for example due to the phenomenon of crosstalk and cross-activation of the same signaling pathways by other networks not included in the model (such crosstalk is well known to occur in plant signaling (27, 28)), then the modeler needs to include additional inputs to the network. For example, in the model of **Fig.** 10.1, the addition of a second input node E, and the modification of the Boolean rule of node D to "D* = B AND (NOT C) OR E," will create a steady state A = 0, B = 0, C = 1, D = 1, E = 1, i.e., A = 0 is not always associated with D = 0 anymore. Changing the Boolean rule of node D to "D* = B AND (NOT C) AND (NOT E)" will create a steady state A = 1, B = 1, C = 0, D = 0, E = 1, i.e., A = 1 is not always associated with D = 1 anymore.

3.9. Assess the Robustness of the Model to Changes in Interactions or in Boolean Rules

A biological system is said to be robust if it maintains the appropriate output in the face of perturbations and fluctuations. One could imagine that robustness is a property that would confer an adaptive advantage under many circumstances and thus would undergo positive selection over evolutionary time. A model is said to be robust if its outcome does not change when the model is subjected to small random perturbations in parameters or assumptions. If a model lacks robustness, this could reflect the true property of the biological system. However, since many cell signaling systems that have been evaluated in this context have been found to indeed exhibit the property of robustness (29–31), if such a system is modeled and the model is found to lack robustness, i.e., is "fragile," then the accuracy and sufficiency (completeness) of the model may be suspect.

One common way to evaluate the robustness of a network model is to determine how susceptible the network outcome is to node rewiring. Rewiring can take the form of randomly switching a positive (activating) edge to a negative (inhibitory) edge or vice versa, randomly adding an edge between two extant nodes in a network, or randomly rewiring pairs of positive or negative edges (19, 32). The robustness of Boolean models to alternative assumptions is assessed by randomly interchanging OR and AND rules (i.e., interchanging the assumption of independent activity with the assumption of conditional dependence). The method of assessing all single node disruptions and then calculating the percentage of cases in which the output is altered as a result probes both the model and the system that is modeled. A model does not have to be robust to every possible change in interactions, assumptions, or parameters to be considered fragile, just as cellular signaling systems are not robust to all perturbations. However, some degree of robustness, e.g., robustness to interchanging a certain percentage of AND and OR rules, is expected from a model.

3.10. Use the Model as Desired to Predict the Outcome When Specific Nodes Are Deleted (Always OFF) or Overexpressed (Always ON), and Use These Outcomes in the Planning of New Wet Bench Experiments. Use the Results from New Wet Bench Experiments to Revise and Extend the Model

Once the model has been shown to accurately simulate known information, it can be used to predict results of experiments that have yet to be conducted. For the simple model of **Fig. 10.1**, we can consider the predictions for knockout (always OFF) or overexpression (with the simplifying assumption that overexpression is equated to "always ON") of the intermediate nodes. Looking at the truth table for node D in our simple network (**Table 10.2**), we see that knockout of node B (node B always set to 0) corresponds to entries 5 and 6, and we see that for these entries the output (node D) is always OFF. Therefore, we predict that if our Boolean rule describing node D is correct, then a genetic or pharmacological knockout of node B will result in a phenotype in which node D is never observed to turn on.

However, what if we actually knocked out node B and found that an output from node D was still sometimes observed? This would be impossible according to the Boolean rule of Eq. [**3**] but, as illustrated in **Table 10.4**, would be expected according to the Boolean rule of Eq. [**4**]. Therefore, such an outcome would lead us to revise the Boolean rule for node D.

Conversely, we can also make predictions for the phenotype of overexpression of node B (node B always set to 1). According to **Table 10.2**, node D would be activated for entry 7 but not for entry 8, while according to **Table 10.4**, node D would always be ON when node B was overexpressed. Thus we can see that the combined results from experiments of knocking out node B and overexpressing node B are likely to be very informative concerning whether Eq. [**3**] or Eq. [**4**] (or neither) is the correct portrayal of the biological system.

Table 10.4
Truth tables for the network illustrated in Fig. 10.1, based on the three Boolean rules: (1) B* = A; (2) C* = NOT A; (3) D* = B OR (NOT C). A 0 signifies that the node is OFF; a 1 signifies that the node is ON. As shown below, the truth table of node D looks quite different from that of Table 10.2 when Eq. [4] is used to define it

Entry	A	B*
1	0	0
2	1	1

Entry	A	C*
3	0	1
4	1	0

Entry	B	C	D*
5	0	0	1
6	0	1	0
7	1	0	1
8	1	1	1

However, there may also be situations where the predictions suggest that certain wet bench experiments will not be informative, and therefore time, effort, and funding should not be spent on them. For example, assume that a genetic knockout of a node is equivalent to setting that node as permanently OFF (=0). Then, according to the Boolean rules of **Table** 10.2, knocking out node C would not be informative, because the status of node D (the output) can be 0 or 1 regardless of whether C is 0 or 1. However, a knockout of B would be informative, because such a knockout would be predicted never to achieve an output of $D^* = 1$.

While in the simple model of **Fig.** 10.1 it is easy to observe the predictions of knockout and overexpression phenotypes, as the model becomes more complex, this will not be the case. For example, in the network for ABA-induction of stomatal closure which we modeled (**Fig.** 10.3; (19)), there are 38 internal nodes. It would be very laborious to write out and evaluate truth tables for all the single, double, and triple knockout combinations of these nodes. However, by modeling this system as described above, we found that, out of all possible double mutant combinations, only 16% were predicted to completely block ABA signaling. We therefore expect that an initial focus on producing and analyzing these double mutants is more likely to prove fruitful than an approach in which double mutant combinations are randomly or arbitrarily chosen for investigation.

Analysis of the network structure can also prove useful in design of future experiments. For example, in our model of ABA-induced stomatal closure, based on information known at the time the network was constructed, we had nine nodes as immediate downstream targets of ABA. Obviously, if it were found that some of these nodes were not independent, but rather participated in the same branch of the signaling pathway, the network structure would be simplified. Therefore, a focus on combinatorial tests of protein–protein interaction between these nine nodes is suggested as an important next experiment.

One of the realities for the modeler is that models become outdated almost as soon as they are published: new nodes are uncovered that must be added to the system and relationships between nodes become better defined. Redrawing the network and revising the Boolean rules take relatively less time than rerunning the entire simulation with the new parameters. Therefore the modeler must decide when there is sufficient new information to justify this task.

4. Conclusions

This chapter has described network analysis and discrete dynamic modeling primarily in the context of cell signaling pathways. It is important to note that this method is not limited to cell signaling but can be applied to any system for which sufficient qualitative information is available concerning the components of the system and their inter-relationships. Some other systems that have been dynamically modeled to date include host–pathogen interactions (33), invertebrate development (26, 34), and floral morphogenesis (35, 36).

In addition, the method can be expanded to include additional qualitative information that goes beyond the simplest ON/OFF formulation of the Boolean model. For example, a model with more parameters than the parameter-free asynchronous Boolean model was used to predict host immune responses to bacteria of the genus *Bordetellae*, the causative agent of whooping cough and related diseases (33). In a discrete model describing segmentation in *Drosophila* embryos, functional products or activities of specific genes were assigned specific integer values that ranged from 0 to 3, and functional threshold values were also assigned to the gene products (37); thus this model was still qualitative but expanded beyond the two-level Boolean format. A similar approach has been applied to the modeling of root hair development (38). Further, a mixed or hybrid model can be developed when partial quantitative information is available. For example, in a further quantification of the Drosophila segmentation network, piecewise linear differential equations were developed to continuously describe the state of each node of the network (26, 39). BooleanNet contains a module for such piece-wise linear modeling.

References

1. Figeys, D., McBroom, L.D., and Moran, M.F. (2001) Mass spectrometry for the study of protein–protein interactions. *Methods* **24**(3), 230–239.
2. Berggard, T., Linse, S., and James, P. (2007) Methods for the detection and analysis of protein–protein interactions. *Proteomics* 7(16), 2833–2842.
3. Walhout, A.J. and Vidal, M. (2001) High-throughput yeast two-hybrid assays for large-scale protein interaction mapping. *Methods* **24**(3), 297–306.
4. Legrain, P. and Selig, L. (2000) Genome-wide protein interaction maps using two-hybrid systems. *FEBS Lett.* **480**(1), 32–36.
5. Fields, S. (2005) High-throughput two-hybrid analysis. The promise and the peril. *FEBS J.* **272**(21), 5391–5399.
6. Obrdlik, P., El-Bakkoury, M., Hamacher, T., et al. (2004) K+ channel interactions detected by a genetic system optimized for systematic studies of membrane protein interactions. *Proc. Natl. Acad. Sci. USA* **101**(33), 12242–12247.
7. Uetz, P., Giot, L., Cagney, G., et al. (2000) A comprehensive analysis of protein–protein interactions in Saccharomyces cerevisiae. *Nature* **403**(6770), 623–627.
8. Schwikowski, B., Uetz, P., and Fields, S. (2000) A network of protein–protein interactions in yeast. *Nat. Biotechnol.* **18**(12), 1257–1261.
9. Rain, J.C., Selig, L., De Reuse, H., et al. (2001) The protein–protein interaction map of Helicobacter pylori. *Nature* **409**(6817), 211–215.
10. Buck, M.J. and Lieb, J.D. (2004) ChIP-chip: considerations for the design, analysis, and application of genome-wide chromatin immunoprecipitation experiments. *Genomics* **83**(3), 349–360.
11. Haring, M., Offermann, S., Danker, T., Horst, I., Peterhansel, C., and Stam, M. (2007) Chromatin immunoprecipitation: optimization, quantitative analysis and data normalization. *Plant Methods* **3**, 11.
12. Hudson, M.E. and Snyder, M. (2006) High-throughput methods of regulatory element discovery. *Biotechniques* **41**(6), 673, 5, 7 passim.
13. Mockler, T.C., Chan, S., Sundaresan, A., Chen, H., Jacobsen, S.E. and Ecker, J.R. (2005)

Applications of DNA tiling arrays for whole-genome analysis. *Genomics* **85**(1), 1–15.

14. de Folter, S., Urbanus, S.L., van Zuijlen, L.G., Kaufmann, K., and Angenent, G.C. (2007) Tagging of MADS domain proteins for chromatin immunoprecipitation. *BMC Plant Biol.* 7, 47.
15. Lee, J., He, K., Stolc, V., et al. (2007) Analysis of transcription factor HY5 genomic binding sites revealed its hierarchical role in light regulation of development. *Plant Cell* **19**(3), 731–749.
16. Peck, S.C. (2006) Phosphoproteomics in Arabidopsis: moving from empirical to predictive science. *J. Exp. Bot.* **57**(7), 1523–1527.
17. Blom, N., Sicheritz-Ponten, T., Gupta, R., Gammeltoft, S., and Brunak, S. (2004) Prediction of post-translational glycosylation and phosphorylation of proteins from the amino acid sequence. *Proteomics* **4**(6), 1633–1649.
18. de la Fuente van Bentem, S. and Hirt, H. (2007) Using phosphoproteomics to reveal signalling dynamics in plants. *Trends Plant Sci.* **12**(9), 404–411.
19. Li, S., Assmann, S.M., and Albert, R. (2006) Predicting essential components of signal transduction networks: a dynamic model of guard cell abscisic acid signaling. *PLoS Biol.* **4**(10), e312.
20. Albert, R., DasGupta, B., Dondi, R., et al. (2007) A novel method for signal transduction network inference from indirect experimental evidence. *J. Comput. Biol.* **14**(7), 927–949.
21. Voit, E.O. (2000) Computational Analysis of Biochemical Systems. Cambridge: Cambridge University Press.
22. Tyson, J.J., Chen, K.C., and Novak, B. (2003) Sniffers, buzzers, toggles and blinkers: dynamics of regulatory and signaling pathways in the cell. *Curr. Opin. Cell Biol.* **15**(2), 221–231.
23. Friedman, C., Kra, P., Yu, H., Krauthammer, M., and Rzhetsky, A. (2001) GENIES: a natural-language processing system for the extraction of molecular pathways from journal articles. *Bioinformatics* **17** Suppl 1, S74–S82.
24. Marcotte, E.M., Xenarios, I., and Eisenberg, D. (2001) Mining literature for protein–protein interactions. *Bioinformatics* **17**(4), 359–363.
25. Jensen, L.J., Saric, J., and Bork, P. (2006) Literature mining for the biologist: from information retrieval to biological discovery. *Nat. Rev. Genet.* 7(2), 119–129.
26. Chaves, M., Albert, R., and Sontag, E.D. (2005) Robustness and fragility of Boolean models for genetic regulatory networks. *J. Theor. Biol.* **235**(3), 431–449.
27. Gazzarrini, S. and McCourt, P. (2003) Cross-talk in plant hormone signalling: what Arabidopsis mutants are telling us. *Ann. Bot.* (Lond) **91**(6), 605–612.
28. Fujita, M., Fujita, Y., Noutoshi, Y., et al. (2006) Crosstalk between abiotic and biotic stress responses: a current view from the points of convergence in the stress signaling networks. *Curr. Opin. Plant Biol.* **9**(4), 436–442.
29. Ingolia, N.T. (2004) Topology and robustness in the Drosophila segment polarity network. *PLoS Biol.* **2**(6), e123.
30. Barkai, N. and Leibler, S. (1997) Robustness in simple biochemical networks. *Nature* **387**(6636), 913–917.
31. von Dassow, G., Meir, E., Munro, E.M., and Odell, G.M. (2000) The segment polarity network is a robust developmental module. *Nature* **406**(6792), 188–192.
32. Li, F., Long, T., Lu, Y., Ouyang, Q., and Tang, C. (2004) The yeast cell-cycle network is robustly designed. *Proc. Natl. Acad. Sci. USA* **101**(14), 4781–4786.
33. Thakar, J., Pilione, M., Kirimanjeswara, G., Harvill, E.T., and Albert, R. (2007) Modeling systems-level regulation of host immune responses. *PLoS Comput. Biol.* **3**(6), e109.
34. Ghysen, A. and Thomas, R. (2003) The formation of sense organs in Drosophila: a logical approach. *Bioessays* **25**(8), 802–807.
35. Mendoza, L. and Alvarez-Buylla, E.R. (1998) Dynamics of the genetic regulatory network for Arabidopsis thaliana flower morphogenesis. *J. Theor. Biol.* **193**(2), 307–319.
36. Espinosa-Soto, C., Padilla-Longoria, P., and Alvarez-Buylla, E.R. (2004) A gene regulatory network model for cell-fate determination during Arabidopsis thaliana flower development that is robust and recovers experimental gene expression profiles. *Plant Cell* **16**(11), 2923–2939.
37. Sanchez, L. and Thieffry, D. (2003) Segmenting the fly embryo: a logical analysis of the pair-rule cross-regulatory module. *J. Theor. Biol.* **224**(4), 517–537.
38. Mendoza, L. and Alvarez-Buylla, E.R. (2000) Genetic regulation of root hair development in Arabidopsis thaliana: a network model. *J. Theor. Biol.* **204**(3), 311–326.
39. Chaves, M., Sontag, E.D., and Albert, R. (2006) Methods of robustness analysis for Boolean models of gene control networks. *Syst. Biol.* (Stevenage) **153**(4), 154–167.

Chapter 11

Quantification of Variation in Expression Networks

Daniel J. Kliebenstein

Abstract

Gene expression microarrays allow rapid and easy quantification of transcript accumulation for almost transcripts present in a genome. This technology has been utilized for diverse investigations from studying gene regulation in response to genetic or environmental fluctuation to global expression QTL (eQTL) analyses of natural variation. Typical analysis techniques focus on responses of individual genes in isolation of other genes. However, emerging evidence indicates that genes are organized into regulons, i.e., they respond as groups due to individual transcription factors binding multiple promoters, creating what is commonly called a network. We have developed a set of statistical approaches that allow researchers to test specific network hypothesis using a priori-defined gene networks. When applied to *Arabidopsis thaliana* this approach has been able to identify natural genetic variation that controls networks. In this chapter we describe approaches to develop and test specific network hypothesis utilizing natural genetic variation. This approach can be expanded to facilitate direct tests of the relationship between phenotypic trait and transcript genetic architecture. Finally, the use of a priori network definitions can be applied to any microarray experiment to directly conduct hypothesis testing at a genomics level.

Key words: Microarray, network, quantitative, systems biology, hypothesis test.

1. Introduction

Phenotypic variation of animals and plants, including disease susceptibility and development, is controlled by quantitative trait loci (QTLs) whose underlying molecular mechanisms are typically studied in QTL mapping experiments (1–3). QTLs are regions of the genome where genetic diversity is associated with phenotypic variation in a specific trait or, if pleiotropic, a suite of traits. These regions may contain genes whose differential expression controls the associated phenotypic variation. Previous methods to link phenotypic variation to its genetic cause required intensive

Dmitry A. Belostotsky (ed.), *Plant Systems Biology, vol. 553*

DOI 10.1007/978-1-60327-563-7_11 Springerprotocols.com

fine-scale mapping experiments. Recently, the genomic technique of microarray-based transcriptomics has been applied to more quickly link phenotypic trait variation with transcriptome variation. This approach uses microarrays to measure global gene expression across a sample of individuals from a natural population. These gene expression values are then used to map expression QTLs (eQTLs) (4–11) or to assess association between transcript variation and phenotypic variation using association mapping style approaches (12–14). These genomics technologies may enable reverse (natural variation) genetics approaches to identify the genetic basis of quantitative traits and facilitate our understanding of network variation within plants (15–19).

The goal of global eQTL analysis is to quickly identify loci controlling the expression variation of gene networks that control distinct biological functions. One approach (4, 6) is to generate a mapping population, assess global gene expression using microarrays, and identify eQTLs controlling the expression of each gene via individual statistical analyses. The eQTL locations for all genes are then summed, "sŭmmation" approach, to identify common regions that control the expression of more genes than expected by random chance, frequently referred to as eQTL hotspots (4, 6, 10, 11, 20–22). This approach is complicated by the potential that individual transcript levels are potentially more variable than the network controlling them. As such, the statistical analysis of individual genes is likely to have significant false-positive and false-negative errors confounding attempts to interpret the biological meaning of any eQTL analysis.

A second complication of the summation is that this requires a posteriori tests to assess whether the genes controlled by an identified eQTL hotspot share a common biological function (e.g., a metabolic pathway, transcriptional co-regulation, similar gene ontology functional annotation) (23, 24). As such, this is descriptive and relies on the presence and absence of individual genes in the list of transcripts significantly controlled by the QTL in question. Hence, we desired to devise a quantitative approach that would allow for the generation of specific hypothesis about transcriptional networks and testing of these hypothesis using microarray analysis of natural genetic variation (25).

In our approach we define the gene networks prior to the statistical analysis allowing quantitative network testing or network eQTL mapping (25). To develop gene networks we rely on existing databases containing either gene co-expression values or predicted metabolic pathways. We define gene networks as a co-regulated set of genes involved in a common biological process. Once we define the networks, we obtain a quantitative measurement of the transcriptional activity of the network by averaging across the individual genes within the network. This single network activity metric can then be used to associate with phenotypic variation or to map eQTL controlling biological networks. A benefit to this approach is that it is possible to predict a network and then identify the loci controlling the network. Further, it allows for rapid hypothesis generation about the biological impact of specific eQTL clusters. A final use of this approach is to apply standard statistical methodologies to test if networks are regulated in response to diverse inputs using standard experimental designs. In this chapter, we describe the approaches and tools required to generate and evaluate transcriptional networks using natural genetic variation.

2. Materials

2.1. *Arabidopsis*

An excellent model plant system for studying quantitative genetics is *Arabidopsis thaliana*. There is a rapidly developing set of both genomics tools and genetic variation populations that greatly aid development and testing of approaches to conduct quantitative network analysis of natural variation.

2.1.1. Natural Genetic Populations

Populations used to study natural genetic variation can be generally classified into structured populations or association populations. Structured populations have known parents allowing for accurate recombination measurements and the application of standard QTL mapping approaches (2). Recently, natural genetic variation in association populations has begun to be queried using linkage disequilibrium mapping approaches (26–28). Structured mapping populations have less genetic variation than association populations but it is unknown if this difference in genetic variation necessarily correlates to levels of phenotypic variation in the two population structures.

2.1.1.1. Structured Populations

In *Arabidopsis*, the main structured populations are made using the recombinant inbred line (RIL) structure where two parents are crossed and the progeny then undergo single seed descent for at least eight generations. After eight generations each resulting line is a homozygous mixture of the two parental genotypes. There are numerous RIL populations in existence in *Arabidopsis*, with the main populations being the Bay-0 × Sha, *Ler* × Col-0 and *Ler* × Cvi (29–33). These populations are of decently large size and have been phenotyped for innumerous diverse phenotypes. In addition to these populations, there are new populations in development or recently released (34–36). Important features of these populations are that they have already been genetically mapped and this information and the lines are or soon will be available from The Arabidopsis Resource Center (www.arabidopsis.org).

2.1.1.2. Association Populations

Recent work is suggesting that association mapping populations are a complementary approach to using structured populations for quantitative analysis of networks (27). These populations consist of large collections of diverse *Arabidopsis* accessions with unknown ancestry (26–28). These populations are designed to contain the vast majority of genetic diversity within *Arabidopsis* providing a rich source of allelic diversity. This is done by sampling a very large population of accessions and then choosing a smaller experimental population that contains the maximal level of diversity within the larger population. The individual accessions have been genotyped at a large number genetic loci using genomics technologies including near complete genome resequencing (37–39) and this sequence or genotyping information is freely available (www.arabidopsis.org). The accessions in these populations are freely available from The Arabidopsis Resource Center (www.arabidopsis.org).

2.1.2. Microarray Data

2.1.2.1. Genetic Variation Data Sets

Microarrays have been utilized to survey transcript accumulation variation in structured *Arabidopsis* populations (10, 11) and small association populations (12–14, 40). The microarray data for the Bay × Sha RIL population and one small association population can be obtained from elp.ucdavis.edu (10, 12, 14). Alternatively, this data can be downloaded from ArrayExpress as data sets E-TABM126 and E-TABM62 (www.ebi.ac.uk/microarray-as/aer/?#ae-main[0]). This database will provide either the raw .CEL files or the normalized gene expression data. Replicated microarray data for another association population can be downloaded from www.weigelworld.org/resources/microarray/AtGenExpress (40). Currently, the microarray data on the *Ler* × Cvi RIL population appears to be available via personal communication with the authors (11).

2.1.2.2. Co-expression Databases

The transcriptomic response of *Arabidopsis* to various environmental, genetic, and developmental perturbations has been intensively queried using microarrays. Most of this data is compiled into databases including www.genevestigator.org, www.Arabidopsis.leeds.ac.uk/ACT, and http://www.atted.bio.titech.ac.jp (41–46). These databases allow the researcher to enter a specific gene or set of genes to identify all other genes that show similar transcriptional variation within the whole database or a subset of the database. This provides an excellent data source for the generation of hypothetical networks as described in **Section 3.1.1**.

2.1.3. Metabolic Network Databases

Biosynthetic pathways are frequently co-regulated at the transcript level and as such are excellent sources of network hypothesis (47, 48). The Aracyc database for *Arabidopsis* contains an extensive list of enzyme encoding genes and their predicted or proven reactions. This database links enzymes and their corresponding genes into predicted or proven metabolic pathways that can be treated as networks (49, 50). This includes both primary and secondary metabolic networks. This database is readily accessible or completely downloadable at the *Arabidopsis* webpage (www.arabidopsis.org) to aid in network generation as described in **Section 3.1.3**.

2.2. Barley

Barley (*Hordeum vulgare*) is the other plant species that has a large existing mapping population that has been intensively analyzed using genomic microarray data. These are both required to enable a network analysis of network eQTL.

2.2.1. Natural Genetic Populations

The main population for quantitative analysis of transcript networks in Barley is a doubled haploid population obtained from a cross of the Steptoe and Morex inbred parents. This doubled haploid population consists of 139 lines that have been high-throughput genotyped to create a dense marker map (51). This population also has extensive phenotypic information available for the lines across multiple environments with significant replication (wheat.pw.usda.gov/ggpages/SxM/phenotypes.html).

2.2.2. Microarray Data

2.2.2.1. Genetic Variation Data Sets

The microarray data for the Steptoe × Morex DH population is available from ArrayExpress as data set E-TABM-112 (http://www.ebi.ac.uk/microarray-as/aer/?#ae-main[0]) (9, 51). This database will provide either the raw .CEL files or the normalized gene expression data.

2.2.2.2. Co-expression Databases

Barleybase (www.barleybase.org) is a database containing numerous microarray experiments from Barley that can allow a researcher to query for co-expressed genes (52, 53). Additionally, microarray data can be downloaded to allow researchers to apply their own co-expression analysis or alter the statistical parameters at their desire. This can be done using a validated batch-learning self-organizing map approach as previously described (54). This provides an excellent data source for the generation of hypothetical networks as described in **Section 3.1.1**.

2.3. Other Species

While barley and *Arabidopsis* are currently the plant species that contain both the genetic populations and microarray analysis to allow for large-scale quantitative analysis of network variation, there are additional projects underway that will assuredly generate similar data for other species. For instance, maize and rice have large mapping populations available that only require the application of microarrays to generate the necessary transcript variation measures (55). Numerous other plants have had targeted microarray analysis of natural genetic variation to address specific questions showing the broad applicability of this technology (8, 56–63).

3. Methods

In the a priori approach to network analysis of gene expression, the hypothetical networks are defined prior to the analysis of the microarray data. The goal of this a priori network approach is to allow the researcher to develop hypotheses about gene sets using prior information and then test these hypotheses utilizing the gene networks and microarray data. For instance, a researcher could hypothesize that a set of genes are critical for defense against a given pathogen. The researcher can then use the following methods to identify pathogen response networks, map eQTL controlling these networks, and compare the resulting data to QTL controlling resistance against the pathogen. Alternatively, these same approaches can be used to directly test if two genotypes that differ in resistance also differ in the expression of their hypothetical defense network. The applications of this approach are only limited to a researcher's ability to generate hypothesis and conduct the experiment.

3.1. Network Assignment

The first step required in this method is to generate groups of genes for which the researcher thinks there is support to presume or hypothesize that the genes within the group are coordinately regulated. The evidence for gene network assignment can be generated from genes having coordinate regulation, having a similar biological function or from numerous existing and developing genomics databases.

3.1.1. Gene Co-expression

Numerous plant species have existing databases containing large collections of microarray analysis which allow for researchers to identify co-regulated genes. These co-regulated genes can function as a priori-defined gene networks that can be used for further analysis. There are two predominant avenues to querying gene expression databases for co-regulated gene networks, the "guide-gene" and "non-targeted" approaches (54, 64).

3.1.1.1. "Guide-Gene" Approach to Co-expression Clustering

The simplest approach to using genomic expression databases for generating co-regulated gene networks is the "guide-gene" approach (54). The guide-gene approach involves researchers identifying their favorite gene, inputting it into the available databases, or using their own statistical analysis to identify all other genes in the genome that show a significant positive correlation across the available microarray data. This positive correlation suggests that these genes are controlled by the same regulatory network with the same directionality. These genes can then be classified as a co-regulated network. *See* **Section 3.1.6** for a discussion of the optimal size of co-regulated gene networks. *See* **Section 3.1.7** for a discussion of correlation thresholds and the potential ramification on the network's utility.

3.1.1.2. "Non-targeted" Approach to Co-expression Clustering

A more intensive and global approach to network definition using co-expression databases is to take the complete data set and compile all gene-to-gene correlations and then utilize this to conduct a complete clustering of all genes based on their correlation (42, 54, 65, 66). This approach will generate massive interconnected gene networks that can be utilized to create putative co-regulated gene networks (66). The genomic network requires dissection into discrete co-regulated gene networks that can then be handled individually. This dissection can be accomplished by deciding upon a correlation threshold required between genes to classify them as a co-regulated network. *See* **Section 3.1.7** for a discussion about correlation directionality and thresholds for calling co-regulated genes. An alternative to the hard correlational threshold is to visually inspect the networks and dissect them based on the density of clustering. Network diagrams typically are comprised of dense local gene clusters that are connected to other clusters via sparser interactions. A researcher could decide that they will dissect clusters based upon the frequency of interconnections within a cluster versus those between clusters. This would not require a hard correlational threshold and may yield more biologically relevant clusters (66).

3.1.2. Metabolic Pathway Network Definition

A useful method to define coordinated biological function is the cooperation of enzymes within a biosynthetic pathway. There are multiple databases containing both validated and predicted metabolic pathways present in *Arabidopsis* and other plant species (49, 50, 67, 68). As biosynthetic pathways exist to optimally transmute a beginning substrate to an end product, the genes in a metabolic pathway are frequently co-regulated (16, 25, 47, 48). As such, metabolic pathways provide an excellent beginning with which to predict coordinate gene expression networks. The available databases can be downloaded to generate a ready network list that can be further modified to the researcher's specific aims.

3.1.3. Protein Interaction Network Definition

Modern genomics technologies are providing a diverse array of data sets to allow gene networks to be defined and then tested. One such genomics data set allowing gene network prediction is protein interaction networks (69–71). These interaction networks predict the presence of protein complexes whose members are likely to be coordinately regulated to provide a common outcome (72, 73). There are two forms of protein interaction networks. In plants, the most common data currently available are for individual protein complexes (73). Another form of data that is coming is massive interactome maps attempting to illustrate all possible protein–protein interactions (69–71). While these interactome maps are highly complex, they do highlight local protein clusters that appear to function in protein complexes (72). A researcher could define the proteins/genes in a local cluster as likely to function in a coordinate fashion and as such be a good candidate for a coordinately regulated gene network. *See* **Section 3.1.6** for a discussion of the optimal size of co-regulated gene networks.

3.1.4. Other Potential Biological Definitions

The above approaches to generating hypothetical gene networks for further testing are not meant to exclude other approaches. In fact, each approach to a priori network definition inherently limits and frames both the questions being tested and the answers obtained. For instance, gene networks defined a priori using metabolic pathways allow a research to test how their experimental variable X controls gene expression for the biosynthetic pathway. Similarly, the proteomics definition limits any test to addressing how the protein complex may be regulated. As such any approach can be used to define the networks and the specific approach to network definition should be chosen to maximize the precision and/or power of the future tests. For instance, if a researcher is interested in using microarray data to address natural variation in trichomes, then a network defined by genes exclusively or predominantly expressed in trichomes will be more powerful than a proteomic or metabolic pathway-defined network. Any data that can allow a researcher to generate a group of genes logically expected to be co-regulated is a valid approach to a priori gene network definition. As the network is simply a tool for hypothesis testing it does not have to be "correct"; future experiments will test the correctness of the original definition.

3.1.5. Duplicated Genes and Optimizing Network Definitions

One complexity of plant genomes is the vast amount of gene duplication that has occurred (74–77). This can lead to the duplication of entire gene networks allowing the duplicated networks to obtain similar but distinct biological functions that may not be co-regulated. For instance, in maize several tryptophan biosynthetic genes have been duplicated and recruited for 2,4-dihydroxy-7-methoxy-1,4-benzoxazin-3-one (DIMBOA) synthesis which is regulated differently from the tryptophan biosynthetic pathway in maize (78). If a researcher's network is defined using protein interaction or metabolic pathways, it is possible that there are duplicated copies of this network, each with its own regulation pattern. As such, the overlapping patterns would diminish the ability to identify a signal of network co-expression.

A simple approach for researchers to test their network for the presence of duplicated networks with opposing expression patterns is to obtain microarray measures of gene expression and conduct a correlational analysis among the genes within a network. If all members of the network are co-regulated, they will show a positive correlation. Genes that constitute a separate network will show no or negative correlation with the other network genes. An illustration of this principle comes from previous work in *Arabidopsis* that utilized metabolic pathway definitions to initially define networks (25). Correlational analysis within these metabolically defined networks showed that each metabolic pathway typically had two different gene networks with opposing gene regulation (25). For instance, the genes predicted for lignin biosynthesis could be separated into two complete lignin sub-pathways that showed a positive correlation within each sub-pathway and negative correlation between the two sub-pathways. This correlational separation of duplicated networks should always be used to maximize the precision of any network definition before proceeding to specific network testing as described in **Section 3.2**.

3.1.6. Number of Genes in a Network

An important consideration in any network definition is how the number of genes within a network may affect future tests of that network. If a network has too few genes, then any statistical test using that network will be sensitive to variation in individual genes. This could create or destroy network significance due to error or variation in an individual gene within the network. Conversely if a network has too many genes, then these genes are likely integrating diverse and independent regulatory inputs and any desired biological specificity may be lost. As such, expression across very large gene networks may act as a measure of the plant's physiological status complicating the ability to resolve and specific biological phenomena (10, 25, 63, 79). Thus, to maximize the statistical power in terms of error potential and to increase the precision on the biological questions being asked, networks must be of a moderate gene membership.

In practice, the minimal gene membership within a network should be no fewer than five, with ten genes being a more optimal limit (12, 16, 25). The upper boundary of a network gene population is harder to define as this is dependent upon the co-regulation among members of a gene network. If the network members are absolutely co-regulated with no other influences separating them, then the network can be of any size. In practice, an analysis of eQTL in *Arabidopsis* showed that gene networks with more than 50 genes typically identified a limited set of eQTL hotspots whereas gene networks of 25 were more specific (Kliebenstein, unpublished data). This suggests that somewhere between 25 and 50 is likely the upper bound of the optimal gene network in *Arabidopsis* for network expression analysis. However, if the physiological measurements are the desired outcome of any analysis, then larger networks are valid uses of this a priori network approach.

3.1.7. Strictness of Network Definition

It is important for the ensuing network analysis that only those genes showing a positive correlation are considered as a co-regulated gene network. Admittedly, many regulatory networks have both positive and negative consequences on gene expression. However, the inclusion of negatively regulated genes would cause the "signal" from the co-regulated gene network to be diminished because these genes' negative changes would erase the positive regulation in the other genes within the co-regulated network. If the researcher feels that the negatively regulated genes are of sufficient interest to merit inclusion the solution is to create a separate negatively co-regulated network for analysis. If in fact the two gene networks are controlled by the same regulatory machinery in different directions, then the two co-regulated gene networks will identify the same factors in the ensuing experiment and can strengthen the researcher's interpretations.

Another important factor in generating gene groups via the co-expression analysis is the level of correlation between the input gene and the other genes that is used as the threshold for calling genes as a co-regulated network. This threshold will impact the results obtained from any network analysis of these genes. While there are no absolute thresholds that can be universally applied, in general the tighter the correlation required to call a group of genes a co-regulated network, the more likely that they will be regulated by a single transcriptional network. The lower the correlation between the genes in the network, the more likely they are regulated by a mixture of transcriptional networks. In this case, the gene network may actually function as more of a measure of some specific physiological condition such as drought or general stress level. Thus, the choice of the correlation level for defining networks by the guide-gene approach will likely alter the results from any network analysis.

3.2. Network Testing with Natural Variation Data

The above approaches to defining gene networks provide the opportunity to test a networks quantitative response to natural genetic variation. This can be in the form of a network eQTL analysis which only requires small changes to the standard single trait methodologies with which most laboratories are familiar. Below, we present a discussion of approaches to analyze a priori networks using eQTL analysis.

3.2.1. Network eQTL Analysis

After previous microarray data from the desired population is obtained (*see* **Sections 2.1.2.2** and **2.2.2.2**) or microarray data from a new population has been generated and gene networks have been defined, the next step is to identify eQTLs controlling these a priori-defined gene networks for which there are two basic approaches readily available to most labs. These are the average z score approach and the multi-trait approach as described below.

3.2.1.1. Average z Score Approach to A Priori Network eQTL

One approach to map network eQTLs for a priori-defined gene networks is to use standard software packages such as QTL Cartographer (80, 81). This requires the generation of a single metric describing the expression of the gene network. In traditional QTL mapping, a single metric for the trait is measured and entered into the QTL algorithm, for example the accumulation of a metabolite. The development of a single metric for a priori-defined gene networks is complicated by the genes having widely varying expression ranges (25). If this difference between genes is not corrected, variation in any single metric for the network will be dominated by those genes with higher expression and defeat the ability of an a priori network to encapsulate the information provided by all genes within the network. One solution to this complication is to conduct a simple mean centering. In this approach, the average expression across the different lines for each gene is set to a preordained value, say 0. The actual value for each gene is independently normalized by subtracting the measured gene expression value in that line by that gene's average expression measured across all lines. This is similar to the RMA adjustment for microarrays where the average gene expression per microarray is set to a constant and the transcript accumulation within each microarray is normalized accordingly (82). While a simple mean-centering approach does normalize the means, it does not compensate for genes with large expression ranges also having larger variances.

Simultaneously compensating for differences in variance and mean expression requires the use of the z scores for each gene within the network (25). This requires standardizing the expression of each gene in each line to its z score. This is accomplished by first subtracting the expression of each gene in each line by the average expression of that gene across all lines. This value is then divided by the standard deviation of that gene's expression across all lines. This forces all genes within the network to have an average expression of 0 and a standard deviation of 1 across all lines. Once the z score for each gene in each line has been determined, the average z score across the genes in the a priori gene network is measured in each line. This provides a single metric or number for the a priori gene networks expression that can be entered into a lab's favorite QTL mapping package to identify network eQTL using all appropriate significance determinations as would be conducted for any other trait (83–86).

3.2.1.2. Multi-trait Approach to A Priori Network eQTL

Multi-trait mapping algorithms provide a second approach to mapping eQTLs for a priori-defined gene networks. These algorithms were initially developed to test for QTLs across multiple environments (87–89). In the standard approach to multi-trait mapping, the same trait is measured in multiple environments and QTLs are mapped in each environment and across the environments. The multi-trait algorithms can be adapted to map gene network QTLs by treating each gene in the network as a separate measure of the gene network's response, hence treating each network as a different "environment" measure of the trait (90). The genes can then be entered into the multi-trait algorithms and eQTLs that map across the genes (environments) are the network eQTL for that specific a priori-defined gene network. An advantage to this approach is that gene-specific eQTLs can be rapidly identified in the ensuing QTL analysis. Additionally if any genes obviously behave differently than the other genes in the network in the multi-trait analysis, they can be dropped from the network and the eQTL analysis repeated to test if this better refines the a priori network. This approach can likely be extended into the more complex Bayesian QTL approaches being developed (90–93).

3.3. Network Testing of Experimental Data

In addition to allowing for analysis of natural variation in gene expression networks, the a priori definition approaches also provide the opportunity to test the network's quantitative response to more traditional experimental variation. This experimental variation could be in the form of environmental or genetic perturbation of the plant. Further, the a priori network analysis only requires small changes to the standard single gene methodologies with which most laboratories are familiar. This approach should be applicable to network testing of metabolomics data (*see* **Notes 1 and 2** for brief discussion).

3.3.1. Experimental Design

If the a priori network is being used to test existing microarray data sets for a network's regulation, then the researcher is limited to what the existing experiments allow. However, the researcher can utilize this a priori network approach to test a network's response to new experimental variables that were not a factor in the network's definition (12, 17). In this case, standard experimental designs should be followed to maximize the statistical power just as if the researcher was focusing on a single gene rather than a network. There is some thought that a network analysis may not require as much replication as an individual gene. However, as the basis of the a priori network approach is that there is a single underlying biological mechanism for the gene's co-regulation, it is possible that the variation present in this biological mechanism is similar to the variation identified in a single gene. This is shown by the lack of increased genetic heritability for the aliphatic glucosinolate network in comparison to the average heritability for the underlying genes (16). Further, individual genes and the networks within which they reside appeared to control similar levels of variation across *Arabidopsis* accessions, suggesting that gene networks and individual genes require similar levels of replication (25). As such, it is advisable to conduct sufficient replication with an experimental design meant to control for and minimize error as much as possible.

3.3.2. Nested ANOVA of Experimental Variables

One key aspect of the a priori network definition is that it facilitates the direct testing of gene network responses to experimental perturbation. This can be done using any standard experimental design meant to query gene expression responses to biotic, abiotic, or genetic perturbations. For this analysis, the gene networks are designed as described (**Section 3.1**), the appropriate experiment conducted, and data collected. The experiment can be a microarray analysis of a wild-type plant versus a mutant, plants grown in normal versus drought conditions, or a factorial experiment combining different experimental factors. An a priori network analysis of this data only requires a modification of the traditional ANOVA that many laboratories already utilize. In this modification, gene and gene network membership for each gene are both entered into the statistical analysis as separate variables. The data are then analyzed as a nested ANVOA whereby gene is nested under the gene network term (25). For instance, genes A, B, and C are considered members of network X and genes D, E, and F are members of network Y. This allows the data for each gene's expression data to be used by the model but only within the specific gene network in which that gene resides. This allows the model to compare expression variation between genes within a network to that between specific networks. For example, variation within the genes A, B, and C for network X is analyzed separately to the variation for genes D, E, and F in network Y. Finally, the variation between network X and Y is analyzed. Additionally, a nested ANOVA can compare the level of variation controlled by each component of the model. For instance, an analysis of natural variation in *Arabidopsis* gene expression suggested that network variation was on a similar order of individual gene variation (25). The ANOVA can then be extended to directly test for effects of different experimental perturbations upon the networks.

An example of this nested ANOVA approach is an analysis of how modifying three MYB transcription factors within *A. thaliana* altered the expression of sulfur utilization networks. In this experiment, WT and the different MYB expression lines were measured with replicated microarrays. The nested ANOVA tested if the introduction of the MYBs into *Arabidopsis* predominantly altered individual genes or the sulfur utilization networks within which the genes reside (17). This found a significant effect of the transcription factors upon the different networks, showing that the MYBs control distinct sulfur networks (17). The nested ANOVA can be easily implemented in any statistical package. However for very large data sets containing numerous genes and networks, the R platform is likely better due to more efficient matrix inversion algorithms. Smaller more discrete tests are feasible in any statistical package.

3.4. Conclusions

Genomics experiments are sometimes thought of as limited to generating hypothesis that are then tested by other methodologies. This leaves a need for developing approaches to allow for hypothesis testing using genomics-scale experiments. In this methods description, we relay one approach to using genomics data, specifically microarray data, to directly test hypothesis and map genetic variation for a priori-defined gene networks. This a priori approach has been mostly used for the analysis of eQTL controlling gene networks but can be extended to nearly any experimental approach. The methods described in this chapter are readily accessible to any laboratory with basic statistical programs such as Excel, R, SAS, or Systat and do not require any special programming. As such, these methods should allow any researcher to being treating gene networks as testable hypothesis using existing or new microarray data. This should allow for an increase in specific biological inference to be derived from transcriptomics data and experiments in any species. Finally, the approaches described here can be adapted to any genomics platform such as metabolomics whereby quantitative measurements of network members can be conducted and networks can be defined.

4. Notes

1. Applying the a priori network approach to metabolomics would be feasible to compare the network responses of biosynthetic pathways, i.e., TCA cycle, to the responses of the individual metabolites within the pathway.
2. A caveat to applying any expression analysis approaches to metabolite analysis is that metabolites can be interconverted from one to another. In contrast, the transcript for one gene cannot be directly converted into the transcript for another gene. As such, this difference in the relationship between metabolites and the relationship between transcripts may generate different variance properties in the two genomics data sets.

Acknowledgments

Funding for this methods development was obtained by a National Science Foundation grants DBI 0642481 to DJK.

References

1. Zeng, Z.-B., Kao, C.-H., and Basten, C.J. (1999) Estimating the genetic architecture of quantitative traits. *Genetic Research* **75**, 345–355.
2. Mackay, T.F.C. (2001) The genetic architecture of quantitative traits. *Annual Review of Genetics* **35**, 303–339.
3. Lander, E.S. and Botstein, D. (1989) Mapping Mendelian factors underlying quantitative traits using RFLP linkage maps. *Genetics* **121**, 185–199.
4. Schadt, E.E., Monks, S.A., Drake, T.A., Lusis, A.J., Che, N., Colinayo, V., Ruff, T.G., Milligan, S.B., Lamb, J.R., Cavet, G.,

Linsley, P.S., Mao, M., Stoughton, R.B., and Friend, S.H. (2003) Genetics of gene expression surveyed in maize, mouse and man. *Nature* **422**, 297–302.

5. Craig, B.A., Black, M.A., and Doerge, R.W. (2003) Gene expression data: the technology and statistical analysis. *Journal of Agricultural Biological and Environmental Statistics* **8**, 1–28.
6. Brem, R.B., Yvert, G., Clinton, R., and Kruglyak, L. (2002) Genetic dissection of transcriptional regulation in budding yeast. *Science* **296**, 752–755.
7. Jansen, R.C. and Nap, J.P. (2001) Genetical genomics: the added value from segregation. *Trends in Genetics* **17**, 388–391.
8. Kirst, M., Basten, C.J., Myburg, A.A., Zeng, Z.B., and Sederoff, R.R. (2005) Genetic architecture of transcript-level variation in differentiating xylem of a eucalyptus hybrid. *Genetics* **169**, 2295–2303.
9. Potokina, E., Druka, A., Luo, Z., Wise, R., Waugh, R., and Kearsey, M. (2007) Gene expression quantitative trait locus analysis of 16 000 barley genes reveals a complex pattern of genome-wide transcriptional regulation. *Plant Journal* doi: 10.1111/j.1365-313X.2007.03315.x.
10. West, M.A.L., Kim, K., Kliebenstein, D.J., van Leeuwen, H., Michelmore, R.W., Doerge, R.W., and St. Clair, D.A. (2007) Global eQTL mapping reveals the complex genetic architecture of transcript level variation in Arabidopsis. *Genetics* **175**, 1441–1450.
11. Keurentjes, J.J.B., Fu, J.Y., Terpstra, I.R., Garcia, J.M., van den Ackerveken, G., Snoek, L.B., Peeters, A.J.M., Vreugdenhil, D., Koornneef, M., and Jansen, R.C. (2007) Regulatory network construction in Arabidopsis by using genome-wide gene expression quantitative trait loci. *Proceedings of the National Academy of Sciences of the United States of America* **104**, 1708–1713.
12. Van Leeuwen, H., Kliebenstein, D.J., West, M.A.L., Kim, K.D., van Poecke, R., Katagiri, F., Michelmore, R.W., Doerge, R.W., and St. Clair, D.A. (2007) Natural variation among *Arabidopsis thaliana* accessions for transcriptome response to exogenous salicylic acid. *Plant Cell* **19**, 2099–2110.
13. Van Poecke, R.M.P., Sato, M., Lenarz-Wyatt, L., Weisberg, S., and Katagiri, F. (2008) Natural variation in RPS2-mediated resistance among Arabidopsis accessions: correlation between gene expression profiles and phenotypic responses. *Plant Cell* **19**, 4046–4060.
14. Kliebenstein, D.J., West, M.A.L., Van Leeuwen, H., Kyunga, K., Doerge, R.W., Michelmore, R.W., and St. Clair, D.A. (2006) Genomic survey of gene expression diversity in *Arabidopsis thaliana*. *Genetics* **172**, 1179–1189.
15. Flint, J., Valdar, W., Shifman, S., and Mott, R. (2005) Strategies for mapping and cloning quantitative trait genes in rodents. *Nature Reviews Genetics* **6**, 271–286.
16. Wentzell, A.M., Rowe, H.C., Hansen, B.G., Ticconi, C., Halkier, B.A., and Kliebenstein, D.J. (2007) Linking metabolic QTL with network and *cis*-eQTL controlling biosynthetic pathways. *PLoS Genetics* **3**, e162.
17. Sønderby, I.E., Hansen, B.G., Bjarnholt, N., Ticconi, C., Halkier, B.A., and Kliebenstein, D.J. (2007) A systems biology approach identifies a R2R3 MYB gene subfamily with distinct and overlapping functions in regulation of aliphatic glucosinolates. *PLoS ONE* **2**, e1322.
18. Hansen, B.G., Kliebenstein, D.J., and Halkier, B.A. (2007) Identification of a flavin-monooxygenase as the S-oxygenating enzyme in aliphatic glucosinolate biosynthesis in Arabidopsis. *Plant Journal* **50**, 902–910.
19. Zhang, Z.-Y., Ober, J.A., and Kliebenstein, D.J. (2006) The gene controlling the quantitative trait locus *EPITHIOSPECIFIER MODIFIER1* alters glucosinolate hydrolysis and insect resistance in Arabidopsis. *Plant Cell* **18**, 1524–1536.
20. Yvert, G., Brem, R.B., Whittle, J., Akey, J.M., Foss, E., Smith, E.N., Mackelprang, R., and Kruglyak, L. (2003) Trans-acting regulatory variation in Saccharomyces cerevisiae and the role of transcription factors. *Nature Genetics* **35**, 57–64.
21. Bystrykh, L., Weersing, E., Dontje, B., Sutton, S., Pletcher, M.T., Wiltshire, T., Su, A.I., Vellenga, E., Wang, J.T., Manly, K.F., Lu, L., Chesler, E.J., Alberts, R., Jansen, R.C., Williams, R.W., Cooke, M.P., and de Haan, G. (2005) Uncovering regulatory pathways that affect hematopoietic stem cell function using 'genetical genomics' *Nature Genetics* **37**, 225–232.
22. Potokina, E., Druka, A., Luo, Z., Wise, R., Waugh, R., and Kearsey, M. (2008) Gene expression quantitative trait locus analysis of 16 000 barley genes reveals a complex pattern of genome-wide transcriptional regulation. *Plant Journal* **53**, 90–101.
23. Subramanian, A., Tamayo, P., Mootha, V.K., Mukherjee, S., Ebert, B.L., Gillette, M.A., Paulovich, A., Pomeroy, S.L., Golub,

T.R., Lander, E.S., and Mesirov, J.P. (2005) Gene set enrichment analysis: a knowledge-based approach for interpreting genome-wide expression profiles. *Proceedings of the National Academy of Sciences of the United States of America* **102**, 15545–15550.

24. Mootha, V.K., Lindgren, C.M., Eriksson, K.F., Subramanian, A., Sihag, S., Lehar, J., Puigserver, P., Carlsson, E., Ridderstrale, M., Laurila, E., Houstis, N., Daly, M.J., Patterson, N., Mesirov, J.P., Golub, T.R., Tamayo, P., Spiegelman, B., Lander, E.S., Hirschhorn, J.N., Altshuler, D., and Groop, L.C. (2003) PGC-1 a-responsive genes involved in oxidative phosphorylation are coordinately downregulated in human diabetes. *Nature Genetics* **34**, 267–273.
25. Kliebenstein, D., West, M., van Leeuwen, H., Loudet, O., Doerge, R., and St. Clair, D. (2006) Identification of QTLs controlling gene expression networks defined a priori. *BMC Bioinformatics* 7, 308.
26. Zhao, K.Y., Aranzana, M.J., Kim, S., Lister, C., Shindo, C., Tang, C.L., Toomajian, C., Zheng, H.G., Dean, C., Marjoram, P., and Nordborg, M. (2007) An Arabidopsis example of association mapping in structured samples. *PLoS Genetics* **3**, e4.
27. Weigel, D. and Nordborg, M. (2005) Natural variation in arabidopsis. How do we find the causal genes? *Plant Physiology* **138**, 567–568.
28. Nordborg, M., Borevitz, J.O., Bergelson, J., Berry, C.C., Chory, J., Hagenblad, J., Kreitman, M., Maloof, J.N., Noyes, T., Oefner, P.J., Stahl, E.A., and Weigel, D. (2002) The extent of linkage disequilibrium in Arabidopsis thaliana. *Nature Genetics* **30**, 190–193.
29. Loudet, O., Chaillou, S., Camilleri, C., Bouchez, D., and Daniel-Vedele, F. (2002) Bay-0 x Shahdara recombinant inbred line population: a powerful tool for the genetic dissection of complex traits in Arabidopsis. *Theoretical and Applied Genetics* **104**, 1173–1184.
30. El-Assal, S.E.D., Alonso-Blanco, C., Peeters, A.J.M., Raz, V., and Koornneef, M. (2001) A QTL for flowering time in Arabidopsis reveals a novel allele of *CRY2*. *Nature Genetics* **29**, 435–440.
31. Koornneef, M., Alonso-Blanco, C., and Vreugdenhil, D. (2004) Naturally occurring genetic variation in *Arabidopsis thaliana*. *Annual Review of Plant Biology* **55**, 141–172.
32. Clarke, J., Mithen, R., Brown, J., and Dean, C. (1995) QTL analysis of flowering time in *Arabidopsis thaliana*. *Molecular and General Genetics* **248**, 278–286.
33. Lister, C. and Dean, D. (1993) Recombinant inbred lines for mapping RFLP and phenotypic markers in *Arabidopsis thaliana*. *Plant Journal* **4**, 745–750.
34. Perchepied, L., Kroj, T., Tronchet, M., Loudet, O., and Roby, D. (2006) Natural variation in partial resistance to *Pseudomonas syringae* is controlled by two major QTLs in *Arabidopsis thaliana*. *PLoS ONE* **1**, e123.
35. Symonds, V.V., Godoy, A.V., Alconada, T., Botto, J.F., Juenger, T.E., Casal, J.J., and Lloyd, A.M. (2005) Mapping quantitative trait loci in multiple populations of *Arabidopsis thaliana* identifies natural allelic variation for trichome density. *Genetics* **169**, 1649–1658.
36. El-Lithy, M.E., Bentsink, L., Hanhart, C.J., Ruys, G.J., Rovito, D.I., Broekhof, J.L.M., van der Poel, H.J.A., van Eijk, M.J.T., Vreugdenhil, D., and Koornneef, M. (2006) New Arabidopsis recombinant inbred line populations genotyped using SNPWave and their use for mapping flowering-time quantitative trait loci. *Genetics* **172**, 1867–1876.
37. Nordborg, M., Hu, T.T., Ishino, Y., Jhaveri, J., Toomajian, C., Zheng, H., Bakker, E., Calabrese, P., Gladstone, J., Goyal, R., Jakobsson, M., Kim, S., Morozov, Y., Padhukasahasram, B., Plagnol, V., Rosenberg, N.A., Shah, C., Wall, J.D., Wang, J., Zhao, K., Kalbfleisch, T., Schulz, V., Kreitman, M., and Bergelson, J. (2005) The pattern of polymorphism in *Arabidopsis thaliana*. *PLoS Biology* **3**, e196.
38. Borevitz, J.O., Hazen, S.P., Michael, T.P., Morris, G.P., Baxter, I.R., Hu, T.T., Chen, H., Werner, J.D., Nordborg, M., Salt, D.E., Kay, S.A., Chory, J., Weigel, D., Jones, J.D.G., and Ecker, J.R. (2007) Genome-wide patterns of single-feature polymorphism in *Arabidopsis thaliana*. *Proceedings of the National Academy of Sciences of the United States of America* **104**, 12057–12062.
39. Clark, R.M., Schweikert, G., Toomajian, C., Ossowski, S., Zeller, G., Shinn, P., Warthmann, N., Hu, T.T., Fu, G., Hinds, D.A., Chen, H.M., Frazer, K.A., Huson, D.H., Schoelkopf, B., Nordborg, M., Raetsch, G., Ecker, J.R., and Weigel, D. (2007) Common sequence polymorphisms shaping genetic diversity in *Arabidopsis thaliana*. *Science* **317**, 338–342.
40. Lempe, J., Balasubramanian, S., Sureshkumar, S., Singh, A., Schmid, M., and Weigel, D. (2005) Diversity of flowering responses

in wild *Arabidopsis thaliana* strains. *PLoS Genetics* **1**, 109–118.

41. Manfield, I.W., Jen, C.H., Pinney, J.W., Michalopoulos, I., Bradford, J.R., Gilmartin, P.M., and Westhead, D.R. (2006) Arabidopsis co-expression tool (ACT): web server tools for microarray-based gene expression analysis. *Nucleic Acids Research* **34**, W504–W509.
42. Obayashi, T., Kinoshita, K., Nakai, K., Shibaoka, M., Hayashi, S., Saeki, M., Shibata, D., Saito, K., and Ohta, H. (2007) ATTED-II: a database of co-expressed genes and cis elements for identifying co-regulated gene groups in Arabidopsis. *Nucleic Acids Research* **35**, D863–D869.
43. Grennan, A.K. (2006) Genevestigator: facilitating web-based gene-expression analysis. *Plant Physiology* **141**, 1164–1166.
44. Jen, C.H., Manfield, I.W., Michalopoulos, I., Pinney, J.W., Willats, W.G.T., Gilmartin, P.M., and Westhead, D.R. (2006) The Arabidopsis co-expression tool (ACT): a WWW-based tool and database for microarray-based gene expression analysis. *Plant Journal* **46**, 336–348.
45. Zimmermann, P., Hennig, L., and Gruissem, W. (2005) Gene-expression analysis and network discovery using Genevestigator. *Trends in Plant Science* **10**, 407–409.
46. Obayashi, T., Okegawa, T., Sasaki-Sekimoto, Y., Shimada, H., Masuda, T., Asamizu, E., Nakamura, Y., Shibata, D., Tabata, S., Takamiya, K.I., and Ohta, H. (2004) Distinctive features of plant organs characterized by global analysis of gene expression in arabidopsis. *DNA Research* **11**, 11–25.
47. Gachon, C.M.M., Langlois-Meurinne, M., Henry, Y., and Saindrenan, P. (2005) Transcriptional co-regulation of secondary metabolism enzymes in Arabidopsis: functional and evolutionary implications. *Plant Molecular Biology* **58**, 229–245.
48. Wei, H.R., Persson, S., Mehta, T., Srinivasasainagendra, V., Chen, L., Page, G.P., Somerville, C., and Loraine, A. (2006) Transcriptional coordination of the metabolic network in Arabidopsis. *Plant Physiology* **142**, 762–774.
49. Zhang, P.F., Foerster, H., Tissier, C.P., Mueller, L., Paley, S., Karp, P.D., and Rhee, S.Y. (2005) MetaCyc and AraCyc. Metabolic pathway databases for plant research. *Plant Physiology* **138**, 27–37.
50. Mueller, L.A., Zhang, P.F., and Rhee, S.Y. (2003) AraCyc: a biochemical pathway database for Arabidopsis. *Plant Physiology* **132**, 453–460.
51. Luo, Z.W., Potokina, E., Druka, A., Wise, R., Waugh, R., and Kearsey, M. J. (2007) SFP genotyping from Affymetrix arrays is robust but largely detects cis-acting expression regulators. *Genetics* **176**, 789–800.
52. Richardson, A., Boscari, A., Schreiber, L., Kerstiens, G., Jarvis, M., Herzyk, P., and Fricke, W. (2007) Cloning and expression analysis of candidate genes involved in wax deposition along the growing barley (Hordeum vulgare) leaf. *Planta* **226**, 1459–1473.
53. Shen, L.H., Gong, J., Caldo, R.A., Nettleton, D., Cook, D., Wise, R.P., and Dickerson, J.A. (2005) BarleyBase – an expression profiling database for plant genornics. *Nucleic Acids Research* **33**, D614–D618.
54. Saito, K., Hirai, M., and Yonekura-Sakakibara, K. (2008) Decoding genes with coexpression networks and metabolomics – 'majority report by precogs'. *Trends in Plant Science* **13**, 36–43.
55. Kearsey, M.J. and Farquhar, A.G.L. (1998) QTL analysis in plants; where are we now? *Heredity* **80**, 137–142.
56. Jordan, M.C., Somers, D.J., and Banks, T.W. (2007) Identifying regions of the wheat genome controlling seed development by mapping expression quantitative trait loci. *Plant Biotechnology Journal* **5**, 442–453.
57. DeCook, R., Lall, S., Nettleton, D., and Howell, S.H. (2006) Genetic regulation of gene expression during shoot development in Arabidopsis. *Genetics* **172**, 1155–1164.
58. Juenger, T.E., Wayne, T., Boles, S., Symonds, V.V., McKay, J., and Coughlan, S.J. (2006) Natural genetic variation in whole-genome expression in *Arabidopsis thaliana*: the impact of physiological QTL introgression. *Molecular Ecology* **15**, 1351–1365.
59. Street, N.R., Skogstrom, O., Sjodin, A., Tucker, J., Rodriguez-Acosta, M., Nilsson, P., Jansson, S., and Taylor, G. (2006) The genetics and genomics of the drought response in Populus. *Plant Journal* **48**, 321–341.
60. An, C.F., Saha, S., Jenkins, J.N., Scheffler, B.E., Wilkins, T.A., and Stelly, D.M. (2007) Transcriptome profiling, sequence characterization, and SNP-based chromosomal assignment of the EXPANSIN genes in cotton. *Molecular Genetics and Genomics* **278**, 539–553.

61. Venu, R.C., Jia, Y., Gowda, M., Jia, M.H., Jantasuriyarat, C., Stahlberg, E., Li, H., Rhineheart, A., Boddhireddy, P., Singh, P., Rutger, N., Kudrna, D., Wing, R., Nelson, J.C., and Wang, G.L. (2007) RL-SAGE and microarray analysis of the rice transcriptome after Rhizoctonia solani infection. *Molecular Genetics and Genomics* **278**, 421–431.
62. Kiani, S.P., Grieu, P., Maury, P., Hewezi, T., Gentzbittel, L., and Sarrafi, A. (2007) Genetic variability for physiological traits under drought conditions and differential expression of water stress-associated genes in sunflower (*Helianthus annuus* L.). *Theoretical and Applied Genetics* **114**, 193–207.
63. Shi, C., Uzarowska, A., Ouzunova, M., Landbeck, M., Wenzel, G., and Lubberstedt, T. (2007) Identification of candidate genes associated with cell wall digestibility and eQTL (expression quantitative trait loci) analysis in a Flint x Flint maize recombinant inbred line population. *BMC Genomics* **8**, 22.
64. Aoki, K., Ogata, Y., and Shibata, D. (2007) Approaches for extracting practical information from gene co-expression networks in plant biology. *Plant and Cell Physiology* **48**, 381–390.
65. Hirai, M.Y., Klein, M., Fujikawa, Y., Yano, M., Goodenowe, D.B., Yamazaki, Y., Kanaya, S., Nakamura, Y., Kitayama, M., Suzuki, H., Sakurai, N., Shibata, D., Tokuhisa, J., Reichelt, M., Gershenzon, J., Papenbrock, J., and Saito, K. (2005) Elucidation of gene-to-gene and metabolite-to-gene networks in Arabidopsis by integration of metabolomics and transcriptomics. *Journal of Biological Chemistry* **280**, 25590–25595.
66. Ma, S.S., Gong, Q.Q., and Bohnert, H.J. (2007) An Arabidopsis gene network based on the graphical Gaussian model. *Genome Research* **17**, 1614–1625.
67. Urbanczyk-Wochniak, E. and Sumner, L.W. (2007) MedicCyc: a biochemical pathway database for Medicago truncatula. *Bioinformatics* **23**, 1418–1423.
68. Caspi, R., Foerster, H., Fulcher, C.A., Hopkinson, R., Ingraham, J., Kaipa, P., Krummenacker, M., Paley, S., Pick, J., Rhee, S.Y., Tissier, C., Zhang, P.F., and Karp, P.D. (2006) MetaCyc: a multiorganism database of metabolic pathways and enzymes. *Nucleic Acids Research* **34**, D511–D516.
69. Li, S.M., Armstrong, C.M., Bertin, N., Ge, H., Milstein, S., Boxem, M., Vidalain, P.O., Han, J.D.J., Chesneau, A., Hao, T., Goldberg, D.S., Li, N., Martinez, M., Rual, J.F., Lamesch, P., Xu, L., Tewari, M., Wong, S.L., Zhang, L.V., Berriz, G.F., Jacotot, L., Vaglio, P., Reboul, J., Hirozane-Kishikawa, T., Li, Q.R., Gabel, H.W., Elewa, A., Baumgartner, B., Rose, D.J., Yu, H.Y., Bosak, S., Sequerra, R., Fraser, A., Mango, S.E., Saxton, W.M., Strome, S., van den Heuvel, S., Piano, F., Vandenhaute, J., Sardet, C., Gerstein, M., Doucette-Stamm, L., Gunsalus, K.C., Harper, J.W., Cusick, M.E., Roth, F.P., Hill, D.E., and Vidal, M. (2004) A map of the interactome network of the metazoan C-elegans. *Science* **303**, 540–543.
70. von Mering, C., Krause, R., Snel, B., Cornell, M., Oliver, S.G., Fields, S., and Bork, P. (2002) Comparative assessment of large-scale data sets of protein–protein interactions. *Nature* **417**, 399–403.
71. Ito, T., Chiba, T., Ozawa, R., Yoshida, M., Hattori, M., and Sakaki, Y. (2001) A comprehensive two-hybrid analysis to explore the yeast protein interactome. *Proceedings of the National Academy of Sciences of the United States of America* **98**, 4569–4574.
72. Geisler-Lee, J., O'Toole, N., Ammar, R., Provart, N.J., Millar, A.H., and Geisler, M. (2007) A predicted interactome for Arabidopsis. *Plant Physiology* **145**, 317–329.
73. Wei, N., Chamovitz, D.A., and Deng, X.W. (1994) Arabidopsis Cop9 is a component of a novel signaling complex mediating light control of development. *Cell* **78**, 117–124.
74. Vision, T.J., Brown, D.G., and Tanksley, S.D. (2000) The origins of genomic duplications in Arabidopsis. *Science* **290**, 2114–2117.
75. Blanc, G., Hokamp, K., and Wolfe, K.H. (2003) A recent polyploidy superimposed on older large-scale duplications in the Arabidopsis genome. *Genome Research* **13**, 137–144.
76. Wolfe, K.H. and Shields, D.C. (1997) Molecular evidence for an ancient duplication of the entire yeast genome. *Nature* **387**, 708–713.
77. Blanc, G. and Wolfe, K.H. (2004) Widespread paleopolyploidy in model plant species inferred from age distributions of duplicate genes. *Plant Cell* **16**, 1667–1678.
78. Ober, D. (2005) Seeing double: gene duplication and diversification in plant secondary metabolism. *Trends in Plant Science* **10**, 444–449.
79. Meyer, R.C., Steinfath, M., Lisec, J., Becher, M., Witucka-Wall, H., Törjék, O., Fiehn, O., Eckardt, A., Willmitzer, L., Selbig, J., and Altmann, T. (2007) The

metabolic signature related to high plant growth rate in Arabidopsis thaliana. *Proceedings of the National Academy of Sciences of the United States of America* **104**, 4759–4764.

80. Basten, C.J., Weir, B.S., and Zeng, Z.-B. (1999) QTL Cartographer, Version 1.13, Department of Statistics, North Carolina State University, Raleigh, N.C.
81. Wang, S., Basten, C.J., and Zeng, Z.-B. (2006) Windows QTL Cartographer 2.5. Department of Statistics, North Carolina State University, Raleigh, NC.
82. Irizarry, R.A., Hobbs, B., Collin, F., Beazer-Barclay, Y.D., Antonellis, K.J., Scherf, U., and Speed, T.P. (2003) Exploration, normalization, and summaries of high density oligonucleotide array probe level data. *Biostatistics* **4**, 249–264.
83. Churchill, G.A. and Doerge, R.W. (1994) Empirical threshold values For quantitative trait mapping. *Genetics* **138**, 963–971.
84. Bogdan, M. and Doerge, R.W. (2005) Biased estimators of quantitative trait locus heritability and location in interval mapping. *Heredity* **95**, 476–484.
85. Doerge, R.W. (2002) Mapping and analysis of quantitative trait loci in experimental populations. *Nature Reviews Genetics* **3**, 43–52.
86. Doerge, R.W. and Churchill, G.A. (1996) Permutation tests for multiple loci affecting a quantitative character. *Genetics* **142**, 285–294.
87. Gilbert, H. and Le Roy, P. (2003) Comparison of three multitrait methods for QTL detection. *Genetics Selection Evolution* **35**, 281–304.
88. Knott, S.A. and Haley, C.S. (2000) Multitrait least squares for quantitative trait loci detection. *Genetics* **156**, 899–911.
89. Ronin, Y.I., Kirzhner, V.M., and Korol, A.B. (1995) Linkage between loci of quantitative traits and marker loci – multi-trait analysis with a single marker. *Theoretical and Applied Genetics* **90**, 776–786.
90. Chen, M. and Kendziorski, C. (2007) A statistical framework for expression quantitative trait loci mapping. *Genetics* **177**, 761–771.
91. Ball, R.D. (2007) Quantifying evidence for candidate gene polymorphisms: Bayesian analysis combining sequence-specific and quantitative trait loci colocation information. *Genetics* **177**, 2399–2416.
92. Hoti, F. and Sillanpaa, M.J. (2006) Bayesian mapping of genotype x expression interactions in quantitative and qualitative traits. *Heredity* **97**, 4–18.
93. Lan, H., Chen, M., Flowers, J.B., Yandell, B.S., Stapleton, D.S., Mata, C.M., Mui, E.T.K., Flowers, M.T., Schueler, K.L., Manly, K.F., Williams, R.W., Kendziorski, C., and Attie, A.D. (2006) Combined expression trait correlations and expression quantitative trait locus mapping. *PLoS Genetics* **2**, 51–61.

Chapter 12

Co-expression Analysis of Metabolic Pathways in Plants

Ann Loraine

Abstract

Co-expression analysis allows experimenters to re-use archived expression microarray data to uncover previously unknown functional relationships between genes. An observation that a group of genes are co-expressed across diverse experimental conditions suggests they may play similar roles in the cell. Several thousand expression microarray experiments performed on samples from *Arabidopsis thaliana* have entered the public domain and it is now possible to use these data to investigate metabolic networks in plants. This chapter explains how to use a Web-based tool (CressExpress) to investigate co-expression of genes involved in metabolic pathways in *Arabidopsis*. Using CressExpress together with desktop visualization and analysis tools, one can easily identify clusters of genes that are co-expressed with one or more genes of interest, making it possible to identify new players in metabolic pathways that are regulated at the level of mRNA abundance.

Key words: Expression array, co-expression, relevance networks, correlation, linear regression.

1. Introduction

More so perhaps than any other genomic technology, expression DNA microarrays have driven a revolution in the conduct of molecular biology. The sheer volume of data even a single well-designed microarray experiment can produce has stimulated molecular biology labs to develop data management and statistical analysis expertise that they can re-deploy in new settings. This chapter will discuss co-expression analysis as one example of this and will describe how researchers can take advantage of archived microarray data to study metabolic pathways, focusing on *Arabidopsis thaliana* and data from the ATH1 microarray from Affymetrix to demonstrate what is possible.

Dmitry A. Belostotsky (ed.), *Plant Systems Biology, vol. 553*

DOI 10.1007/978-1-60327-563-7_12 Springerprotocols.com

The ATH1 array from Affymetrix consists of around 500,000 25-bp oligonucleotide probes that are grouped into probe sets, where each probe set includes 11 perfect match and 11 mismatch or control probes. All the probes in a probe set are selected from a 200 to 300 base region near the three prime end of a target transcript, and the probes may or may not overlap, depending on the transcript. The ATH1 array contains 22,814 probe sets, including control probe sets useful for assessing sample quality.

One noteworthy aspect of the ATH1 array is that the target transcript sequences were based on a set of gene model annotations that were partially hand-annotated as part of the *Arabidopsis* Genome Initiative (1). Other array designs from Affymetrix, especially the mammalian arrays, have targeted expressed sequences harvested from the dbEST, Genbank, and Unigene databases. Quality issues with EST data, such as the impossibility in many cases of identifying the transcribed strand, create technical challenges in probe set design and often result in probe sets that yield problematic data. This means that, relatively speaking, the ATH1 data may be unusually high quality when compared to data from other Affymetrix 3-prime arrays. It will be interesting to test this idea as informatics methods improve and we can more easily access data in high-throughput fashion.

Another less unusual feature of the ATH1 array is that many probe sets are promiscuous in the sense that they may hybridize with target transcripts arising from multiple locations in the genome. This is reflected in the probe set nomenclature, which uses suffixes to indicate when a probe set uniquely identifies its target (suffix "_at") or may recognize targets from multiple genes (suffix "_x_at"). However, users of the data should note that these names were assigned years ago, and the cross-hybridizing probe sets received their "x_at" designations based on only those sequences that were provided to the array design pipeline in the initial design phase. Thus, some probe sets with "_at" designations may cross-hybridize with other targets that were unknown at the time the ATH1 array was created. Furthermore, different groups have produced probe set to target gene annotations, including both Affymetrix and The Arabidopsis Information Resource (TAIR), and so it is important to note that sometimes these annotations disagree. For this reason, a careful investigator should always keep track of probe set names during an analysis and should also note which probe set annotation is being used.

As of early 2008, the Gene Expression Omnibus contained over 3,000 ATH1 sample (GSM) records, each one corresponding to one array hybridization (2). The availability of such a large amount of data in an accessible location makes it tempting to experiment with meta-analysis methods that compare samples from completely different experiments. Some care must be taken in this, however. For example, users should beware of attempting to compare expression across samples from different sources, since any observed differences could be due to the fact that samples come from different labs, not because of any interesting biological variation.

However, it is valid to examine how probe set readings vary across multiple samples and experiments relative to each other, and this is the crux of why co-expression analysis is a useful technique. Consider **Fig. 12.1**, which shows a scatter plot of expression values for two probe sets from 1,771 ATH1 array hybridizations. Each point represents a single array; the x and y co-ordinates are the expression values for the x-axis and y-axis probe sets, respectively. Note that when the y-axis probe set values are high, the x-axis probe set values are also high, and the reverse is true. There is a linear relationship between expression values from the two probe sets, and this relationship is so tight that one can use the value of one variable to predict the value of the other with relatively good confidence. Based on this, we can say that the targets for these two probe sets vary in concert, and this co-variation appears across the vast majority of arrays represented on the plot throughout the expression ranges of each probe set.

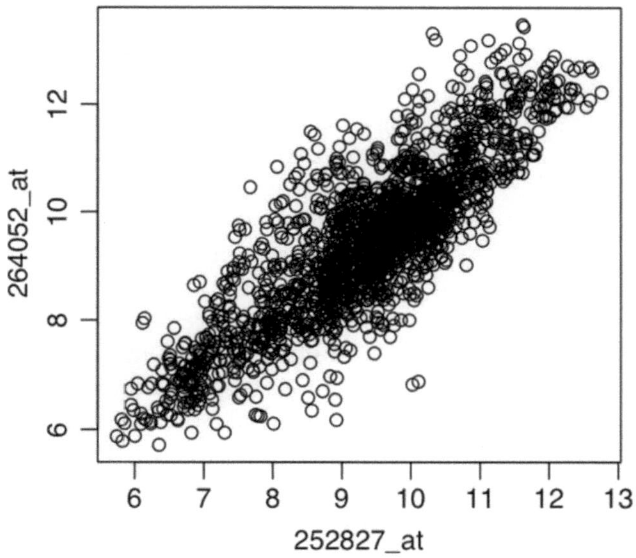

Fig. 12.1. Positive co-expression between target genes for two ATH1 array probe sets.

To quantify the closeness of this relationship, we can compute a correlation coefficient (Pearson's r) from the data. Pearson's r ranges from –1 to 1, and values closer to zero indicate less association, while values closer to 1 or –1 indicate a tighter clustering about a line. Performing a linear regression of y on x yields a linear model one might use to predict the value of y given x. In co-expression analysis, we typically use linear regression as a method for obtaining an r^2 value, which is square of Pearson's r and expresses the percentage of variation in y that is explained by variation in x. Higher r^2values that are closer to one indicate a tighter relationships between variables, while the slope of the regression line (positive or negative) indicates the directionality of the relationship.

Methods exist for assigning probabilities (p values) to regressions and correlations; in general, plots with more points clustering more tightly around the regression line yield smaller p values. For example, consider a pairwise comparison between two genes that yields an r^2 value of 0.75 and a p value of 0.001. This means that the probability of obtaining an r^2 value equal to or greater than 0.75 purely by chance is 0.001. Stated another way, we expect that only 1 in 1,000 of randomly created plots would exhibit an r^2 as large as what we observe in the data.

If we are interested in comparing only two genes, and only in performing one such comparison, a p value of 0.001 is very unlikely, given that we could have chosen any pair of genes. However, for a large-scale, data-mining experiment in which we are examining the entire genome, we might compute a linear regression for every possible pair of probe sets. For the ATH1 array, this means that we could consider 22,810 choose two distinct probe set plots, over 260 million combinations. At a significance level of 0.001, that means that under the assumption of random (non-linear) relationships between probe set expression values, we might expect that around 260,000 comparisons (0.001×260 million) would yield r^2 values that appear to be significant at a pre-designated alpha level of 0.001. In other words, we would expect to obtain 260,000 false positives, i.e., pairs of genes that appear to be co-expressed but are not.

We can improve our chance of avoiding such a large number of false positives by changing our p value threshold for deciding significance (the alpha level) by lowering it from 0.001 to some other value. But how much should we lower it? According to the Bonferroni adjustment, which is one of the most conservative approaches, we can achieve a 0.001 probability of having no false positives among all 260 million tests, equivalent to a family-wise error rate of 0.001, by dividing 0.001 by the number of tests we perform and using the result as the new alpha threshold for each test. (The derivation of this calculation is relatively straightforward and is explained on a number of different Web sites.) For the ATH1 array, this calculation yields 3.8×10^{-12}, which may seem like a ridiculously small number, but, in practice, p values for linear regressions involving expression data from 200 or more arrays frequently achieve this and smaller p values. Indeed, regressions involving the plots shown in **Fig. 12.2** have p values that are many orders of magnitude smaller than this Bonferroni-adjusted threshold.

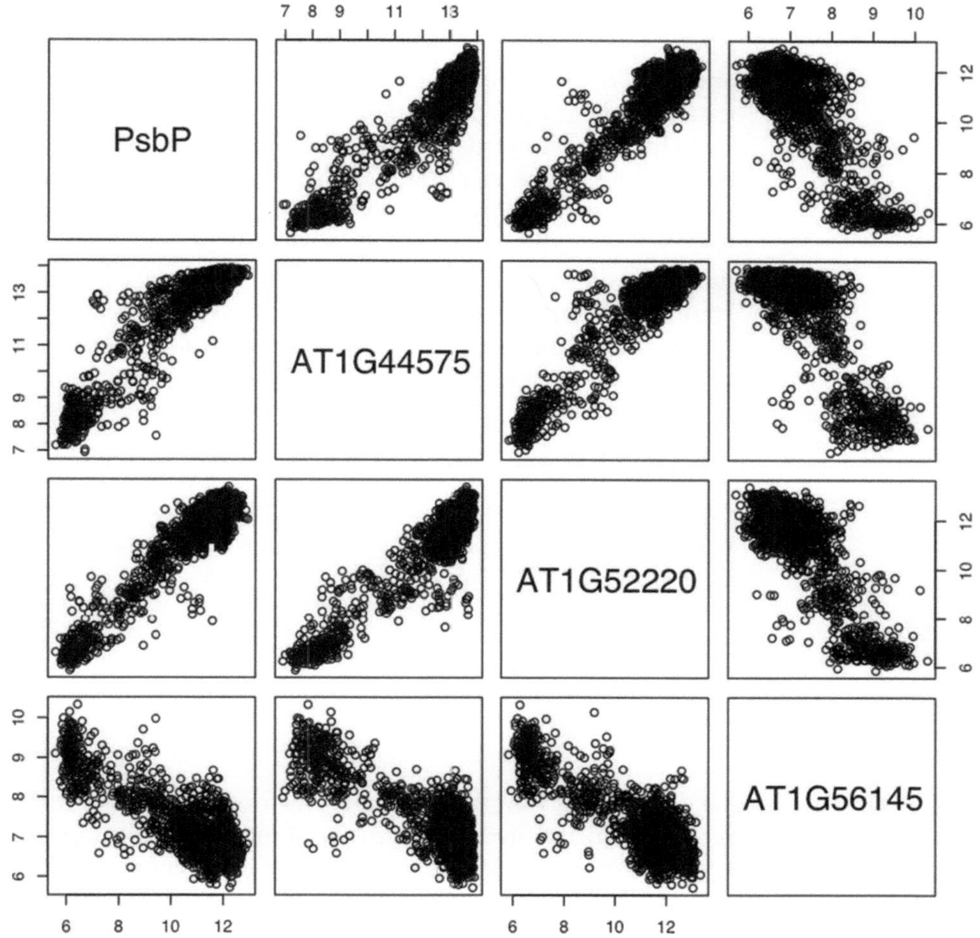

Fig. 12.2. Positive and negative co-expression between AT1G56145 (predicted protein kinase) and three genes encoding predicted and known photosystem II components. Probe sets (*left* to *right*) interrogating these genes were 251784_at, 245213_at, 259838_at, and 262093_at, according to annotations from TAIR. In general, negative correlation (rightmost column) is less common than positive correlation.

As of January 2008, the GEO contained results from almost 200,000 individual microarray hybridizations. Clearly, there are abundant data available for co-expression analysis, and there will likely be more in the future as more journals make public release of the data a prerequisite for publication. As a result, these data are finding a second life as raw material for co-expression analysis.

GEO makes these data available in formats that make harvesting and mining the data relatively easy. The GEO Web site interface is sometimes confusing and difficult to use, but the ftp site accompanying GEO is relatively well organized and presents few problems for computational scientists who need to get the data in bulk. Other resources that store and distribute plant array data include the Nottingham Arabidopsis Stock Center AffyWatch and NASCArrays service (3) which offers access to Arabidopsis data, and PlexDB (4), which aims to provide access to the expression data from 17 species of plants and plant pathogens.

This trend toward greater portability and accessibility of the data will no doubt accelerate, and the ease of harvesting the data has already helped many groups recycle the data, using it to populate Web-accessible data-mining tools that re-deploy the data in new forms for data-mining experiments. But how can individual laboratories interested in specific pathways or processes utilize these data in their research? This chapter will demonstrate one example of how this can work, using the CressExpress tool hosted at http://www.cressexpress.org, and two flexible and powerful desktop data analysis programs: R (www.r-project.org) and TableView (5). This chapter will describe the use of large-scale co-expression analysis to characterize a pathway and identify potential new players in the pathway, which can then be tested in the laboratory.

2. Methods

2.1. Step 1: Identify Probe Sets (Array Elements) That Interrogate the Pathway Genes

To demonstrate how this works, we will focus on a single pathway in secondary metabolism: biosynthesis of indolic glucosinolates from tryptophan. Glucosinolates are sulfur- and nitrogen-containing compounds that are synthesized from amino acid precursors and are found in a number of *Brassica* species, including *Arabidopsis* (6, 7). Glucosinolate breakdown products are responsible for the pungent flavors of horseradish and wasabi mustard, and some have been studied for their potential to defend against cancer. In the plant, glucosinolate compounds are believed to play roles in pathogen resistance, defense against herbivory, and as chemical attractants. Our goal in this co-expression analysis will be to identify candidate genes that may be involved in glucosinolate biosynthesis, using the expression patterns of genes that are already known to be a part of the pathway.

The first step in the analysis is to identify genes involved in the pathway of interest and then identify the ATH1 probe sets that interrogate these genes. To start, we turn to the AraCyc database, an online database of metabolism in *Arabidopsis*, available via links from the TAIR Web site (8). AraCyc reports that the indolic glucosinolates biosynthesis pathway involves five reactions, catalyzed by six gene products (**Table 12.1**).

We can use the gene names or their AGI codes to look up the associated ATH1 probe sets on the TAIR Web site. **Figure 12.3A** shows screen captures from the TAIR Web site showing how to

Table 12.1
Glucosinolate biosynthesis from tryptophan. Gene symbols, probe set to gene target annotations, and annotations are from TAIR and AraCyc version 3.5

AGI code	Gene symbol	Probe set (ATH1)	Enzyme
At2g22330	CYP79B3	264052_at	Cytochrome p450
At4g39950	CYP79B2	252827_at	Cytochrome p450
At4g31500	CYP83B1, ATR4, RED1, RNT1, SUR2	253534_at	Cytochrome p450
At2g20610	SUR1, ALF1, HLS3, RTY, RTY1, SUPERROOT 1	263714_at	Transaminase activity
At1g24100	UGT74B1	264873_at	UDP-glucosyl transferase
At1g74100	F2P9.3, F2P9_3	260387_at	Sulfotransferase

look up probe sets from the ATH1 array for CYP79B2. First, we run a search using the search box on the TAIR home page. This retrieves a list of results, including a locus named for the AGI (Arabidopsis Genome Initiative) code name for CYP79B2. Clicking the locus link opens a locus-level page for CYP79B2, which reports array elements associated with the gene.

Many genes have multiple probe sets from one or both of two Affymetrix arrays, including ATH1 and the older AG (Arabidopsis Genome) array. Usually, the probe set with the longer name is from ATH1. Clicking a probe set name opens a new page describing it. To find out if it is on the ATH1 array, click the "+" icon next to the text "See list of array designs."

Many probe sets are promiscuous, i.e., they have multiple targets. It is unclear how using these probe sets will affect results, and so anyone performing a co-expression analysis should be aware when a probe set interrogates multiple genes. As of this writing, TAIR displays the identity of alternative probe set targets in the space above the page heading "Array Element." An example is shown in **Fig. 12.3B**.

2.2. Step 2: Are They Co-expressed with Each Other?

The next step of the analysis will be to determine the extent to which these six genes are co-expressed with each other. If we learn that all six exhibit a high degree of co-expression with each other, we may then be able to use them as a kind of computational bait to identify other candidate genes that may also play a role in the biosynthesis of glucosinolates from tryptophan.

To determine whether the genes are co-expressed with each other, we will use R, a freely available, open source tool for statistical analysis, together with a simple Web service that delivers

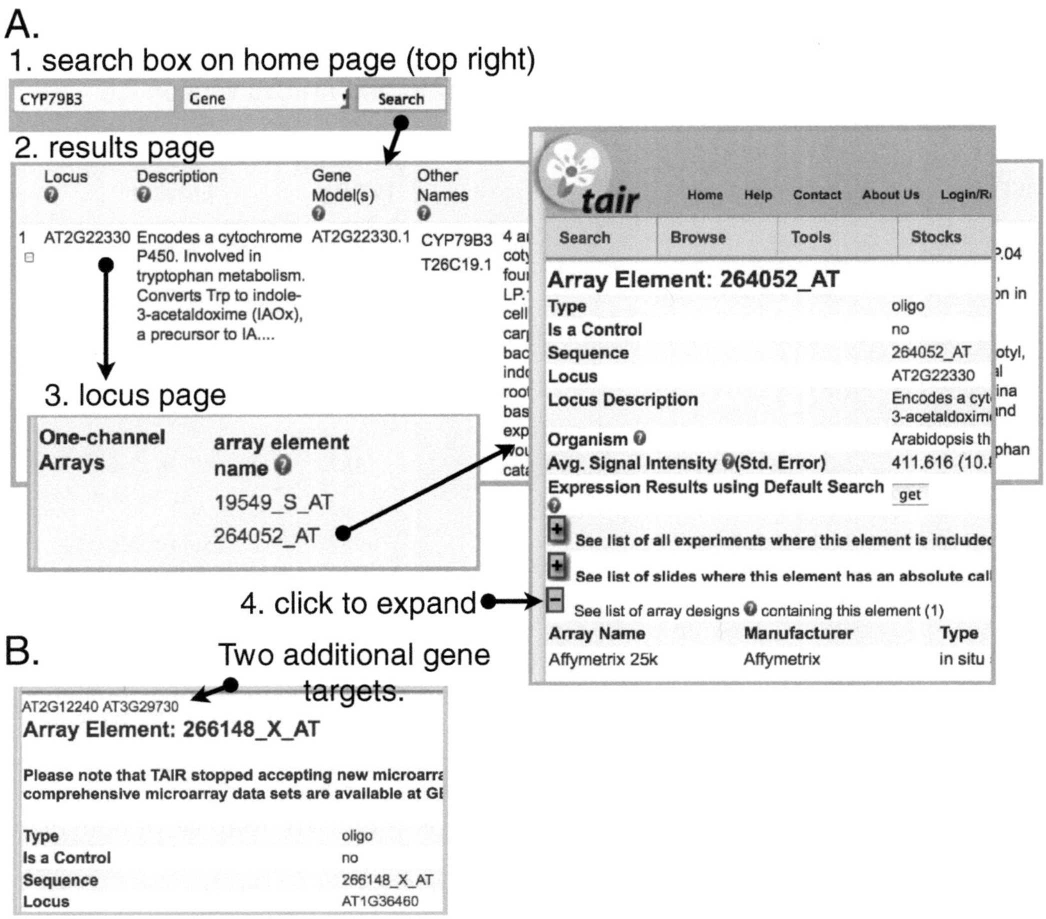

Fig. 12.3. Flowchart illustrating how to use TAIR to find probe sets that uniquely recognize a target gene. **A**. Looking up the ATH1 probe set for CYP79B3. **B**. Where to look on a probe set page to find out if a probe set interrogates multiple target genes.

expression data in simple tabular or comma-delimited formats. (To find out more about R, visit the R project home page at http://www.r-project.org.) We will use R to access expression values for the six glucosinolate probe sets, compute Pearson's correlation coefficient for each pair of genes, and display a scatter plot similar to **Fig. 12.2**.

First, we launch a program called the R interpreter, a program into which we type commands to read files, manipulate data, perform statistical tests, etc. Note that commands that are typed into the interpreter can also be typed into and then saved in plain text file and then run in the R interpreter by typing the source command and the name of the file. This capability is why R is considered to be both a programming language one can use to write scripts (commands that should be performed in a sequence) and an interactive tool for analyzing data. In practice, most people use R in both ways, since it is often convenient to re-use a sequence of analysis steps many times. Note however that if you want to write scripts for R to run again and again, you must save them in plain text formats. R cannot read Word documents or any other file that is not a plain text (e.g., a ".txt") file.

Figure 12.4 shows the commands typed into the interpreter as part of an R session, along with output from each command. Commands typed into the R interpreter appear next to the R

```
R is a collaborative project with many contributors.
...
Type 'q()' to quit R.

> base = "http://www.cressexpress.org/cgi-bin/getExpVals.py"
> probesets = c('264052_at','252827_at','253534_at','263714_at','264873_a
+   '260387_at')
> pss = paste(probesets,collapse=',')
> url = paste(c(base,'?version=3_0&file-
format=comma&pss=',pss),sep='',collapse='')
> url
[1] "http://obiwan.ssg.uab.edu:8080/coexpression/cgi-
bin/getExpVals.py?version=3_0&file-
format=comma&pss=264052_at,252827_at,253534_at,263714_at,264873_at,260387
> dat = read.delim(url,sep=',',header=TRUE)
> dim(dat)
[1] 1771   11
> heads = c('cel','CYP79B3','CYP79B2','CYP83B1','SUR1','UGT74B1','At1g741
+   'exp','slide','ks','url')
> names(dat)=heads
> cor(dat[,2:7])
            CYP79B3   CYP79B2   CYP83B1      SUR1   UGT74B1 At1g74100
CYP79B3   1.0000000 0.8429500 0.7924926 0.6376598 0.6604806 0.7220173
CYP79B2   0.8429500 1.0000000 0.7970052 0.6100147 0.6322989 0.7642592
CYP83B1   0.7924926 0.7970052 1.0000000 0.7764918 0.7659748 0.8430911
SUR1      0.6376598 0.6100147 0.7764918 1.0000000 0.8319885 0.7611434
UGT74B1   0.6604806 0.6322989 0.7659748 0.8319885 1.0000000 0.7814670
At1g74100 0.7220173 0.7642592 0.8430911 0.7611434 0.7814670 1.0000000
> plot(dat[,2:7])
> model = lm(dat$CYP79B2~dat$SUR1)
> summary(model)
Call:
lm(formula = dat$CYP79B2 ~ dat$SUR1)

Residuals:
     Min       1Q   Median       3Q      Max
-3.64168 -0.84895 -0.04724  0.78613  4.18958

Coefficients:
            Estimate Std. Error t value Pr(>|t|)

dat$SUR1     0.94919    0.02931  32.379   <2e-16 ***
---
Signif. codes:  0 '***' 0.001 '**' 0.01 '*' 0.05 '.' 0.1 ' ' 1

Residual standard error: 1.144 on 1769 degrees of freedom
Multiple R-Squared: 0.3721,    Adjusted R-squared: 0.3718
F-statistic:  1048 on 1 and 1769 DF,  p-value: < 2.2e-16
```

Fig. 12.4. A sample session with the R interpreter.

prompt (>>>) and outputs appear on the next lines. First, we define a variable (url) that represents a Web address for data associated with the six probe sets listed in **Table 12.1**. The URL has several parameters, or components, that specify the data we want to retrieve. The first parameter (version) specifies the data release version. More information about specific data releases can be found on the CressExpress Web site in the FAQ section. In this case, release 3_0 corresponds to version 3.0, which includes around 1,770 arrays that were processed using the RMA algorithm. The second URL parameter (file-format) specifies how the data should be formatted. In this case, we request that commas be used as a field separator. The last parameter (pss) gives a comma-separated list of probe sets whose expression data we would like to retrieve. This list can include 1–50 probe set names, but requesting larger numbers of probe sets will result in slower response time. Note that every parameter except the first one must be proceeded by an "&" character, and all parameters must be followed by an "=" sign and then the requested value.

To see how the data are formatted, and to save it to a local file, you can enter the URL in a Web browser's Navigation Toolbar and then save the resulting "page" as a plain text file. You should then be able to open it in Excel or any other program that can read tabular data (*see* **Fig. 12.5**). Depending on the browser, you may need to experiment with the "save as" options to get the data into the proper format. (In Firefox, use the "All Files" option.)

glucosinolates.txt

	A	B	C	D	E	F	
1	cel	264052_at	252827_at	exp	slide	ks	url
2	SHM002_ATH1	9.7994	9.3265	10	Murray_A	0.02	http:/
3	SHM002_ATH1	9.4053	9.2606	10	Murray_A	0.041	http:/
4	SHM002_ATH1	9.9792	10.042	10	Murray_A	0.037	http:/
5	MC001_ATH1_	7.912	8.6416	14	Campb-32	0.096	http:/
6	MC001_ATH1_	8.8488	8.249	14	Campb-33	0.039	http:/
7	MC001_ATH1_	8.8192	9.0895	14	Campb-32	0.073	http:/
8	MC001_ATH1_	9.7531	9.4786	14	Campb-32	0.043	http:/
9	MC001_ATH1_	8.3206	8.9945	14	Campb-32	0.016	http:/
10	MC001_ATH1_	10.054	9.534	14	Campb-31	0.052	http:/
11	MC001_ATH1_	8.7295	8.751	14	Campb-32	0.061	http:/
12	MC001_ATH1_	9.7983	9.6517	14	Campb-31	0.049	http:/

Fig. 12.5. Expression data from the CressExpress Direct Access Web service. Columns B and C contain RMA-processed data for two probe sets that interrogate glucosinolate pathway genes.

Within the R interpreter, we retrieve the data from the Web service using the read.delim command and store the output to a variable called dat, which is an R object called a data frame. We can then pass dat to various functions that use the data stored in dat to compute correlations, regressions, make plots of the data, and many other possibilities. Note that to find out more information about any given function, you can type help(cmd) where cmd is the name of the function.

The read.delim command can access data either from a local file or a URL Web address, and it can accept several options that modify how it reads the data. The sep=',' option indicates that the field delimiter for the incoming data is a comma character, and the header=TRUE option indicates that the first row of the data contains column titles. Once we have retrieved the data, we can change the column headings to gene names using the names command to make it easier to reference columns we want. To compute pairwise correlations, we use cor, and pass it a part of the data set that contains the expression values (dat[,4:7]) corresponding to columns 4 through 7. To view a multi-scatter plot similar to **Fig. 12.3**, we use the command plot and again pass it columns 4 through 7. To compute a linear regression (saved as variable model), we use the lm (linear model) command and then pass it to summary as shown. In this case, we find that the p value associated with comparing CYP79B3 with CYP79B2 is less than 10^{-16}, a significant result.

2.3. Finding Other Genes That Are Co-expressed

It is clear that the six genes are highly co-expressed with each other, which suggests that co-expression analysis may be able to help us find other genes that play a role in the pathway or in related pathways. The CressExpress tool will allow us to select specific experiment we would like to include in a co-expression analysis, and so before we get started, we might want to identify experiments where the six genes' expression levels are relatively high. Although the genes appear to be highly co-expressed throughout their range, it is possible that other genes that play a role in the pathway exhibit co-expression with the "bait" pathway genes only in high-expression situations. We could use R to identify these experiments using R commands like subset or sort; however, for the purposes of this chapter, we will demonstrate how to do this using a different tool: TableView.

TableView is data exploration and analysis tool that is freely available and open source, developed originally at the University of Minnesota *(5)*. It has a number of useful visualizations and functions, but here we will focus on just one function: its ability to display interactive, clickable scatter plots. To launch TableView, visit the Web site at http://igb.bioviz.org/links.shtml and follow the links that lead to a JavaWebStart page (file extension JNLP) that, once loaded, will trigger download and launch of the application. For more information about Java Web Start and launching TableView, see the CressExpress tutorial on using TableView to visualize expression data available under the "Visualization" tab accessible from the CressExpress home page.

To identify experiments and conditions where the six indolic glucosinolates biosynthesis genes are highly expressed, we launch TableView and select the Load Table option under the File menu, which opens a second window where we enter the same data direct access URL as before (*see* **Fig. 12.6**). Once the data appear in the

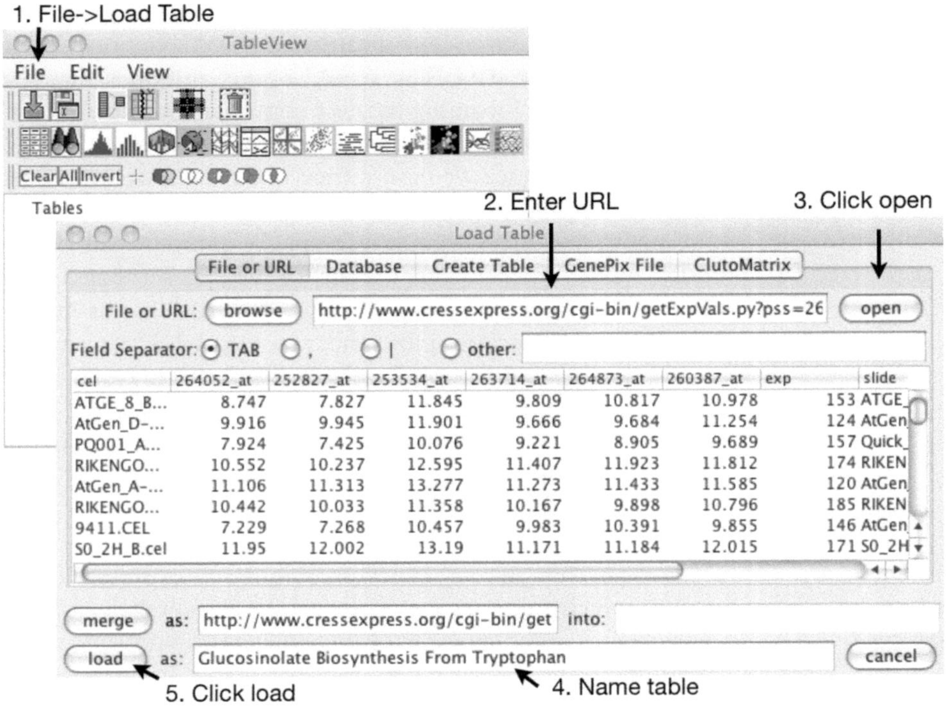

Fig. 12.6. Retrieving expression values from the CressExpress Direct Access Web service using TableView. The URL used to retrieve these data was http://www.cressexpress.org/cgi-bin/getExpVals.py?version=3_0&file-format=tab&pss=264052_at,252827_at,253534_at,263714_at,264873_at,260387_at.

second window, we click the "Load" button to load the data as a new table into TableView. Note that in the example (**Fig. 12.6**), we have named the table "Glucosinolate Biosynthesis From Tryptophan" for convenience.

Once the data are loaded as a new table data set, we can select it in the main window and display it using the viewing options represented by the buttons at the top of the display (**Fig. 12.7**). If we click the Table icon, TableView will show a spreadsheet view of the data. To make working with the data more convenient, we first click the column heading for the "exp" column to sort and group the arrays based on their experimental affiliation (these numbers are assigned by NASCArrays). Doing this causes the spreadsheet to display all arrays from the same lab and same experiment in consecutive rows, which will be useful later when we view scatter plots between probe set expression values. Next, we click the multi-scatter plot icon and then select individual cells within the display to view pairwise scatter plots between probe set pairs. To select the arrays with larger values, we can click-drag over the upper right quadrant of a plot. Note that doing this causes the corresponding rows in the spreadsheet view to be selected. Also, if we click another cell in the multi-scatter plot

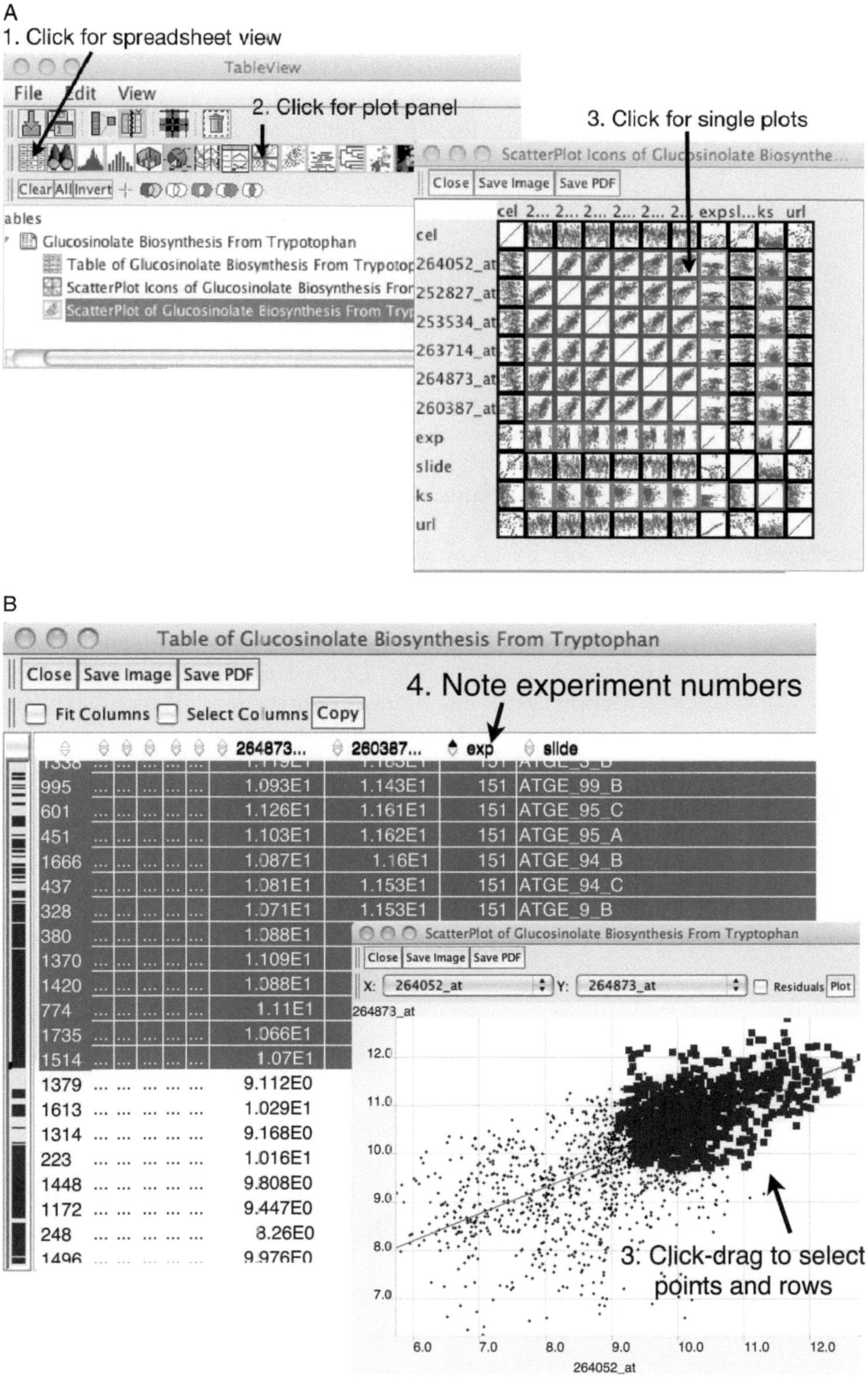

Fig. 12.7. Using TableView to identify arrays where glucosinolate probe sets indicate high expression values. (A) Multiple scatter plots (B) Interactive scatter plots and tabular view.

view and open another scatter plot between a different pair of probe sets, points corresponding to the same selected rows will also be highlighted.

We can then copy the selected rows into a new spreadsheet program (e.g., Excel) or just scroll up and down to find the experiments for which a majority of arrays appear in the high range for both probes sets. (These are easy to spot as blocks of contiguous, highlighted rows. For the glucosinolates, experiments with id numbers 337, 335, 330, 319, 192,191, 190, 188, 187, 186, 185, 180, 181, 179, 177, 171, 169, 168, 167, 166, 162, 155, 151, 150, 149, 147, 145, 144, 143, 142, 141, 140, 139, 137, 136, 132, 124, 123, 120, 103, 81, 79, 71, 60, 53, 46, and 26 have a majority of arrays with high-expression values.) To find out the tissue types and conditions associated with these samples, we can cut and paste the Web addresses (from the column labeled "url") into a Web browser and read the description of the experiment at the Nottingham Arabidopsis Stock Center Web site.

2.4. CressExpress Pathway-Level Co-expression

Now that we know which experiments contain samples that yield relatively high expression values for the six glucosinolate biosynthesis genes, we will use the CressExpress Web tool to query these same experiments and identify additional genes that are highly co-expressed with all six genes (*see* **Fig. 12.8**). This will give us some clues as to what other genes may be involved in pathway function and regulation.

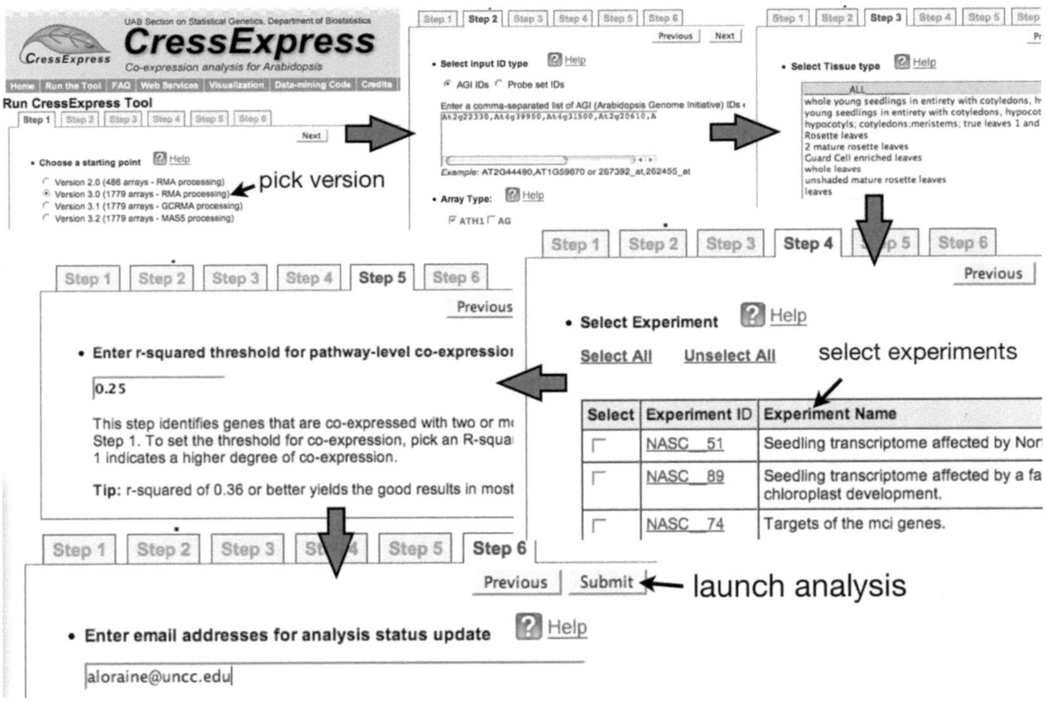

Fig. 12.8. Using CressExpress to identify genes that are co-expressed with glucosinolate biosynthesis pathway genes.

However, it is important to note that sub-selecting these experiments may not necessarily improve the results in every case. Our goal here is to assemble a good list of candidate genes that may play a role in glucosinolate function and/or biosynthesis, and we hope that sub-selecting based on high expression may help identify co-expression relationships that would otherwise be obscured by a higher degree of variation in the lower ranges of expression. To test whether sub-selecting based on expression level really does improve co-expression for this pathway, we would need to perform some additional analyses. For example, using R, we could re-calculate correlation using just the expression data from the higher ranges, using data both from the glucosinolate query genes and genes identified from a whole-genome co-expression analysis, which will identify using the CressExpress pathway-level co-expression analysis described below.

To perform the whole-genome co-expression analysis, we visit the CressExpress Web site (http://www.cressexpress.org) and click the link labeled "Run the Tool." We then step through a series of screens in which we set up and run the whole-genome co-expression analysis. Using CressExpress, we will perform linear regression between the six glucosinolate biosynthesis genes and all other genes represented on the ATH1 microarray using data from experiments we identified in TableView. Note that it would be difficult, if not impossible, to perform an analysis involving all 22,000 ATH1 probe sets using the desktop tools described thus far. The CressExpress Web site serves as an interface to a more powerful system that performs the same types of calculations as we did using R but on a much grander scale.

To start, we choose a data release (Step One), enter the AGI codes for the six genes (Step Two), and select the ATH1 array as the data source (Step Two). The next screen (Step Three) presents a listing of available tissues; here, we accept "All," which is the default option. Then, in Step Four, we check the experiments whose ids we identified using TableView.

The next screen (Step Five) allows us to set up a pathway-level co-expression (PLC) experiment. This part of the analysis will search the genome for genes that are co-expressed with two or more of the six query genes we entered in Step One, where two genes are considered to be co-expressed when their pairwise linear regression r^2 value is equal to or exceeds the designated threshold. In this case, we enter an r^2 threshold of 0.25, corresponding to a Pearson's correlation coefficient of 0.5. This means that genes that are co-expressed with two or more of the query genes with r^2 value of 0.25 or better will be reported. Next (Step Six), we enter an email address and launch the analysis. (For additional details on how PLC works, readers should consult a paper describing pathway-level co-expression analysis of metabolic pathways described in the AraCyc database (9).)

CressExpress runs a whole-genome co-expression analysis that compares the query genes' probe sets to all the other probe sets on the ATH1 array using arrays from the experiments designated in Step Four. Once the analysis completes, CressExpress sends an email message to the address entered in Step Six. The email contains a link to a compressed package of results files (a "zip" file) stored on the CressExpress site. Once we download and unpack the file, we can use a Web browser to view a results file named PLCResults.html that lists all genes that were co-expressed with two or more of the original query genes.

Figure 12.9 presents a screen capture of the PLC results Web page showing several genes that are co-expressed with glucosinolate queries. (In this case, we used r^2 threshold of 0.35.) The Web page presents a table that lists the co-expressed genes (and their probe sets) in the first column, together with Gene Ontology annotations describing their known or predicted functions. Each gene name links to the corresponding locus page at TAIR, and the probe set names link to a page at Affymetrix' NetAffx Web site (10). The next column in the table lists the query genes with which they were co-expressed and their pairwise r^2 values.

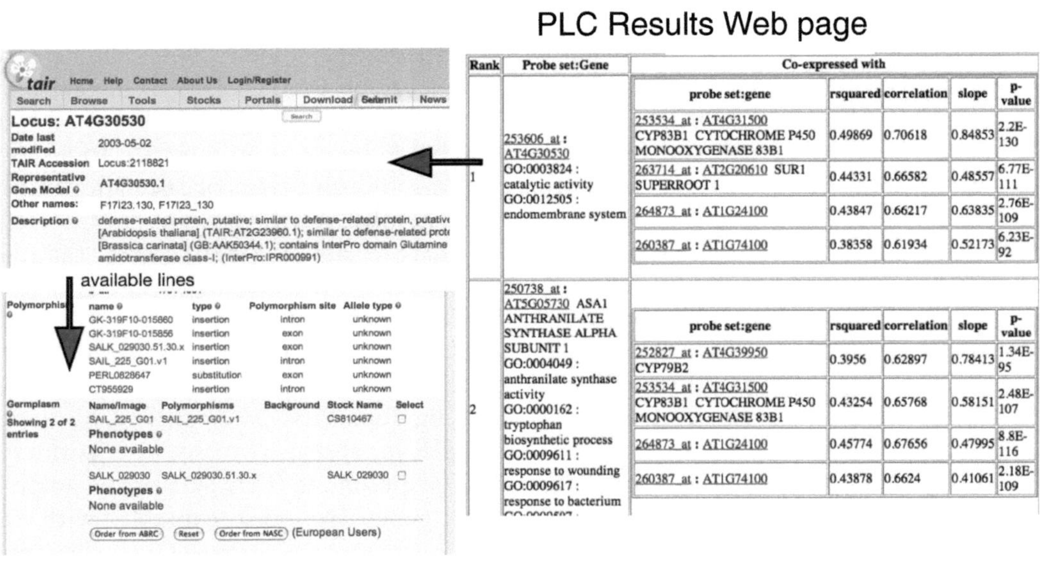

Fig. 12.9. Genes co-expressed with the indole glucosinolate pathway in Arabidopsis.

In this analysis, we find that genes annotated with functions related to tryptophan biosynthesis (e.g., ASA1 encoding anthranilate synthase beta subunit) appear high on the list, along with genes of unknown function. At this point, we have reached the limit of what this particular co-expression analysis tool can do; the next step, clearly, would be to identify T-DNA or other mutant lines that contain lesions in these genes and then test their effects on glucosinolate biosynthesis, pathogen defense, and/or other related phenotypes. By following the links in the PLC Results Web pages, we can easily identify publicly available seeds stocks believed to contain lesions in these top-ranked co-expressed genes, order the lines, and then test them for glucosinolate-related phenotypes.

3. Conclusion

This chapter provides an introduction to computational methods one can use to identify candidate players in metabolic pathways, assuming that genes that play related roles in the cell require some form of coordinate regulation. If this regulation is carried out at the level of mRNA abundance, then it is very likely that expression microarrays can detect it and that these relationships will be evident in the vast storehouses of expression microarray data currently available in the public domain. In this chapter, we explain one of the most straightforward and accessible approaches to identify co-expression: linear regression and correlation. However, it should be noted that methods used to combine expression data from many sources are still being developed, and experimental biologists need ways to incorporate these methods into their research workflow. This chapter aims to provide a roadmap for how this can work using freely available tools that have uses in many settings far beyond what is presented here.

References

1. Redman, J.C., Haas, B.J., Tanimoto, G., and Town, C.D. (2004) Development and evaluation of an Arabidopsis whole genome Affymetrix probe array. *Plant J.* **38**, 545–561.
2. Wheeler, D.L., Barrett, T., Benson, D.A., Bryant, S.H., Canese, K., Chetvernin, V., Church, D.M., DiCuccio, M., Edgar, R., Federhen, S., Geer, L.Y., Helmberg, W., Kapustin, Y., Kenton, D.L., Khovayko, O., Lipman, D.J., Madden, T.L., Maglott, T.R., Ostell, J., Pruitt, K.D., Schuler, G.D., Schriml, L.M., Sequeira, E., Sherry, S.T., Sirotkin, K., Souvorov, A., Starchenko, G., Suzek, T.O., Tatusov, R., Tatusova, T.A., Wagner, L., and Yaschenko, E. (2006) Database resources of the National Center for Biotechnology Information. *Nucleic Acids Res.* **34**, D173–D180.
3. Craigon, D.J., James, N., Okyere, J., Higgins, J., Jotham, J. and May, S. (2004) NASCArrays: a repository for microarray data generated by NASC's transcriptomics service. *Nucleic Acids Res.* **32**, D575–D577.
4. Wise, R.P., Caldo, R.A., Hong, L. Shen, L., Cannon, E., and Dickerson, J.A. (2007) BarleyBase/PLEXdb: A unified expression profiling database for plants and plant pathogens. *Methods Mol. Biol.* **406**, 347–364.
5. Johnson, J.E., Stromvik, M.V., Silverstein, K.A., Crow, J.A., Shoop, E., and Retzel, E.F. (2003) TableView: portable genomic data visualization. *Bioinformatics* **19**, 1292–1293.
6. Grubb, C.D. and Abel, S. (2006) Glucosinolate metabolism and its control. *Trends Plant Sci.* **11**, 89–100.
7. Halkier, B.A. and Gershenzon, J. (2006) Biology and biochemistry of glucosinolates. *Annu. Rev. Plant Biol.* **57**, 303–333.
8. Zhang, P., Foerster, H., Tissier, C.P., Mueller, L., Paley, S., Karp, P.D., and Rhee, S.Y. (2005) MetaCyc and AraCyc: Metabolic pathway databases for plant research. *Plant Physiol.* **138**, 27–37.

9. Wei, H., Persson, S. Mehta, T., Srinivasasainagendra, V., Chen, L., Page, G.P., Somerville, C., and Loraine, A. (2006) Transcriptional coordination of the metabolic network in Arabidopsis. *Plant Physiol.* **142**, 762–774.

10. Liu, G., Loraine, A.E., Shigeta, R., Cline, M., Cheng, J., Valmeekam, V., Sun, S., Kulp, D., and Siani-Rose, M.A. (2003) NetAffx: Affymetrix probesets and annotations. *Nucleic Acids Res.* **31**, 82–86.

Chapter 13

Integration of Metabolic Reactions and Gene Regulation

Chen-Hsiang Yeang

Abstract

Metabolic reactions and gene regulation are two primary processes of cells. In response to environmental changes cells often adjust the regulatory programs and shift the metabolic states. An integrative investigation and modeling of these two processes would improve our understanding of the cellular systems and may generate substantial impacts in medicine, agriculture, environmental protection, and energy. We review the studies of the various aspects of the crosstalk between metabolic reactions and gene regulation, including models, empirical evidence, and available databases.

Key words: Gene regulation, metabolic reactions.

1. Introduction

Metabolic reactions and gene regulation are two essential and tightly coupled processes of life. On the one hand metabolism serves the chemical functions of living organisms such as producing energy, synthesizing elementary materials, and removing toxic wastes. On the other hand gene regulation serves the controlling function by modulating RNA and protein syntheses. Since both processes are essential and fundamental to all the living organisms, their order of appearance has been a contentious issue in evolutionary biology (e.g., (1, 2)). Regardless of their origins, the two processes have become inseparable. In response to environmental changes cells often adjust regulatory programs and shift metabolic states. The shifts of metabolic states result from the regulation of enzyme gene expression and activities. Conversely, gene expression is often modulated directly or indirectly by metabolites.

Dmitry A. Belostotsky (ed.), *Plant Systems Biology, vol. 553*

DOI 10.1007/978-1-60327-563-7_13 Springerprotocols.com

Decades of studies have accumulated a large amount of knowledge and data regarding metabolism, gene regulation, and their crosstalk. This chapter reviews some recent modeling work, empirical evidence, databases, and data sets regarding the crosstalk between the two processes. While the studies of the crosstalk of metabolism and gene expression in plant systems are only beginning (please see the chapter by Loraine and colleagues in this volume), we will focus here for the most part on unicellular organisms such as *Escherichia coli* and *Saccharomyces cerevisiae* because of much more extensive availability of data and lower complexity of the processes in such systems. However, the same general principles should be applicable to plants as well.

1.1. Effects of Gene Regulation on Metabolic Reactions

As an essential aspect of life, all species invest substantial amount of genomic resources on metabolism. About half of the genes in *E. coli* and *S. cerevisiae* serve metabolic functions. Naturally the regulation of those genes modulates the metabolic reactions. Individual genes regulating the enzymes and transporters of some metabolic processes have been identified. Moreover, certain principles pertaining to the evolution of metabolic states have also been proposed and experimentally validated. These genes and models provide the information about the effects of gene regulation on metabolic reactions at both local and global levels.

1.2. General Theoretical Considerations of Determination and Measurements of Metabolic Fluxes

The activity of a metabolic reaction is characterized by its metabolic flux. The flux of a metabolic reaction denotes the rate of production or consumption of its substrates. The law of mass conservation requires that the net production or consumption rate of a metabolite equals to fluxes producing the metabolite minus the fluxes consuming it. In a matrix representation,

$$\frac{dx}{dt} = Av \qquad [1]$$

where $\frac{dx}{dt}$ denotes a vector of production or consumption rates of all metabolites, v is a vector of metabolic fluxes, and A is the matrix of stoichiometric coefficients of all reactions. Each row of A denotes the balance equation of a metabolite. In a steady state, $\frac{dx}{dt} = 0$. Hence any feasible set of metabolic fluxes must satisfy the balance equations:

$$Av = 0. \qquad [2]$$

In a metabolic network, a metabolite often participates in multiple reactions. Hence, the number of reactions typically exceeds the number of metabolites, and equation [2] is underdetermined. Additional constraints are therefore needed in order to uniquely determine the metabolic fluxes in a cell.

Flux balance analysis (FBA, (3–5)) tackles this problem by imposing a specific set of constraints. The lower and upper bounds of each flux can be specified by the thermodynamic properties of the reaction and the physiological conditions of the cell:

$$\alpha_i \leq v_i \leq \beta_i \qquad [3]$$

where α_i and β_i denote the lower and upper bounds of flux v_i. Furthermore, FBA assumes that the cell adjusts the metabolic fluxes within the physical constraints in order to optimize the growth. The growth rate of biomass is a linear function of metabolic fluxes based on biomass composition (6):

$$r = \mathrm{g}^T \cdot v. \qquad [4]$$

The growth optimization assumption yields a unique set of metabolic fluxes. The optimal $\hat{v}$ should maximize equation [4], subject to the constraints of equations [2] and [3]. This is a standard linear programming problem and can be solved by many mathematical tools (e.g., (7, 8)).

Flux balance analysis is a simple yet powerful tool for predicting metabolic fluxes. The measured fluxes of wild-type *E. coli* strains were shown to match the FBA predictions (9). However, the limitation of FBA is the requirement for growth optimality. In a real physiological system, there may exist many suboptimal flux modes that help the organism to adapt to specific environmental conditions. This idea leads to an alternative formulation of the flux balance equations – the elementary mode analysis (10–12).When the metabolic fluxes do not have to maximize the growth objective function (equation [4]), valid solutions of equations [2] and [3] constitute a convex set C. Each element in C can be expressed as a linear combination of multiple basis vectors:

$$C = \left\{ v | v = \sum_{i=1}^{k} w_i p_i, w_i \geq 0 \right\}. \qquad [5]$$

Each p_i is a vector of metabolic fluxes for each reaction. A set of p_i vectors consist of what is defined as *elementary modes* if none of them is a linear combination of others (i.e., they are linearly independent).

An elementary mode can be viewed as a collection of metabolic reactions. The linear independence of two modes suggests the two sets of reactions are not coupled. In real metabolic networks, an elementary mode often corresponds to a biologically meaningful pathway. For instance, the elementary modes of the carbon metabolism and amino acid synthesis include such pathways as the TCA cycle, glyoxylate shunt, and glutamate synthesis (10).

Elementary modes represent all physiologically viable fluxes including the optimal flux sets. They are hence applicable to a wider range of problems. For example, *E. coil* knockout mutants destroying all elementary modes were shown to be lethal, and the growth efficiency of fluxes averaged over elementary modes was correlated with the expression levels of the corresponding enzymes (12).

Measuring the flux of a given metabolic reaction directly is challenging, since its reactants often participate in other metabolic reactions in the cell. Instead, chemists often choose to label the key substrates/metabolites with stable isotopes, such as the commonly used C^{13} and observe the distribution of the tagged atoms in the cell (13). Mass spectrometry or nuclear magnetic resonance (NMR) can be used to monitor the distribution of the isotopically labeled compounds, and the production or consumption rates of certain metabolites can be delineated from the resulting spectra. This information adds more constraints to the flux balance equations, and the fluxes of certain reactions can be inferred with such additional constraints.

Current data sets of metabolic fluxes are concentrated on microbes such as *E. coli* (e.g., (14–16)) and *S. cerevisiae* (e.g., (17–19)). Most data sets probe the well-known glucose metabolism network (glycolysis, TCA cycle, pentose phosphorylation).

1.3. Gene Regulation Constraints on Metabolic Fluxes

Gene regulation can modulate metabolic reactions by changing the levels of enzymes. An immediate extension of FBA is to incorporate the expression levels of metabolic enzymes as additional constraints on metabolic fluxes (20, 21). The fluxes of reactions catalyzed by enzymes with low expression levels are set to 0, while an up-regulated enzyme does not exert additional constraints. However, the effects of gene levels and regulation on metabolic fluxes are much subtler than the simple binary switch. Metabolic fluxes are tightly regulated in order to maintain homeostasis of organisms. Alterations of external or internal conditions often induce only minor changes of metabolism. For instance, Ishii et al. found that the metabolite levels of *E. coli* are robust against disruptions of enzymes in the central carbon metabolism (22).

The robustness of metabolic fluxes is partially attributed to the redundancy of enzymes (as well as to the fact that they are often but not always synthesized in excess of what is minimally required for optimal function, H Kacser and JA Burns, Genetics 97:639–666, 1981). Moreover, many reactions can be catalyzed by multiple isozymes (23). These isozymes often function under different metabolic conditions or, in multicellular organisms, in different tissues or cell types. The effect of deleting one enzyme can often be rescued by its complementary isozymes. In addition to redundancy, the cells also demonstrate robustness at a global level. For example, measurements of metabolic fluxes in E. coli knockout strains demonstrate substantial deviations from FBA predictions of the knockout strains, where the fluxes of the reactions catalyzed by the perturbed enzymes are set to zero. Instead of rearranging many fluxes to optimize growth under the new constraints, cells tend to adopt a minimum adjustment from the wild type. Segre et al. introduced an algorithm that performs a minimization of metabolic adjustment (MOMA) upon gene deletion (24, 25). Instead of the linear objective function in FBA (equation [4]), MOMA minimizes the square of the distance (or squared distance) between the unperturbed flux v_w and the new flux v:

$$r = (v - v_w)^T (v - v_w). \qquad [6]$$

The constraints are the linear equalities/inequalities in FBA (equations [2] and [3]) plus the zero constraints on perturbed fluxes. Minimization of the quadratic objective function (equation 6) with the linear constraint functions (equations 2 and 3) can be solved by many mathematical tools (e.g., see 26).Predictions from MOMA fit the empirical flux measurements of knockout strains more closely than FBA (24).

The results of MOMA indicate the inertia of cells against changes. The gene expression and other regulatory apparatus in wild-type strains have evolved over millions of years of evolution to fit the growth conditions that are most often found in nature. The perturbed system is hence unlikely to shift quickly to the new optimal state. For unicellular organisms, readjustment of metabolic fluxes to optimum may be achieved only through evolution. Indeed, the empirical evidence indicates that cells reach the growth optimum predicted by FBA after certain number of generations of evolution (27).

Regulation of enzyme activities adds an additional layer of complexity to the regulation of metabolism. Metabolites can modulate the catalytic efficiencies of enzymes by binding to proteins' allosteric (i.e., regulatory, non-active) sites (23). For instance, in glycolysis the conversion of fructose-6-phosphate into fructose-1,6-bisphosphate (EC\# 2.7.1.11) is inhibited by phosphoenolpyruvate, downstream of reaction 2.7.1.11 (28), and the conversion of phosphoenolpyruvate into pyruvate (EC\# 2.7.1.40) is activated by fructose-1,6-biphosphate, upstream of reaction 2.7.1.40 (29). Without requiring the expression changes of enzymes, this feedback control allows the pathways to quickly respond to metabolic conditions.

1.4. General Principles Governing the Global Distribution of Metabolic Fluxes

The global distribution of metabolic fluxes is the product of a complex control system that underwent many millions of years of evolution. It is tuned to optimize the fitness of the organisms in their specific environmental niches. However, the definition of fitness is sometimes elusive and must be considered in the context of a need to balance several contradictory goals. Here we discuss two important factors that shape the metabolic flux distribution.

- *Growth optimality.* Accumulating biomass with efficiency is one of the major functions of metabolism. It is thus reasonable to assume that metabolic fluxes are globally distributed to maximize growth. Flux balance analysis derived from the growth optimality criterion already successfully fits the flux data of wild-type bacterial strains (9, 27). The poor fit of FBA on knockout strains (24) further supports the optimality criterion in evolution, since the expressions and functions of genes are tuned to optimize growth in a specific environment, and the loss of certain genes often deviates the cells from the optimal configuration. Furthermore, results of artificial evolution indicate that a strong selective pressure and fast mutation rates of bacteria can force the knockout strains to rapidly evolve toward achieving the new optimal condition (27, 30).
- *Robustness.* Robustness of metabolic fluxes is essential for homeostasis. Alterations of fluxes may accumulate or deplete certain metabolites and generate toxic effects. Robustness of metabolic fluxes can be achieved by redundancy and a tight feedback control. Many reactions are catalyzed by multiple isozymes. In addition, redundancy also exists at the pathways level, because the synthesis or degradation of the key metabolites and the production of energy can often be achieved by multiple pathways. For instance, both glycolysis and pentose phosphate pathways convert glucose-6-phosphate into glyceraldehyde-3-phosphate. Moreover, the enzyme kinetics as well as the expression of the respective genes encoding them are controlled by multiple feedback loops. In many metabolic pathways, the enzymes that catalyze the upstream reactions are down-regulated by the products of the downstream reactions, and conversely, the enzymes responsible for the downstream reactions are positively regulated by the products of the upstream reactions (e.g., (28, 29)). In addition, the expression of enzymes is also directly or indirectly regulated by metabolites (see the next section).

The balance between growth efficiency and robustness may not only govern the distribution of metabolic flux modes but also influence the network topology of metabolic reactions. The connectivity of metabolic networks, similar to other biological networks, follows a power law distribution (31). There exists a small number of "key metabolites" that appear in many reactions, while most metabolites appear in only one or a few reactions. The power law distribution allows an efficient allocation of metabolic resources, as the cells can concentrate protein synthesis on the enzymes for the "hub" reactions. It is also robust against random removal of nodes (metabolites) or edges (reactions) in the network, since most nodes have low connectivity (31). However, it is also fragile in the same way that the Internet is vulnerable to removal of the hubs in the network (31, 32). Whether the power law distribution of the metabolic network is an evolutionary consequence of the selection for efficiency and robustness or the product of other mutational processes is still under debate.

1.5. Key Regulators of Metabolic Enzymes

In addition to the global properties of the regulation of metabolic fluxes, the key regulators of many metabolic enzymes in different organisms have been identified. Here we give a brief overview of several well-understood examples of regulators of the major metabolic pathways.

ArcA and ArcB form a two-component regulatory system for respiratory control in *E. coli* (33). ArcB is a membrane-bound sensor kinase and ArcA is the cognate response regulator. Under the conditions of oxygen deficiency, ArcB phosphorylates ArcA, which then represses the expression of many enzymes involved in aerobic respiration. Some of the enzymes thus regulated include those that catalyze the TCA cycle and the glyoxylate shunt, such as *gltA*, *acnAB*, *icdA*, *sucABCD*, *sdhCDAB*, *fumA*, *mdh*, and *aceB* (33–40).

CRP is a transcriptional dual regulator in *E. coli* (41). CRP binds to the promoters of operons involved in glucose metabolism, lactose metabolism, electron transfer, and many others (42, 43). The CRP–cAMP complex is the best characterized system for catabolite repression of bacteria (41). The dimeric CRP–cAMP complex binds to promoters and activates transcription. Exogenous glucose both inhibits cAMP synthesis and stimulates the efflux of cAMP from cytoplasm (44), which therefore reduces the CRP–cAMP complex levels and causes glucose repression.

Another global regulator for carbon metabolism of bacteria is Cra (FurR) (41). Cra represses some enzymes in the central carbon pathway such as *pfkA*, *pykF*, *zwf*, and *edd-eda* and activates others such as *ppsA*, *fbp*, *pckA*, *icd*, *aceA*, and *aceB* (45). Experiments on Cra knockout strains of *E. coli* indicate that Cra activates glucogenesis, while represses the catabolic pathways of glucose such as glycolysis (40, 45).

In addition to global regulators such as CRP and Cra, the pathways of *E. coli* amino acid synthesis are also regulated by the amino acid-specific regulators. For instance, the operon of genes *leuA-D* in leucine biosynthesis is regulated by the *LeuO* transcription activator (46). Various genes involved in methionine synthesis are regulated by *MetJ* transcriptional repressor (47). Genes involved in the biosynthesis and transport of aromatic amino acids are controlled by *TyrR* transcriptional dual regulator (48).

Many transcription factors in the budding yeast *S. cerevisiae* are known to regulate genes involved in metabolism. *Gal4* and *Gal80* constitute the well-studied antagonistic pair of factors that regulate the genes of galactose metabolism. *Gal4* binds to the promoters of enzymes and transporters of galactose utilization and activates their expression. When galactose is deficient, the repressor *Gal80p* binds to *Gal4p* and inhibits its interaction with the general transcription apparatus, hence repressing the transcription of galactose metabolism genes. When galactose is plentiful, *Gal80* disassociates from *Gal4p* and those genes are activated (49, 50).

In yeast, *Gcn4p* is a master transcription factor controlling the genes that are involved in the synthesis of amino acids. Upon amino acid starvation, *Gcn4p* activates many genes involved in amino acid synthesis (51). Large-scale chromatin immunoprecipitation (ChIP-Chip) assays also indicate that the promoters interacting with *Gcn4p* are enriched for the amino acid synthesis genes (52). In addition to *Gcn4p*, enzymes of different amino acid synthesis pathways are also activated by different transcription factors. Examples include *Bas1* for histidine and arginine synthesis (53), *Leu3* for leucine synthesis (54), and *Cbf1* for methionine synthesis (55).

2. Effects of Metabolic Reactions on Gene Regulation

Changes of external or internal metabolic conditions often induce pronounced changes in the expression of a large number of genes. Combined with the regulation of the enzyme activity, gene expression responses are necessary to maintain cellular homeostasis under different metabolic conditions. With the recent progress of high-throughput assays, many genome-wide expression data sets for various metabolic conditions and organisms have become accessible. The mechanisms of some of these gene regulatory effects have been studied in detail. Furthermore, certain patterns relating the regulation of metabolic enzymes and their positions in the metabolic network have now emerged. These global and local information data sets provide the basis for further inquiry into the effects of metabolic reactions on gene regulation.

2.1. Gene Expression Responses to Metabolic Shifts

Besides the fast responses via the modulation of enzyme activities, metabolic shifts also activate or repress the expression of many genes. Notwithstanding its relative slowness, regulation of gene expression confers numerous selective advantages (this is particularly true for sessile organisms such as plants, because they cannot evade their changing environment and therefore must flexibly adjust to it). It is easier to control a large number of genes or operons by inserting or creating the binding sites on promoters than altering the structure and modification of each individual protein (56). Regulation via gene expression is thereby more effective and malleable for evolution.

It has long been recognized that a particular carbon source in the growth medium of microbes can inhibit the synthesis of enzymes involved in the metabolism of alternative carbon sources (41). One of the best studied catabolic repression is so called the glucose effect. In the presence of glucose, the expression of enzymes involved in other sugar metabolic pathways, such as galactose and glycerol, is repressed (41, 57). Conversely, in a glucose-limiting medium, the genes involved in the TCA cycle, NADH dehydrogenase, and electron transfer are up-regulated relative to the amino acid-limiting medium (41, 58). Other genes involved in carbon metabolism and energy biosynthesis are likewise differentially expressed under different carbohydrate conditions (e.g., 15, 57, 59)).

Another example of differential expression of metabolic enzymes under different carbon sources is the diauxic growth of microbes (60). In a mixture of glucose and lactose, *E. coli* experiences two sequential exponential growth phases, where each phase reflects the utilization of one carbon source. The levels of enzymes and transporters for one carbon source are repressed when the cells utilize the other carbon source (61).

With the rapid progress of microarray technology, global gene expression responses of many organisms and tissues under various metabolic conditions are becoming available. Examples include the supply of different carbon sources such as acetate (59), lactose (62), sucrose (57), amino acid starvation (63), nitrogen supply (64), salt (65), and environmental stress (66). Overall metabolic perturbations often induce global expression responses. Some general rules of gene regulation in metabolic shifts will be discussed in the later section.

In addition to alterations of metabolic conditions, the expression profiles of knocking out the various metabolic enzymes are also reported. Examples include enzymes in the glycolysis pathway in *E. coli* (22, 67), the galactose metabolism pathway in yeast (68), and a global compendium of knockout assays in yeast (69). Remarkably, deletion of genes along the central carbon metabolism often induces only mild responses along these pathways (22, 67), illustrating the robustness of cells at the level of gene expression, as well. On the other hand, some gene knockouts demonstrate phenotypes only in specific growth medium, while other gene knockouts have lethal phenotypes (e.g., (70)).

2.2. Mechanisms of Inducing Expression Responses upon Metabolic Shifts

To alter expressions under the conditions of metabolic shifts, a cell needs to establish feedback mechanisms to sense the metabolites and regulate the transcription and/or translation apparatus. Below we introduce a simple classification of the feedback mechanisms and give a few examples in each class.

- *Metabolites can directly interact with the transcriptional apparatus.* In microbes, some metabolites can physically bind to *cis* -regulatory elements or interact with transcription factors and directly regulate transcription. One of the most well-known cases is the *lac* operon (61) in *E. coli*. In the absence of lactose, a repressor binds to its promoter and blocks the transcription of *lac* genes. Lactose metabolites bind to the repressor, facilitates its dissociation from the promoter, and allows the RNA polymerase to initiate transcription. In yeast, galactose also regulates the transcription of enzymes and transporters by directly interacting with *Gal4p* and *Gal80p* transcription factors (49, 50). In the absence of galactose, repressor *Gal80p* binds to the promoter and blocks transcription. Galactose binds to *Gal80p*, removes it from the promoter, and allows the activator *Gal4p* to occupy the promoter and initiate transcription.

- *Feedback regulation mediated through signal transduction pathways.* The most common pathways of regulating transcription by metabolites are the signal transduction cascades. Metabolites, small molecules, and other environmental changes are detected by receptors on the cellular surface. The signals are transduced through a series of protein modifications (e.g., phosphorylation, ubiquitination) and eventually regulate transcription factors or other proteins. In yeast, glucose-induced expression responses are triggered by various signal transduction pathways (71). Glucose levels are detected by Snf3 or Rgt2 sensors and modulate the activity of G proteins (Ras and Gpa2), which bind independently to adenylase cyclase (Cyr1) and stimulate cAMP production. cAMP binds to the protein kinase A (PKA) tetramer and facilitates the dissociation of the kinase subunit (TPK). TPK is translocated into nucleus and regulates transcription factors. Similarly, the glucose responses in *E. coli* are regulated by the CRP–cAMP complex (41). Glucose inhibits the synthesis of cAMP by adenylase cyclase, thus reducing the level of the CRP–cAMP complex and repressing the transcription of many genes.
- *Riboswitches.* A recently discovered mechanism for metabolite-driven gene regulation, the riboswitches are parts of mRNAs that bind to small molecules and modulate the translation of mRNA in *cis* (72). The binding of small molecules often inhibits translation by forming early termination hairpins, blocking the ribosome binding sites, or inducing self-cleavage. Different classes of riboswitches respond to different metabolites such as thiamine derivatives (73), vitamin B12 (74), purine (75), and other metabolites.

2.3. Governing Principles of Metabolic Gene Regulation

Except for some well-studied systems in a small number of model organisms, the mechanisms of regulation of many metabolic genes are still yet to be revealed. However, the nearly complete characterization of metabolic networks in a few species as well as the rapidly accumulating volume of high-throughput gene expression data facilitates the studies of the relations between gene expression and their functions in the metabolic system. High-level rules that govern such relationships have been proposed, and some of them are being empirically tested. These studies provide valuable insight at systems level and useful guidelines for further experiments.

The apparent rule of homeostasis often prevails when the organisms undergo metabolic shifts or environmental stress. When glucose is supplied, cells begin to produce enzymes required for glycolysis and aerobic respiration while the enzymes metabolizing other carbon sources such as lactose or acetate are repressed (57, 59). The opposite responses occur upon glucose starvation (15). Under amino acid starvation, *E. coli* and yeasts up-regulate genes involved in amino acid synthesis (15, 51). Many metabolic shifts, such as nutrient starvation or intoxication, or drastic changes of osmotic pressure also induce stress responses of the cells (e.g., (66)). General responses to stress include cell cycle arrest, halt of biomass accumulation, down-regulation of mRNA and protein synthesis, over-expression of stress response genes, sporulation for unicellular organisms, and apoptosis for multicellular organisms.

To induce coordinated expression responses, genes must possess coordinated regulatory structures. In bacteria, operons serve as basic units of co-expression. In *E. coli* and many other species, genes located in the same operons often function in the same metabolic pathways (e.g., (61)). Minimization of the quadratic objective function (equation 6) with the linear constraint functions (equations 2 and 3) can be solved by many mathematical tools (e.g., see 26). Thus, additional rules are needed to explain the relationships of metabolic gene regulation and functions. Below, we summarize a few rules that emerged from the recent studies.

- *Proximity*. The operon structures suggest genes with similar functions in the metabolic network are co-regulated. One way to define functional similarity is the distance between two enzymes in the metabolic network. Therefore, one may expect that close enzymes in the metabolic network are co-regulated. Kharchenko et al. examined the expression data in *E. coli* and found that positive co-expression decreases with distance in the network, whereas negative co-expression increases with distance, up to a certain threshold (76). Other studies also suggest that proximal genes in the metabolic networks are co-expressed (e.g., (77)). In addition to obvious co-regulation, the genes along a metabolic pathway may exhibit subtler regulatory relationship. One such example is the "just-in-time" regulatory system of arginine biosynthesis in *E. coli* identified by Zaslaver et al. (78). In this system, the genes in a linear metabolic pathway are sequentially activated, such that the enzyme of a reaction is synthesized just in time to process the substrates generated in the previous step. These genes are regulated by the same set of transcription factors, and their sequential activation results from differential strengths of the binding motifs.

- *Network topology*. The rule of co-regulation may not apply when genes are in different linear pathways. Two metabolic pathways may be converged to or diverged from a third pathway. Several authors have studied the coexpression of genes in convergent and divergent pathways. Kharchenko et al. showed that in the three-gene motifs of the S. cerevisiae metabolic network enzymes catalyzing divergent reactions are significantly co-expressed compared to enzymes catalyzing convergent reactions (76). In contrast, Ihmels et al. showed that in yeast, co-expression occurs only along one of the two divergent pathways (79). Moreover, they showed that isozymes are often separately co-expressed with distinct processes. These conflicting results suggest that the relationships between co-regulation and metabolic network topology require further in-depth study.
- *Metabolic fluxes*. As the distribution of metabolic fluxes is, at least in part, the consequence of gene expression, the rules of metabolic flux distribution should be correlated with the rules governing gene regulation. Several authors correlated properties of metabolic fluxes with gene expression. Stelling et al. constructed an index of "control-effective flux" (CEF) of a substrate from metabolic mode analysis and used it to measure the efficiency of biomass and energy production using a specific substrate (12). The CEF scores were used to predict the relative mRNA levels under varying substrate availability conditions. The expression data for 50 genes on acetate versus glucose as carbon source showed good agreement with the scores. Furthermore, Bilu et al. found a strong correlation between the flexibility of metabolic fluxes and the diversity of expression levels and an anticorrelation between flux flexibility and promoter conservation (80). In addition, they also showed genes active in many optimal metabolic fluxes tend to have conserved sequences. Both results suggest that the expression of genes is tied to the diversity and importance of metabolic fluxes.
- *Feedback*. Some of the rules of feedback regulation are discussed in earlier sections, e.g., the supply of an input substrate often enables its metabolic pathway while disabling the competing pathways metabolizing alternative input substrates. Moreover, the over-production of an output substrate of a pathway often turns the pathway off. These changes of a metabolic pathway can be achieved by altering the activities of enzyme or by modulating the expression of genes encoding them. One example of such feedback system is the regulation

of enzymes in the tetralin degradation pathway of *E. coli* (81). The input substrate (tetralin) induces the enzyme gene expression along the pathway, whereas an intermediate product (reduced ferredoxin) inhibits transcription of the gene(s) encoding the key the enzyme(s). The composite effects of feedback through both gene regulation and allosteric regulation maintain the homeostasis of metabolism.

3. Integration of Metabolic Reactions and Gene Regulation

In the post-genomic era, focused investigations of the global properties and component interactions in the biological systems become more and more important as well as powerful because of huge amount of data that are becoming available. Integration of metabolic reactions and gene regulation is a testing ground for systems biology. Decades of biochemical studies have already mapped the complete metabolic networks (or a large portion of) in multiple organisms. Genome sequences of many organisms have already been published, a large amount of microarray mRNA expression data are already available, and more recently, large volumes of metabolic flux data have been generated. Furthermore, other advanced technologies have produced various large-scale data sets probing different aspects of cellular processes such as protein expression levels, protein–DNA interactions, protein–protein interactions, and protein or DNA modification. These data allow us to considerably deepen our understanding of the integrated system of metabolic reactions and gene regulation. Databases integrating various metabolic and regulatory information have already been established, and some integrated models that take advantage of these databases have been proposed. Results from some of integrative research projects of this kind demonstrate great potential for applications in agriculture, medicine, as well as environmental and energy sciences.

3.1. Integrated Databases

One of the most comprehensive databases of metabolic information is BioCyc ((82), http://.biocyc.org). It is a collection of 371 pathway/genome databases covering the metabolic reactions, substrates, enzymes, genes, operons, and transcription factors from 374 species. The database is organized by species and pathways. Pathways are hierarchically classified and visualized with information of different levels of details. The operons of genes as well as their respective transcription factors are also displayed. In addition, users can choose to compare the pathways across multiple species.

BioCyc contains both curated information from the literature and computationally derived information with either moderate or no manual curation. As subsets of BioCyc, MetaCyc ((83), http://metacyc.org) and EcoCyc ((84), http://ecocyc.org) are intensively curated databases which contain only information reported from the literature. MetaCyc contains the metabolic network information from many species but no operon information. EcoCyc contains both metabolic and operon information for *E. coli* alone.

BioCyc contains primarily the data for prokaryotes organisms, although human, *S. cerevisiae* and *S. pombe* are also covered. Recently, a global reconstruction of the human metabolic network has become publicly available (85). The authors reconstructed the metabolic network from genome sequences and annotations, pathway databases, and manual curation of network components from the "bibliome" of >1500 published articles over the period of more than 50 years. The reconstruction database contains the information on metabolites, reactions and enzymes, compartmentalization of reactants, description of gene–protein relationships, and confidence scores and references on various pathways. Despite its comprehensiveness and accuracy on the human metabolic network, the reconstruction database does not contain information about the gene regulation.

Genomic, regulatory, and metabolic information about plants remains relatively under-represented. AraCyc contains the metabolic information of *Arabidopsis thaliana* ((86), http://www.arabidopsis.org/biocyc/index.jsp). Several detailed databases of yeast gene function, regulation, and metabolism are available, including the Saccharomyces Genome Database (SGD, (87), http://www.yeastgenome.org/), a proprietary Yeast Proteome Database (YPD, https://www.proteome.com/proteome/), and the Munich Information Center for Protein Sequences (MIPS, (88), http://mips.gsf.de/). MIPS also contains the information on other species such as human as well as on *A. thaliana*.

Kyoto Encyclopedia of Genes and Genomes (KEGG) contains information about genes, proteins, reactions, and pathways of 761 species ((89), http://www.genome.jp/kegg/kegg1.html). The curated pathways in KEGG include both metabolic and signal transduction pathways. Unlike BioCyc, KEGG covers many eukaryotic species, especially animals, yet it does not contain the regulatory information.

3.2. Integrated Models

Various methods have been proposed to integrate the information of metabolic and regulatory networks for different types of data-sets. The expression levels of enzymes can be treated as additional constraints in flux balance analysis. For instance, Covert et al. imposed zero constraints of linear programming on the metabolic fluxes where the enzyme expression levels are low (20, 21). This method is extended to predict both metabolic fluxes and gene expression under gene knockout or metabolic perturbations (20, 90–92). The expression levels of enzyme genes are constrained by the known relationships between metabolite concentration and transcription factors.

Alternatively, Stelling et al. predicted gene expression levels from the information of metabolic fluxes alone (12). They quantified the effectiveness of a substrate as the weighted sum of the relative substrate uptake over all viable metabolic flux modes. This control-effective flux index was used to predict the mRNA levels of the catalyzing enzymes.

As opposed to the metabolic flux analysis, many gene expression models adopt information of the metabolic network or utilize experiments under metabolic perturbations to build the gene regulatory models. For instance, Gat-Viks et al. have built a factor graph model to incorporate both known regulatory functions and noisy measurements from multiple sources and extended the known model by learning the augmented graph structures (93). They used part of the metabolic network in the initial network model. In addition, many such studies incorporate data from the relevant knockout mutants. Ideker et al. compared single and double knockout gene expression data with galactose supplu to infer the causal dependencies of enzymes involved in galactose metabolism in yeast (68), with the assumption that if two genes are in the same pathway, then knocking out the downstream gene should manifest the same phenotype as the double knockout (i.e., this would incur no additional fitness cost).

The coupling of metabolic and regulatory networks is bidirectional. On the one hand expressions of enzyme mRNAs/proteins can modulate the reaction activities and metabolic fluxes. On the other hand metabolite concentrations and metabolic fluxes can regulate gene expressions. The models described above emphasize the coupling along either direction. Yeang et al. developed a probabilitic graphical model to study the coupling of gene regulatory and metabolic networks in both directions (94). They used pathways in the joint network to explain the changes of gene expression and metabolic fluxes under knockout or metabolite perturbations. The forward links from enzyme gene expressions to the fluxes of metabolic reactions are determined by functional annotations of enzymes. The feedback links between metabolites and transcription factors are learned from the data to maximize the number of explained pertubration effects.

The models described above are discrete and stationary. To explicate the quantitative and dynamic systems properties, various dynamic system models of metabolic reactions and gene regulation have been proposed. Examples include the models of diauxic shift (95), pheromone response pathways (96), and general metabolic reactions (97).

Besides predicting metabolic fluxes or gene expression changes under perturbations, some programs can also identify the target genes that would maximize the flux of desired products if knocked out (98,99). Furthermore, empirical studies of metabolic flux analysis have been applied in metabolic engineering (see the review (100)).

4. Conclusion

The studies on the integration of gene regulation and metabolic reactions have been progressing with accelerating pace, thanks to new experimental technologies, massive amount of data, and development of novel bioinformatic methods/tools. This progress sets a stage toward a deeper understanding of integration of cellular subsystems and its potential applications in biotechnology, medicine, as well as environmental and energy sciences. There are many open problems and new directions being actively pursued by researchers from multiple disciplines. Here we list some of them which are not covered in this review.

While most studies referred in this review focus on *E. coli* and yeast, the whole-genome sequences, metabolic networks, and gene regulation information of many other species (especially microbes) are already available, and plants, particularly *A. thaliana*, are not far behind. Comparative studies of these types of information across multiple organisms can reveal the evolution of the integrated gene regulatory and metabolic systems. Comparisons of genome sequences, regulatory and metabolic networks are already active research fields in computational biology. Integration of these comparisons will be the next step toward a system-level understanding of cells.

The communication between gene regulatory and metabolic systems in plants and other multicellular organisms is intrinsically much more complex than in unicellular systems. In multicellular organisms, many metabolic pathways are active only in specialized tissues/organs, and their inputs and products are transported from and to other tissues. The crosstalks between metabolism and gene regulation are thus likely tissue specific and are subject to more complex control beyond the simple rules stated above. Investigation of the integrated systems in multicellular organisms requires more tissue-specific data (and especially metabolic flux data), as well as better understanding of the tissue-specific regulatory programs.

In the natural environment, physiological activities of the inhabitants are related. It is hence possible to study coevolving or symbiotic relations in the metabolic or regulatory networks of the inhabitant species. For instance, sequencing the aggregate samples of microbial genomes from the environment – metagenomics – has already led to many discoveries regarding the symbiotic relations (101). With the aid of fast sequencing technologies and comparative studies, it will be possible to study the interactions of species at genetic/molecular levels. The study of the interacting metabolic networks of the inhabitants in a specific environment – "metametabolomics" – would possibly reveal the circulations of carbons, nitrogens, toxic compounds and energy in the environment.

Knowledge of the communications between gene regulatory and metabolic systems can lead to a wide range of applications. Genetic engineering and metabolic engineering already have substantial impacts in agriculture and pharmaceutical industry. With the progress in systems biology and synthetic biology, it is becoming possible to re-program or synthesize organisms to produce or consume optimal quantities of molecules in specific timing or environmental cues. These new organisms can contribute both to the basic knowledge and to the improvements in production of biofuels, antibiotics, proteins, agricultural yield, as well as to cleanup of environmental wastes.

References

1. Dyson, R. (1999) Origins of life. Cambridge University Press, New York, USA.
2. Maynard Smith, J. and Szathmary, E. (1999) The origins of life: from the birth of life to the origin of language. Oxford University Press, New York, USA.
3. Varma, A. and Palsson B.O. (1994) Stoichiometric flux balance models quantitatively predict growth and metabolic by-product secretion in wild type *Escherichia coli* W3110. *Appl. Environ. Microbiol.* **60**(10), 3724–3731.
4. Bonarius, H.P.J., Schmid, G., and Tramper, J. (1997) Flux analysis of underdetermined metabolic networks: the quest for the missing constraints. *Trends Biotechnol.* **15**, 308–314.
5. Edwards, J.S. and Palsson, B.O. (1998) How will bioinformatics influence metabolic engineering? *Biotechnol. Bioeng.* **58**, 162–169.
6. Varma, A. and Palsson B.O. (1993) Metabolic capabilities of *Escherichia coli*. II. Optimal growth patterns. *J. Theor. Biol.* **165**, 503–522.
7. Danzig, G.B., Orden, A., and Wolfe, P. (2003) The generalized simplex method for minimizing a linear form under linear inequality restraints. In The basic George B. Danzig. Stanford University Press, California, USA.
8. Karmarkar, N. (1984) A new polynomial-time algorithm for linear programming. *Combinatorica.* **4**(4), 373–395.
9. Edwards, J.S., Ibarra, R.U., and Palsson, B.O. (2001) In *silico* prediction of *Escherichia coli* metabolic capabilities are consistent with experimental data. *Nat. Biotechnol.* **19**, 125–130.
10. Schuster, S., Dandekar, T., and Fell, D.A. (1999) Detection of elementary flux modes in biochemical networks: a promising tool for pathway analysis and metabolic engineering. *Trends Biotechnol.* **17**, 53–60.
11. Schuster, S., Fell, D.A., and Dandekar, T.A. (2000) A general definition of metabolic pathways useful for systematic organization and analysis of complex metabolic networks. *Nat. Biotechnol.* **18**, 326–332.
12. Stelling, J., Klamt, S., Bettenbrock, K., Schuster, S., and Gilles, E.D. (2002) Metabolic network structure determines key aspects of functionality and regulation. *Nature.* **420**, 190–193.
13. Wiechert, W. (2001) C13 metabolic flux analysis. *Metab. Eng.* **3**, 195–206.
14. Emmerling, M., Dauner, M., Ponti, A., Fiaux, J., Hochuli, M., Szyperski, T., Wuthrich, K., Bailey, J.E., and Sauer, U. (2002) Metabolic flux responses to pyruvate kinase knockout in *Escherichia coli*. *J. Bacteriology.* **184**(1), 152–164.
15. Hua, Q., Yang, C., Baba, T., Mori, H., and Shimizu, K. (2003) Response of the central metabolism in *Escherichia coli* to phosphoglucose isomerase and glucose-6-phosphate dehydrogenase knockouts. *J. Bacteriol.* **185**(24), 7053–7067.
16. Fischer, E., Zamboni, N., and Sauer, U. (2004) High-throughput metabolic flux analysis based on gas chromatography-mass spectrometry derived C13 constraints. *Analy. Biochem.* **325**, 308–316.
17. Velagapudi, V.R., Wittmann, C., Schneider, K., and Heinzle, E. (2007) Metabolic flux screening of *Saccharomyces cerevisiae* single knockout strains on glucose and galactose supports elucidation of gene function. *J. Bacteriol.* **132**(4), 395–404.

18. Costenoble, R., Muller, D., Barl, T., van Gulik, W.M., van Winden, W.A., Reuss, M., and Heijnen, J.J. (2007) 13C-Labeled metabolic flux analysis of a fed-batch culture of elutriated *Saccharomyces cerevisiae*. *FEMS Yeast Res.* 7(4), 511–526.
19. Kleijn, R.J., Geertman, J.M., Nfor, B.K., Ras, C., Schipper, D., Pronk, J.T., Heijnen, J.J., van Maris, A.J., and van Winden, W.A. (2007) Metabolic flux analysis of a glycerol-overproducing *Saccharomyces cerevisiae* strain based on GC-MS, LC-MS and NMR-derived C-labelling data. *FEMS Yeast Res.* 7(2), 216–231.
20. Covert, M., Schilling, C., and Palsson, B.O. (2001) Regulation of gene expression in flux balance models of metabolism. *J. Theor. Biol.* **213**, 73–78.
21. Covert, M. and Palsson, B.O. (2003) Constraints-based models: regulation of gene expression reduces the steady-state solution space. *J. Theor. Biol.* **221**, 309–325.
22. Ishii, N., Nakahigashi, K., Baba, T., Robert, M., Soga, T., et al. (2007) Multiple high-throughput analyses monitor the response of *E. coli* to perturbations. *Science.* **316**, 593–597.
23. Lehninger, A.L. (1982) Principles of biochemistry. Worth Publishers, New York, USA.
24. Segre, D., Vitkup, D., and Church, G. (2002) Analysis of optimality in natural and perturbed metabolic networks. *Proc. Natl. Acad. Sci. USA.* **99**(23), 15112–15117.
25. Segre, D., Zucker, J., Katz, J., Lin, X., D'haeseleer, P., et al. (2003) From annotated genomes to metabolic flux models and kinetic parameter fitting. *OMICS.* 7(3), 301–316.
26. Bertsekas, D. (1995) Nonlinear programming. Athena Scientific, Belmont, MA, USA.
27. Ibarra, R.U., Edwards, J.S., and Palsson, B.O. (2002) *Escherichia coli* K-12 undergoes adaptive evolution to achieve in silico predicted optimal growth. *Nature.* **420**, 186–189.
28. Uyeda, L. (1979) Phosphofructokinase. *Adv. Enzymol. Relat. Areas Mol. Biol.* **48**, 193–244.
29. Waygood, E.B., Mort, J.S., and Sanwal, B.D. (1976) The control of pyruvate kinase of *Escherichia coli*. Binding of substrate and allosteric effectors to the enzyme activated by fructose 1,6-bisphosphate. *Biochemistry.* **15**(2), 277–282.
30. Stolovicki, E., Dror, T., Brenner, N., and Braun, E. (2006) Synthetic gene recruitment reveals adaptive reprogramming of gene regulation in yeast. *Genetics.* **173**(1), 75–85.
31. Jeong, H., Tombor, B., Albert, R., Oltvai, Z.N., and Barabási, A.L. (2000) The large-scale organization of metabolic networks. *Nature.* **407**(6804), 651–654.
32. Carlson, J.M. and Doyle, J. (2002) Complexity and robustness. *Proc. Natl. Acad. Sci.* **99**(suppl. 1), 2538–2545.
33. Iuchi, S. (1993) Phosphorylation/dephosphorylation of the receiver module at the conserved aspartate residue controls transphosphorylation activity of histidine kinase in sensor protein ArcB of *Escherichia coli*. *J. Biol. Chem.* **268**(32), 23972–23980.
34. Iuchi, S. and Lin, E.C. (1988) ArcA (dye), a global regulatory gene in *Escherichia coli* mediating repression of enzymes in aerobic pathways. *Proc. Natl. Acad. Sci.* **85**(6), 1888–1892.
35. Lynch, A.S. and Lin, E.C. (1996) Transcriptional control mediated by the ArcA two-component response regulator protein of *Escherichia coli* : characterization of DNA binding at target promoters. *J. Bacteriol.* **178**(21), 6238–6249.
36. Park, S.J., McCabe, J., Turna, J., and Gunsalus, R.P. (1994) Regulation of the citrate synthase (gltA) gene of *Escherichia coli* in response to anaerobiosis and carbon supply: role of the arcA gene product. *J. Bacteriol.* **176**(16), 5086–5092.
37. Park, S.J., Cotter, P.A., and Gunsalus, R.P. (1995) Regulation of malate dehydrogenase (mdh) gene expression in *Escherichia coli* in response to oxygen, carbon, and heme availability. *J. Bacteriol.* **177**(22), 6652–6656.
38. Park, S.J., Tseng, C.P., and Gunsalus, R.P. (1995) Regulation of succinate dehydrogenase (sdhCDAB) operon expression in *Escherichia coli* in response to carbon supply and anaerobiosis: role of ArcA and Fnr. *Mol. Microbiol.* **15**(3), 473–482.
39. Park, S.J., Chao, G., and Gunsalus, R.P. (1997) Aerobic regulation of the sucABCD genes of *Escherichia coli*, which encode alpha-ketoglutarate dehydrogenase and succinyl coenzyme A synthetase: roles of ArcA, Fnr, and the upstream sdhCDAB promoter. *J. Bacteriol.* **179**(13), 4138–4142.
40. Perrenoud, A. and Sauer, U. (2005) Impact of global transcriptional regulation by ArcA, ArcB, Cra, Crp, Cya, Fnr, and Mlc on glucose catabolism in *Escherichia coli*. *J. Bacteriol.* **187**(9), 3171–3179.
41. Saier, M.H., Ramseier, T.M., and Reizer, J. (1996) Regulation of carbon utilization. In

Escherichia coli and *Salmonella* : cellular and molecular biology. Edited by Neidhardt, F.C., et al. ASM Press, Washington, DC, USA.

42. Zheng, D., Constantinidou, C., Hobman, J.L., and Minchin, S.D. (2004) Identification of the CRP regulon using in vitro and in vivo transcriptional profiling. *Nucleic Acids Res.* **32**(19), 5874–5893.
43. Grainger, D.C., Hurd, D., Harrison, M., Holdstock, J., and Busby, S.J. (2005) Studies of the distribution of *Escherichia coli* cAMP-receptor protein and RNA polymerase along the *E. coli* chromosome. *Proc. Natl. Acad. Sci. USA.* **102**(49), 17693–17698.
44. Makman, R.S. and Sutherland, E.W. (1965) Adenosine 3',5'-phosphate in *Escherichia coli. J. Biol. Chem.* **240**, 1309–1314.
45. Saier, M.H. and Ramseier, T.M. (1997) The catabolite repressor/activator (Cra) protein of enteric bacteria. *J. Bacteriol.* **178**, 3411–3417.
46. Henikoff, S., Haughn, G.W., Calvo, J.M., and Wallace, J.C. (1988) A large family of bacterial activator proteins. *Proc. Natl. Acad. Sci. USA.* **85**(18), 6602–6606.
47. Su, C.H. and Greene, R.C. (1971) Regulation of methionine biosynthesis in *Escherichia coli* : mapping of the metJ locus and properties of a metJ plus-metJ minus diploid. *Proc. Natl. Acad. Sci. USA.* **68**(2), 367–371.
48. Pittard, J., Camakaris, H., and Yang, J. (2005) The TyrR regulon. *Mol. Microbiol.* **55**(1), 16–26.
49. Griggs, D. and Johnston, M. (1991) Regulated expression of Gal4 activator gene in yeast provides a sensitive genetic switch for glucose repression. *Proc. Natl. Acad. Sci. USA.* **88**(19), 8597–8601.
50. Lohr, D., Venkov, P., and Zlatanova, J. (1995) Transcriptional regulation in the yeast Gal gene family: a complex genetic network. *FASEB J.* **9**, 777–787.
51. Natarajan, K., Meyer, M.R., Jackson, B.M., Slade, D., Roberts, C., Hinnebusch, A.G., and Marton, M.J. (2001) Transcriptional profiling shows that Gcn4p is a master regulator of gene expression during amino acid starvation in yeast. *Mol. Cell Biol.* **21**(13), 4347–4368.
52. Lee, T., Rinaldi, N., Robert, F., Odom, D., Bar-Joseph, Z., et al. (2002) A transcriptional regulatory network map for *Saccharomyces cerevisiae. Science.* **298**, 799–804.
53. Denis, V., Boucherie, H., Monribot, C., and Daignan-Fornier, B. (1998) Role of the Myb-like protein Bas1p in *Saccharomyces cerevisiae* : a proteome analysis. *Mol. Microbiol.* **30**(3), 557–566.
54. Xiao, W. and Rank, G. (1990) Branched chain amino acid regulation of the ilv2 locus in *Saccharomyces cerevisiae. Genome.* **33**(4), 596–603.
55. O'Connel, K., Surdin-Kerjan, Y., and Baker, R. (1995) Role of the *Saccharomyces cerevisiae* general regulatory factor cp1 in methionine biosynthetic gene transcription. *Mol. Cell Biol.* **15**, 1879–1888.
56. Carroll, S.B. (2005) Evolution at two levels: on genes and form. *PLoS Biol.* **3**(7), e245.
57. Barrangou, R., Azcarate-Peril, M.A., Duong, T., Conners, S., Kelly, R.M., and Klaenhammer, T.R. (2006) Global analysis of carbohydrate utilization by *Lactobacillus acidophilus* using cDNA microarrays. *Proc. Natl. Acad. Sci. USA.* **103**(10), 3816–3821.
58. Hua, Q., Yang, C., Baba, T., Mori, H., and Shimizu, K. (2004) Analysis of gene expression in *Escherichia coli* in response to changes of growth-limiting nutrient in chemostat cultures. *Appl. Env. Microbiol.* **70**(4), 2354–2366.
59. Oh, M.K. and Liao, J. (2000) Gene expression profiling by DNA microarrays and metabolic fluxes in *Escherichia coli. Biotechnol. Prog.* **16**, 278–286.
60. Monod, J.D. (1947) The phenomenon of enzymatic adaptation and its bearing on problems of genetics and cellular differentiation. *Growth.* **11**, 223–289.
61. Jacob, F. and Monod, J. (1961) Genetic regulatory mechanisms in the synthesis of proteins. *J. Mol. Biol.* **3**, 318–356.
62. Smeianov, V.V., Wechter, P., Broadbent, J.R., Hughes, J.E., Rodriguez, B.T., et al. (2007) Comparative high-density microarray analysis of gene expression during growth of *Lactobacillus helveticus* in milk versus rich culture medium. *Appl. Environ. Microbiol.* **73**(8), 2661–2672.
63. Durfee, T., Hansen, A.M., Zhi, H., Blattner, F.R., and Jin, D.J. (2008) Transcription profiling of the stringent response in *Escherichia coli. J. Bacteriol.* **190**(3), 1084–1096.
64. Gutierrez, R.A., Lejay, L.V., Dean, A., Chiaromonte, F., Shasha, D.E., and Coruzzi, G.M. (2007) Qualitative network models and genome-wide expression data define carbon/nitrogen-responsive molecular machines in Arabidopsis. *Genome Biol.* **8**(1), R7.

65. Ma, S., Gong, Q., and Bohnert, H.J. (2006) Dissecting salt stress pathways. *J. Exp. Bot.* **57**(5), 1097–1107.

66. Gasch, A.P., Spellman, P.T., Kao, C.M., Carmel-Harel, O., Eisen, M.B., Storz, G., Botstein, D., and Brown, P.O. (2000) Genomic expression programs in the response of yeast cells to environmental changes. *Mol. Biol. Cell.* **11**(12), 4241–4257.

67. Siddiquee, K.A., Arauzo-Bravo, M.J., and Shimizu, K. (2004) Effect of a pyruvate kinase (pykF-gene) knockout mutation on the control of gene expression and metabolic fluxes in *Escherichia coli. FEMS Microbiol. Lett.* **235**(1), 25–33.

68. Ideker, T., Thorsson, V., Ranish, J.A., Christmas, R., Buhler, J., et al. (2001) Integrated genomic and proteomic analyses of a systematically perturbed metabolic network. *Science.* **292**(5518), 929–934.

69. Hughes, T.R., Marton, M.J., Jones, A.R., Roberts, C.J., Stoughton, R., et al. (2000) Functional discovery via a compendium of expression profiles. *Cell.* **102**(1), 109–126.

70. Giaever, G., Chu, A.M., Ni, L., Connelly, C., Riles, L., et al. (2002) Functional profiling of the *Saccharomyces cerevisiae* genome. *Nature.* **418**(6896), 387–391.

71. Santangelo, G.M. (2006) Glucose signaling in *Saccharomyces cerevisiae. Microbiol. Mol. Biol. Rev.* **70**(1), 253–282.

72. Edwards, T.E., Klein, D.J., and Ferre-D'Amare, A.R. (2007) Riboswitches: small-molecule recognition by gene regulatory RNAs. *Curr. Opin. Struct. Biol.* **17**(3), 273–279.

73. Winkler, W., Nahvi, A., and Breaker, R.R. (2002) Thiamine derivatives bind messenger RNAs directly to regulate bacterial gene expression. *Nature.* **419**(6910), 952–956.

74. Nahvi, A., Sudarsan, N., Ebert, M., Zou, X., Brown, K.L., and Breaker, R. (2002) Genetic control by a metabolite binding mRNA. *Chem. Biol.* **9**(9), 1043–1049.

75. Kim, J.N. and Breaker, R.R. (2008) Purine sensing by riboswitches. *Biol. Cell.* **100**(1), 1–11.

76. Kharchenko, P., Church, G.M., and Vitkup, D. (2005) Expression dynamics of a cellular metabolic network. *Mol. Syst. Biol.* **1**, 2005.0016 (online).

77. Ge, H., Liu, Z., Church, G.M., and Vidal, M. (2001) Correlation between transcriptome and interactome mapping data from *Saccharomyces cerevisiae. Nat. Genet.* **29** (4), 482–486.

78. Zaslaver, A., Mayo, A.E., Rosenberg, R., Bashkin, P., Sberro, H., Tsalyuk, M., Surette, M.G., and Alon, U. (2004) Just-in-time transcription program in metabolic pathways. *Nat. Genet.* **36**(5), 486–491.

79. Ihmels, J., Levy, R., and Barkai, N. (2004) Principles of transcriptional control in the metabolic network of *Saccharomyces cerevisiae. Nat. Biotechnol.* **22**(1), 86–92.

80. Bilu, Y., Shlomi, T., Barkai, N., and Ruppin, E. (2006) Conservation of expression and sequence of metabolic genes is reflected by activity across metabolic states. *PLoS Comput. Biol.* **2**(8), e106.

81. Martinez-Perez, O., Lopez-Sanchez, A., Reyes-Ramirez, F., Floriano, B., and Santero, E. (2007) Integrated response to inducers by communication between a catabolic pathway and its regulatory system. *J. Bacteriol.* **189**(10), 3768–3775.

82. Karp, P.D., Ouzounis, C.A., Moore-Kochlacs, C., Goldovsky, L., Kaipa, P., et al. (2005) Expansion of the BioCyc collection of pathway/genome databases to 160 genomes. *Nucleic Acids Res.* **33**(19), 6083–6089.

83. Caspi, R., Foerster, H., Fulcher, C.A., Hopkinson, R., Ingraham, J., et al. (2006) MetaCyc: a multiorganism database of metabolic pathways and enzymes. *Nucleic Acids Res.* **34**, D511–D516.

84. Karp, P.D., Keseler, I.M., Shearer, A., Latendresse, M., Krummenacker, M., et al. (2007) Multidimensional annotation of the *Escherichia coli* K-12 genome. *Nucleic Acids Res.* **35**(22), 7577–7590.

85. Duarte, N.C., Becker, S.A., Jamshidi, N., Thiele, I., Mo, M.L., Vo, T.D., Srivas, R., and Palsson, BO. (2007) Global reconstruction of the human metabolic network based on genomic and bibliomic data. *Proc. Natl. Acad. Sci.* **104**(6), 1777–1782.

86. Mueller, L.A., Zhang, P., and Rhee, S.Y. (2003) AraCyc: a biochemical pathway database for Arabidopsis. *Plant Physiol.* **132**(2), 453–460.

87. Cherry, J.M., Ball, C., Weng, S., Juvik, G., Schmidt, R., et al. (1997) Genetic and physical maps of *Saccharomyces cerevisiae. Nature.* **387**(6632 Suppl), 67–73.

88. Mewes, H.W., Frishman, D., Gruber, C., Geier, B., Haase, D., et al. (2000) MIPS: a database for genomes and protein sequences. *Nucleic Acids Res.* **28**(1), 37–40.

89. Kanehisa, M. and Goto, S. (2000) KEGG: Kyoto Encyclopedia of Genes and Genomes. *Nucleic Acids Res.* **28**, 27–30.

90. Covert, M.W., Knight, E.M., Reed, J.L., Herrgard, M.J., and Palsson, B.O. (2004) Integrating high-throughput and

computational data elucidates bacterial networks. *Nature.* **429**(6987), 92–96.

91. Herrgard, M.J., Fong, S.S., and Palsson, B.O. (2006) Identification of genome-scale metabolic network models using experimentally measured flux profiles. *PLoS Comput. Biol.* **2**(7), e72.
92. Shlomi, T., Eisenberg, Y., Sharan, R., and Ruppin, E. (2007) A genome-scale computational study of the interplay between transcriptional regulation and metabolism. *Mol. Syst. Biol.* **3**, 101.
93. Gat-Viks, I., Tanay, A., and Shamir, R. (2004) Modeling and analysis of heterogeneous regulation in biological networks. *J. Comput. Biol.* **11**(6), 1034–1049.
94. Yeang, C.H. and Vingron, M. (2006) A joint model of regulatory and metabolic networks. *BMC Bioinformatics.* **7**, 332.
95. Narang, A. (2006) Comparative analysis of some models of gene regulation in mixed-substrate microbial growth. *J. Theor. Biol.* **242**(2), 489–501.
96. Kofahl, B. and Klipp, E. (2004) Modelling the dynamics of the yeast pheromone pathway. *Yeast.* **21**(10), 831–850.
97. Varner, J.D. (2000) Large-scale prediction of phenotype: concept. *Biotechnol. Bioeng.* **69**(6), 664–678.
98. Burgard, A.P., Pharkya, P., and Maranas, C.D. (2003) Optknock: a bilevel programming framework for identifying gene knockout strategies for microbial strain optimization. *Biotechnol. Bioeng.* **84**(6), 647–657.
99. Patil, K.R., Rocha, I., Forster, J., and Nielsen, J. (2005) Evolutionary programming as a platform for in silico metabolic engineering. *BMC Bioinformatics.* **6**, 308.
100. Kim, H.U, Kim, T.Y., and Lee, S.Y. (2008) Metabolic flux analysis and metabolic engineering of microorganisms. *Mol. Bio. Syst.* **4**, 113–120.
101. Tringe, S.G. and Rubin, E.M. (2005) Metagenomics: DNA sequencing of environmental samples. *Nat. Rev. Genet.* **6**(11), 805–814.

Chapter 14

Applying Word-Based Algorithms: The IMEter

Ian F. Korf and Alan B. Rose

Abstract

Important patterns can be found in strings of characters such as nucleotides in a DNA sequence by examining the frequency of occurrence of specific character combinations or words. The abundance of words can reveal the presence of underlying trends governing the order of characters, even if the biological reasons for those trends remain mysterious. As an example of one way in which word frequencies have provided insight, we describe the IMEter, a word-based algorithm for analyzing introns and their effect on gene expression. The IMEter demonstrates that introns located near the beginning of genes are compositionally distinct from later introns and that these differences are closely related to the ability of some introns to increase gene expression. This word-based approach has proven more successful than deletion analysis at identifying the sequences responsible for elevating expression because they are dispersed throughout stimulatory introns.

Key words: Markov model, word based, nucleotide frequency, odds ratios, intron, gene expression, motif, intron-mediated enhancement, IMEter.

1. Sequences as Markov Models

Whether the medium is conversation, chalkboards, journal articles, or computers, biological sequences are often represented as text. While we know that DNA, RNA, and protein molecules are dynamic entities with physical and chemical properties that depend on their shape and environment, representing biological sequences as one-dimensional *strings* is convenient and allows one to take advantage of analysis techniques pioneered in Cryptography and Natural Language Processing.

One of the simplest analyses one can perform with any text is to count the frequencies of individual symbols. **Table 14.1** shows the frequency of each letter in Darwin's Origin of Species. Not

Dmitry A. Belostotsky (ed.), *Plant Systems Biology, vol. 553*

DOI 10.1007/978-1-60327-563-7_14 Springerprotocols.com

Table 14.1
Letter frequencies in Origin of Species

Symbol	%	Symbol	%
A	7.98	N	7.17
B	1.69	O	7.21
C	3.50	P	1.89
D	3.70	Q	0.09
E	13.18	R	6.27
F	2.78	S	6.88
G	1.82	T	9.00
H	4.99	U	2.56
I	7.43	V	1.19
J	0.07	W	1.60
K	0.37	X	0.24
L	4.19	Y	1.64
M	2.51	Z	0.05

surprisingly, the most common letter is "e", which accounts for approximately 13% of all letters. Similarly, one can count the letters from the *Arabidopsis thaliana* genome and observe that the genome is approximately 36% GC (**Table 14.2**). Although these analyses are very simple, they provide useful information. For example, the letter frequencies in Darwin's book closely resemble the letter frequencies in English in general, and one could use such information to deduce that Origin of Species was published in

Table 14.2
Genomic nucleotide frequencies

	A. thaliana					*D. radiodurans*
			Intron			
Symbol	Genome	Exon	All	Proximal	Distal	Genome
A	32.00	29.85	26.73	25.7	27.3	16.54
C	18.02	20.14	15.46	14.8	16.2	33.51
G	18.01	20.16	17.16	17.0	15.8	33.45
T	31.97	29.84	40.64	42.5	40.7	16.49

English without ever reading the book. Similarly, one can examine the nucleotide compositions of various genomes and recognize that each organism has a characteristic (though not necessarily unique) composition, and one might use these frequencies to look for horizontal gene transfer or contamination events.

While simple letter counting can be useful, it throws away the context of each letter. Let us first consider the immediate context of a letter, which is defined by adjacent letters. It may seem strange, but in text and sequence analysis, only the preceding letter is used as the context. The reason for this is that we model sequences as the products of Markov processes. A useful way to think of a Markov model (or chain) is as a machine that randomly generates plausible observations. Let us consider a Markov model for the daily weather with three states called *Sunny*, *Cloudy*, and *Rainy*. We must define transition probabilities between the various states that determine how often and in what order the observations are generated. For example, we might define the probability of transitioning between Sunny and Cloudy to be greater than Sunny and Rainy because clouds usually precede a rain storm. **Figure 14.1** shows such a model. If we want the weather model to mimic actual weather patterns, the transition probabilities should match local weather observations. Given the model in **Fig. 14.1**, we can start in a state, such as Sunny and then create a new day of weather by "rolling dice" to determine if tomorrow will be Sunny, Cloudy, or Rainy. It should be obvious that whatever tomorrow's weather is depends only on today, not last year. This property where the future is independent of the distant past is known as the *Markov property*.

Getting back to biology, let us now consider that genomes are products of Markov processes. The simplest Markov model one can make is that each nucleotide is generated without respect to context. That is, we can simply define probabilities for A, C, G, and T and draw these at random to generate a sequence. Let us say we do this for the *A. thaliana* genome and the first three nucleotides

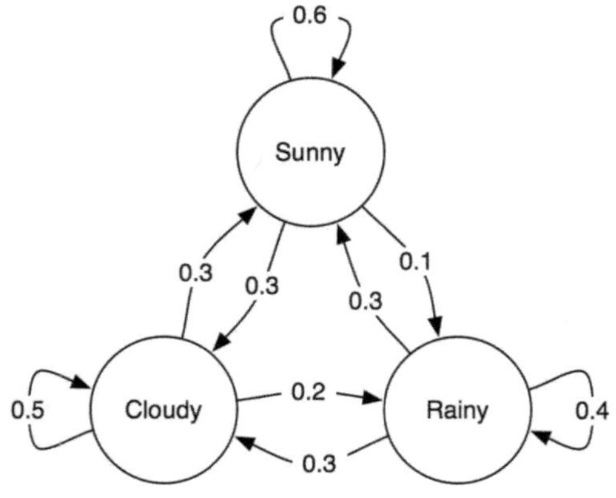

Fig. 14.1. Markov model for daily weather.

generated are GCT. The probability with which this particular sequence was generated by the model is approximately 0.010 ($0.18 \times 0.18 \times 0.32$). To take nearby context into account, we must record the conditional probability of each letter given the preceding letter(s). That is, we need to know the probability of generating a T given that we have already seen a T (this is similar to the weather example above). If we are concerned with only a single preceding letter, this is called a first-order Markov chain. If we are concerned with two letters of context, for example to model the probability of generating a G given that we have already seen an A followed by a T, this is a second-order Markov chain. In general terms, biological sequences are often modeled by an *n*th-order Markov chain where *n* is some non-negative integer. In practice, the value of *n* is commonly in the range of 1–5.

The letter frequencies from **Tables 14.1** and **14.2** can now be understood as 0th-order Markov models. They predict that the probability of observing CG is simply the probability of C multiplied by the probability of G. While you might believe this to be true of the *A. thaliana* genome, you certainly would not believe this of English since you have never seen any words where C precedes G (because there are none). The immediate context of a letter is clearly very important in language. Is this also true of biological sequences? **Table 14.3** shows a first-order Markov model for the *A. thaliana* genome. There are some interesting properties. For example, the probabilities of the homodimers (AA, CC, GG, TT) are greater than expected. Also, the probabilities are not symmetric. For example, the probability of C followed by G is not the same as G followed by C. For this reason, when modeling biological sequences, it is generally a good idea to go beyond 0th-order Markov chains. Exactly how far beyond zero depends on several factors. One important consideration is the size of the training set. A 15th-order Markov chain requires approximately 1 billion (4^{15}) contexts. If such a model was applied to the human genome (approximately 3 billion bp), each observation would receive less than one count on average, which does not lead to a very useful model.

Table 14.3
First-order Markov chain from A. thaliana

	Preceding symbol			
Symbol	**A**	**C**	**G**	**T**
A	36.18	35.24	35.58	23.98
C	16.36	18.81	16.65	20.02
G	18.57	12.99	18.74	19.86
T	28.89	32.97	29.03	36.14

In addition to the immediate context of a symbol, there are also larger contexts. Chromosomes do not have uniform compositions throughout their lengths, and a single Markov model does not capture regional variations very well. **Table 14.2** shows nucleotide frequencies for *A. thaliana* exons and introns. The differences are obvious even at the 0th-order level. The compositional difference between coding and non-coding sequences has been used by many researchers to identify protein-coding genes in genomic DNA. Early efforts (1) used second-order Markov models to report regions that were likely to be coding but did not attempt to define exon–intron boundaries. Today, gene-finding programs are much more sophisticated (2) and take into account such biological features as splice sites, promoters, and poly-A sites. Still, at the root of these gene prediction algorithms, the sequence is generally represented as an *n*th-order Markov model.

Previously we noted that GCT would be generated by an *A. thaliana* 0th-order Markov model with a probability of approximately 0.010. If the model had been derived from *Deinococcus radiodurans* (*see* **Table 14.2**), the sequence would be generated with a probability of approximately 0.019 (0.335 × 0.335 × 0.165). If we are given just the sequence GCT and asked to guess its origin, the odds are almost twice as great that it came from *D. radiodurans* than from *A. thaliana*. Similarly, given models for coding and non-coding sequence, it is very simple to examine an unknown sequence (or segment of a sequence) to determine which model is more likely to have generated the sequence. In essence, this is how gene-finding algorithms work, though they use higher order Markov chains.

Probabilities and odds ratios are typically calculated as logarithms in bioinformatics applications. One reason for this is that genome data is large, and if you repeatedly multiply numbers, you may overflow or underflow the numeric representation in the computer (e.g., if you keep squaring a number on a calculator, you eventually reach a limit as the value approaches infinity or zero). The base of the logarithm is generally 2 or *e*, and the corresponding units are *bits* or *nats*. In general, log-odds values are called *scores*. For example, in sequence alignment, the score for any two amino acids is the log-odds ratio of the observed and expected pairings, where the observed pairing is calculated from multiple alignments of related proteins and the expected pairing is the random expectation from individual amino acid frequencies. A high score indicates the amino acids are found more often than by chance. For example, the score for valine and leucine is positive because these are chemically similar amino acids that can often substitute for one another without compromising protein function. In gene finding, the score of a predicted exon reflects the log-odds ratio of having been generated by a coding vs. non-coding model. A positive score indicates the region is probably coding while a negative score indicates it is probably non-coding.

2. Words as Functional Units

Both in language and in sequence analysis, the concept of a *word* is very useful. In text, it is not letters but words that convey information. Similarly, in biology, specific strings such as the amino acids RGD (extra-cellular matrix attachment) or the nucleotides GAATTC (*Eco*RI restriction site) have known functional roles. In both language and sequence, the meaning of a word often depends on its context.

Identifying words in text is trivial if the language uses spaces as delimiters, but in languages without delimiters, such as ancient Roman, one must use context to parse letters into words. The words and meanings in biology are much more difficult to parse than natural languages for several reasons: (a) there are only four letters in DNA; (b) there are no delimiters or punctuation; (c) there is no dictionary of legal words; (d) we do not know the rules of the language; and (e) words can often be misspelled and retain their biological meaning.

In sequence analysis we use a simplified definition of *word* that does not require knowing its meaning (function). A word is simply a sequence of length k where k is some positive integer. Words are also called k-tuples or k-mers or even oligos. To identify all the words in a sequence, one simply moves a window of length k one letter at a time along the entire sequence (without making truncated words at either end). Words created this way are therefore very similar to nth-order Markov models.

Like letter frequencies, word frequencies can be informative in text and sequence. **Table 14.4** shows the top 20 words from the Origin of Species. As you might expect, *the* tops the list and *species* is quite common. If we construct a quasi-genomic version of the book by removing all the spaces and punctuation, the text becomes one long chromosome-like string that is difficult to interpret.

malsonecatforinstancetakingtocatchratsanothermiceonecata
ccordingtomrstjohnbringinghomewingedgameanotherhareso
rrabbitsandanotherhuntingonmarshygroundandalmostnigh
tlycatchingwoodcocksorsnipesthetendencytocatchratsrathertha
nmiceisknowntobeinheritednowifanyslightinnatechangeofha
bitorofstructurebenefitedanindividualwolfitwouldhavethebe
stchanceofsurvivingandofleavingoffspringsomeofitsyoungwo
uldprobablyinheritthesamehabitsorstructureandbytherepeti
tionofthisprocessanewvarietymightbeformedwhichwouldeith
ersupplantorcoexistwiththeparentformofwolforagainthewol
vesinhabitingamountai

Since we no longer know where the word boundaries are, we can use the sequence analysis concept of a word to identify all possible words of some size k and determine their frequencies.

Table 14.4
Word frequency analysis of Origin of Species

Rank	Word	3-mer	4-mer	5-mer	6-mer	7-mer	8-mer
1	the	the	tion	ofthe	specie	species	differen
2	of	and	nthe	ation	pecies	thesame	havebeen
3	and	ion	ofth	speci	softhe	thatthe	especies
4	in	ing	fthe	pecie	hesame	natural	selectio
5	to	tio	thes	ecies	thesam	differe	election
6	a	nth	ther	inthe	thatth	ifferen	varietie
7	that	ent	that	which	differ	havebee	arieties
8	as	tha	othe	tions	hatthe	avebeen	naturals
9	have	oft	atio	other	ations	especie	lselecti
10	be	her	spec	ction	ofthes	lection	alselect
11	is	sof	peci	onthe	natura	selecti	characte
12	on	fth	have	natur	atural	electio	haracter
13	species	hes	inth	softh	ection	varieti	aturalse
14	by	for	cies	atthe	tionof	arietie	turalsel
15	which	ati	ecie	andth	genera	rieties	uralsele
16	or	int	hich	thesa	especi	fromthe	ralselec
17	are	ere	whic	tothe	eofthe	animals	distinct
18	it	hat	sand	esame	iffere	aturals	ifferent
19	for	oth	tthe	hesam	havebe	lselect	conditio
20	with	ies	with	their	fferen	ication	ondition

Table 14.4 shows that even without knowing the true word boundaries, it is possible to identify common words and phrases. Now imagine if the text was written in a different language and was interspersed with lots of seemingly random letters. This is the problem we have to deal with in analyzing biological sequences.

Even though genomes are complex entities and our knowledge of genome biology is still in its infancy, it is possible to make significant advances using methods as simple as word frequency analysis. As an example, we describe our research on how introns affect gene expression. While our work utilized the *A. thaliana* genome (3), and this work would have been more difficult without the entire sequence, the sine qua non was not the genome, but rather knowing what to model. In other words, the key was understanding the biology.

3. The Biology of Intron-Mediated Enhancement

Shortly after introns were discovered, it was noted that several genes were expressed very poorly when their introns were removed. Conversely, inserting introns into reporter genes that lacked them, including bacterial genes such as lacZ or GUS, often increased the expression of those genes in transgenic organisms. Most of the introns characterized could only affect expression from within transcribed sequences and in their natural orientation, indicating that they operated by a still undefined mechanism that is distinct from transcriptional enhancer elements. This intron-mediated enhancement (IME) (4) has been observed in a broad diversity of organisms including mammals, fungi, nematodes, insects, and plants, suggesting that it is an ancient and fundamental feature of eukaryotic gene regulation.

A puzzle arose from attempts to identify the sequences within introns that are responsible for elevating expression. Some efficiently spliced introns clearly stimulate expression much more than others do, suggesting the presence of enhancing sequences within some introns. However, motifs that are conserved between stimulatory introns could not be found by conventional homology searches due to the large heterogeneity in intron sequence and size coupled with the relatively small number of introns known to stimulate expression. Attempts to locate enhancing regions by deletion analysis also have failed because no unique sequences are individually required for the intron to increase mRNA accumulation (5). That is, introns that contain large internal deletions but are still spliced usually stimulate mRNA accumulation as much as does the full-length intron, even when the combined deletions span the entire intron. How can differences between introns be sequence based if no unique sequences are involved? One possible explanation is that the enhancing sequences are redundant and distributed throughout introns. This idea was confirmed using hybrid introns constructed from parts of enhancing and non-enhancing introns (6). The dispersed nature of the expression-affecting sequences contrasts with more familiar regulatory elements such as enhancers, promoters, RNA secondary structures such as stem loops, binding sites for proteins or small RNAs, or aptamers, whose functions depend on discrete and localized individual sequences.

Three lines of evidence suggest that introns near the start of their genes are more likely than other introns to stimulate expression. First, virtually all of the introns known to boost expression are first introns, although not all first introns have this ability. Second, introns that elevate expression when located in the 5′-UTR lose this effect when they are moved to the 3′-UTR. Third, an enhancing intron that is placed progressively farther downstream in a gene starts to lose its enhancing ability at about 500 bp from the promoter and has no effect ~1 kb or more from the 5′ end (7).

Defining the sequences required for IME is desirable because it would provide a toehold for the biochemical isolation of trans-acting factors that bind to those sequences, which could be an important path to understanding the novel mechanism through which introns affect expression. In addition, identifying the enhancing sequences would provide a means to predict whether or not an untested intron is likely to elevate expression. Eventually, with the knowledge of what sequences enhance expression, it may be possible to design synthetic introns that are more powerful than any naturally occurring one, which would be very useful for transgenic applications seeking to maximize gene expression.

4. The IMEter

The key to modeling IME is the hypothesis that IME signals are enriched in introns located near the promoter (proximal) compared to those farther down the transcript (distal). Starting with simple letter counting, we can look at the sequence composition of proximal and distal introns. **Table 14.2** shows the single letter frequencies where proximal introns are defined as those that begin within 500 bp of the promoter and distal introns are defined as those that begin more than 500 bp from the promoter. The fact that the compositions are not identical suggests that there may be important differences between proximal and distal introns. Even if the compositions were identical, however, there may be higher order differences not visible at the 0th order.

Our program for predicting IME, the IMEter (6), is very similar to algorithms for predicting coding regions. Instead of comparing arbitrary genomic regions to the compositions of coding and non-coding sequences, we compare arbitrary introns to the compositions of proximal and distal introns. A positive score indicates an intron is more similar to proximal introns than distal introns and is therefore more likely to contain elements responsible for IME. Since we do not consider the splice donor and acceptor sequences to be generated from the same model as the body of the intron, the IMEter omits these regions. Before training the IMEter, we must choose several parameters: (a) the word size, (b) the cutoff for proximal introns, (c) the cutoff for distal introns, (d) the length of the splice donor site to omit, and (e) the length of the splice acceptor site to omit. For simplicity, the cutoff for both proximal and distal can be a single value, such as the 500 bp we used for **Table 14.2**.

The IMEter scoring function can be described by the following equation:

$$S = \sum_{i=1+D}^{i \leq L-K-A} \log\left(\frac{P_{w_i}}{Q_{w_i}}\right)$$

where S is the IMEter score, L is the length of the intron, K is the word size, A is the length of the splice acceptor consensus, D is the length of the splice donor consensus, w_i is a word of length K at position i, and P and Q are frequency distributions for words of length K in proximal and distal introns.

Training the IMEter requires a set of genes where the position of the 5′ end of the transcript and the positions of introns are known. Fortunately, the *Arabidopsis* genome annotation contains thousands of experimentally identified examples due to the efforts of the full-length cDNA sequencing project (8). While we utilized thousands of genes, we find that it is also possible to train the IMEter on a few hundred conserved genes. As part of our training procedure, we removed highly paralogous genes (to limit over-training on large gene families) and those genes with suspect features (e.g., very short or GC-rich introns) that may indicate genome annotation errors.

To test whether IMEter scores have biologically meaningful values, we trained the IMEter with an educated guess at the parameters (word size 5, proximal/distal cutoff at 400 bp, 5 bp donor site, 10 bp acceptor site) and examined the scores of introns whose effect on gene expression was already known. The only data set appropriate for this analysis comes from experiments in *Arabidopsis* where the enhancing ability of different introns has been tested with the same reporter gene in single-copy lines. Even though the quantitative data set was small, representing just six introns, it was the largest known for any organism. Furthermore, the data are very reproducible, as indicated by the small amount of variation in expression, presumably because only single-copy transgenic plants were analyzed. When IMEter scores were compared to the expression values, a very strong linear correlation was found between an intron's IMEter score and the degree to which that intron stimulates mRNA accumulation (**Fig. 14.2** filled circles). The tight correlation suggested that IMEter scores might be able to predict the enhancing ability of previously uncharacterized introns. To test this, six additional introns were chosen, and all enhanced expression to the degree expected from their IMEter scores (**Fig. 14.2** open

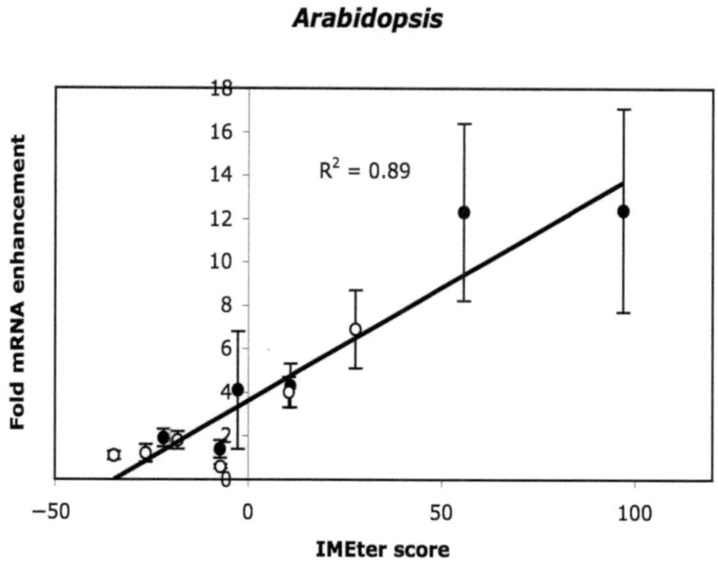

Fig. 14.2. IMEter score is correlated with enhancement.

circles). Further evidence supporting the connection between IMEter scores and enhancing ability comes from the 21 other *Arabidopsis* introns reported in the literature to boost expression of different genes. All but one of these introns have a positive IMEter score, and 18 have scores in the top 5% of all *Arabidopsis* introns.

The IMEter can be optimized by changing various parameters. **Figure 14.3** shows how variations in word size and the proximal/distal cutoff affect performance as measured by the R^2 value. A word size of 1, corresponding to a 0th-order Markov model, is not very useful. Larger word sizes perform much better, but as the word size gets over 8, the R^2 value drops off. This is especially apparent when the proximal/distal cutoff is low. This is probably due to the smaller amount of sequence available and the larger number of words. In *Arabidopsis*, a variety of parameter combinations perform approximately equivalently. This may not be true in other genomes or other sequence analysis scenarios, so it is a good idea to survey the parameter space as we have done.

The observation that promoter-proximal and distal introns gave different k-mer profiles indicated that introns are structurally unequal depending on the location of those introns in their genes. To explore genome-wide differences in intron composition, the entire set of *Arabidopsis* introns was randomly divided into two equal groups. The introns in one group were used to train the IMEter, which was then used to analyze the introns in the other. The distribution frequency of IMEter scores forms a bell-shaped curve centered near zero. When only the first introns from genes are considered, the distribution shifts to the right (mean score = 10.6),

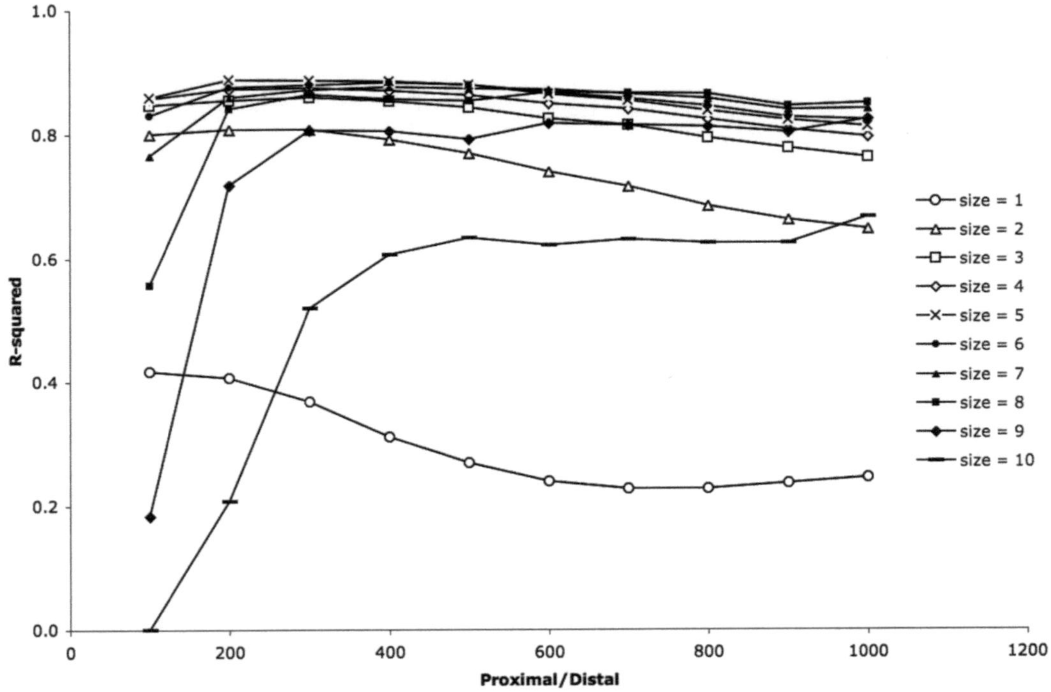

Fig. 14.3. Optimizing IMEter parameters.

and virtually all of the highest-scoring introns in the genome are first introns. The relationship between IMEter scores and location can be seen more clearly by plotting the scores of introns against their distance from the start of transcription (**Fig. 14.4**). Average

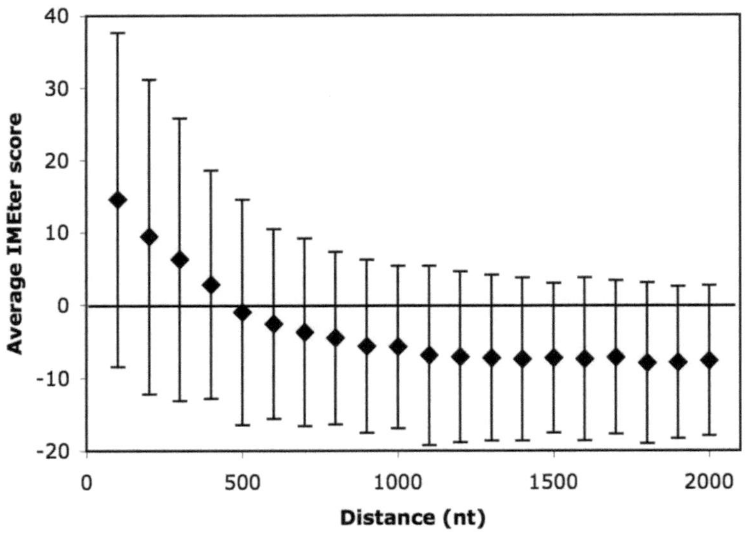

Fig. 14.4. IMEter score of introns as a function of distance from their promoter.

IMEter scores are highest in introns near the start and decline with distance, and very few introns more than 1000 nt from the start have a positive score. This pattern is in striking agreement with the ability of an intron to stimulate mRNA accumulation, which also declines with distance from the promoter until it is lost entirely between 550 and 1100 nt from the start of transcription.

4.1. Identifying Enhancing Sequences

One drawback of analyzing word frequencies is that the biological signals that give rise to high scores are not immediately apparent. To identify candidate sequences involved in IME, we employed a motif-finding algorithm, NestedMica (9), to find sequence patterns that are over-represented in the 100 introns with the highest IMEter scores. Several motifs were found, and these were ranked by how well they correlated with the set of introns with known effects on expression. This analysis was therefore very similar to that shown in **Fig. 14.2**, except that a combined motif score was used in place of the IMEter score. The motif that was most correlated with observed enhancement is shown in **Fig. 14.5**, and we call this the IME motif.

Fig. 14.5. IME motif.

Rather than evaluating an entire intron, one can also look for regions of high IMEter score in genomic context by calculating IMEter score in a fixed, sliding window. **Figure 14.6** shows that high scores are most abundant in the intron and occur in the same regions as high IME motif density. While there is a great deal of variation from gene to gene, the general pattern is for IME signals to be concentrated in proximal introns and virtually absent from other regions of the genome.

4.2. IME Signals in Other Species

The IMEter can be applied as described to any organism where there are known exon–intron structures for a few hundred genes or more. Unfortunately, there are no organisms aside from *Arabidopsis* where the IMEter can be quantitatively evaluated because rigorous intron-swapping experiments have not been performed elsewhere. It remains to be established whether or not promoter proximity is relevant to IME in all organisms. Furthermore, the IMEter may be ineffective in species in which introns are very large in size (as in mammals) or small in number (as in *Saccharomyces cerevisiae*). Given the aforementioned caveats, we have examined

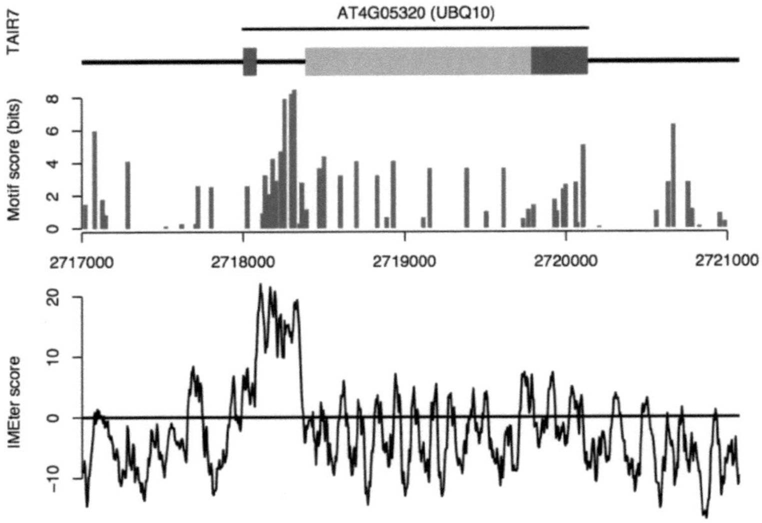

Fig. 14.6. IMEter score and IME motif density in the UBQ10 region. The genomic region of the UBQ10 gene is shown. The exon–intron structure is shown at the top. The dark regions are untranslated, and the light regions are coding. The middle panel shows the score of the IME motif. Higher bars indicate a better match to the consensus. The lower panel shows IMEter score in a 50 bp window.

IME signals in rice and find that a rice IMEter behaves similarly to the *Arabidopsis* IMEter (data not shown, see (6)). We also find that there is a good correlation (R^2 0.74) between rice-trained IMEter scores and *Arabidopsis* expression values, which indicates that the IME machinery may be very similar in these organisms.

5. Summary

The IMEter illustrates some of the strengths and weaknesses of word-based algorithms. On the positive side, the IMEter revealed previously unsuspected differences in the composition of introns, a large collection of very diverse elements. No prior assumptions about the length or positions of the relevant sequences were required. However, word frequency analysis is not the appropriate method for all sequence elements. IME signals are both dispersed and redundant, so there was a good fit between the biological signals and the statistical model. If we had been looking for an isolated signal where position was an important factor, for example the TATA box, one would not expect word frequency analysis to be very useful.

Perhaps the most serious weakness of word-based analyses is the difficulty in identifying the functional elements that are being recognized from among the entire dictionary of words. A number of statistical measures can be employed, but ultimately the biological significance of any candidates must be evaluated experimentally. Despite the inherent limitations in word-based analyses, they can be very useful tools for the systems biologist because they provide a means to detect previously unrecognized patterns in complex sets of data, thereby revealing new connections. While it is expected that more sophisticated statistical models (e.g., hidden Markov models) and experimental molecular biology (e.g., gene expression studies, proteomics) are required to identify the biological entities involved, word-based analyses can provide a critical first step for the journey ahead.

References

1. Staden, R. and McLachlan, A.D. (1982) Codon preference and its use in identifying protein coding regions in long DNA sequences. *Nucleic Acids Res.* **10**, 141–156.
2. Brent, M.R. (2008) Steady progress and recent breakthroughs in the accuracy of automated genome annotation. *Nat. Rev. Genet.* **9**, 62–73.
3. Arabidopsis thaliana Consortium (2000) Analysis of the genome sequence of the flowering plant Arabidopsis thaliana. *Nature.* **408**, 796–815.
4. Mascarenhas, D., Mettler, I.J., Pierce, D.A., and Lowe, H.W. (1990) Intron-mediated enhancement of heterologous gene expression in maize. *Plant Mol. Biol.* **15**, 913–920.
5. Rose, A.B. and Beliakoff, J.A. (2000) Intron-mediated enhancement of gene expression independent of unique intron sequences and splicing. *Plant Physiol.* **122**, 535–542.
6. Rose, A.B., Elfersi, T., Parra, G., and Korf, I. (2008) Promoter-proximal introns in Arabidopsis thaliana are enriched in dispersed signals that elevate gene expression. *Plant Cell.* **20**, 543–551.
7. Rose, A.B. (2004) The effect of intron location on intron-mediated enhancement of gene expression in Arabidopsis. *Plant J.* **40**, 744–751.
8. Seki, M., Narusaka, M., Kamiya, A., Ishida, J., Satou, M., Sakurai, T., Nakajima, M., Enju, A., Akiyama, K., Oono, Y., Muramatsu, M., Hayashizaki, Y., Kawai, J., Carninci, P., Itoh, M., Ishii, Y., Arakawa, T., Shibata, K., Shinagawa, A., and Shinozaki, K. (2002) Functional annotation of a full-length Arabidopsis cDNA collection. *Science.* **296**, 141–145.
9. Down, T.A. and T.J. Hubbard (2005) NestedMICA: sensitive inference of over-represented motifs in nucleic acid sequence. *Nucleic Acids Res.* **33**, 1445–1453.

Section III

Imaging

Chapter 15

Live-Imaging and Image Processing of Shoot Apical Meristems of *Arabidopsis thaliana*

G. Venugopala Reddy and A. Roy-Chowdhury

Abstract

The shoot apical meristem (SAM) of higher plants represents a dynamic network of different cell types which exhibit distinct patterns of gene expression and cellular behaviors. The regulation of distinct patterns of gene expression and cellular behaviors is mediated by cell–cell communication networks. Live-imaging of spatiotemporal dynamics of cell–cell communication networks, gene expression patterns, and cellular behaviors is critical to deduce principles that underlie SAM growth and maintenance. In this chapter, we describe live-imaging methods, fluorescent reagents, and image processing protocols that have been developed to visualize the regulatory dynamics of SAM growth in *Arabidopsis thaliana*.

Key words: Fluorescence, stem cells, real-time imaging, image segmentation, 3D reconstruction, cell lineage, cell tracking.

1. Introduction

Pattern formation and stem-cell maintenance in the shoot apical meristems (SAMs) of higher plants involve co-ordinated regulation of gene expression and growth (1). In *Arabidopsis thaliana*, the SAM stem-cell niche consists of approximately 500 cells (5 μM each in size) organized into three clonally distinct layers of cells. The SAM is further subdivided into distinct functional domains (2). The central zone (CZ) is located at the tip and it contains a set of stem cells. The progeny of stem cells enters into differentiation pathways when they enter the surrounding peripheral zone (PZ). The CZ also supplies cells to the rib meristem (RM) located beneath the CZ and the RM cells differentiate and become part of the stem. Thus, cell fate specification within SAMs is a dynamic

Dmitry A. Belostotsky (ed.), *Plant Systems Biology, vol. 553*

DOI 10.1007/978-1-60327-563-7_15 Springerprotocols.com

process in which transient changes in gene activation/repression and changes in growth patterns are tightly coupled in both space and time. Therefore, visualization of gene expression and growth, in real time, by employing live-imaging methods may provide new insights into the dynamic interaction between growth and cell fate specification mediated by cell–cell communication (3). Recent studies have attempted to understand the dynamic spatiotemporal contours of cell–cell communication networks and that of patterns of gene expression and cell behaviors in living SAMs (4–9). This effort has involved the development of new live-imaging methods, new fluorescent reporter lines, and image processing protocols. In this chapter we discuss the live-imaging methods and image processing methods that have been developed to study the SAMs of *A. thaliana*.

2. Materials

1. Ziess 310 or Zeiss 510 upright confocal microscope with multi-channel imaging capability
2. 63X achroplan water dipping objective lens (0.95 NA; Zeiss)
3. Clear plastic boxes (Part #: DG-0720; http://www.durphypkg.com/)
4. A pair of fine tweezers (#5 INOX; Dumont)
5. 1.5% agarose
6. Sterile water
7. Transgenic plants with appropriate fluorescent constructs
8. Fluorescent dyes such as FM 1-43/FM4-64 (50 μg/mL; Molecular probes) series.
9. Microscopy platforms: Care should be taken to preserve the integrity of the specimen and at the same time acquire images at sufficiently high signal-to-noise ratio to achieve the required spatial and temporal resolution. The cell-type specification and cell divisions in the *A. thaliana* SAM occur over a period of several hours and can therefore be reliably reconstructed by imaging at intervals of hours using conventional confocal scanning microscopes (4–7). The 3D nature of SAMs imposes a severe limitation on achieving the required spatial resolution from images that are taken at deeper layers. Multi-photon imaging systems have been shown to yield better spatial resolution than confocal systems; however, the two-photon imaging is surprisingly toxic to SAM cells (10, 11). Objective lenses with higher light-gathering ability can

improve spatial resolution; for example, 63X water dipping achroplan lens with 0.95 numerical aperture (Zeiss), which has a working distance of 2 mm, can be employed to acquire better quality images. Taken together, SAMs can be imaged by using a Zeiss 310 or Zeiss 510 upright confocal microscope fitted with a 63X water dipping achroplan lens.

10. Plant growth and plant care: Plants are germinated on MS-agar plates and allowed to grow for 10 days before transferring them into clear plastic boxes containing MS-agar. The plants are maintained in aseptic conditions until bolting. Upon bolting, when the shoot apex emerges out of the rosette, the plants are prepared for live-imaging (*see* next section). Clear plastic boxes (Part #: DG-0720; http://www.durphypkg.com/) are filled with water to submerge the plant prior to each imaging session, which can last for 30s–1 min. The water is then discarded and the plants are returned to normal growth conditions. This process is repeated for different imaging intervals as required.

11. Cellular dynamics markers: The main tools for fluorescence imaging of SAM cells and gene expression include proteins with intrinsic fluorescence such as green fluorescent protein (GFP) and its derivatives and analogous chromophore-containing proteins. Several guides are available to choose appropriate fluorescent proteins, based on their brightness, photostability, rate of protein maturation, oligomerization, and for the use in multiple labeling experiments (12, 13).

    Several fluorescent protein chimeras have been described to follow cellular dynamics within the SAMs. Plasma membrane-localized YFP (35S::YFP29-1) is an EYFP fused to a protein tag which targets the protein chimera to the plasma membrane and therefore it has been used to follow cell expansion and cell division patterns (4, 14). Protein chimera consisting of mYFP fusion to Histone2B (35S::H2B:mYFP) localizes to chromatin and hence it has been used to monitor nuclear divisions (15). CyclinB1;1:GFP is a chimeric protein between GFP and a mitotic cyclin expressed under native cyclin promoter. The fusion protein is expressed in cells that undergo G2-M transition, and it can be used as a marker to follow cells that are about to enter or in the process of division (4). Alternatively, the water-soluble lipophilic dyes such as FM1-43 and FM4-64 can be directly applied to the SAMs to highlight the cellular boundaries and thereby to follow cell division dynamics (5) *(see* **Note1***)*. However, the major limitation in using theses dyes is that they are progressively taken up by cells and therefore it becomes impossible to follow cellular outlines beyond 12–24 h. Among all the markers described, the plasma membrane marker is an ideal choice because both cell expansion and the cell division events can be scored long after the actual event has occurred. In addition, an extensive collection of fluorescent protein tags, which target different intracellular compartments of plant cells, have been described and can be found at http://deepgreen.stanford.edu/index.html.

12. Markers for gene expression dynamics: The choice of type of fluorescent protein construct is primarily based on the biological questions being investigated. Fluorescent proteins expressed from native promoter elements of individual genes form the basis for live-imaging of gene expression and cellular identity transition within the SAM (6, 7). Alternatively, protein dynamics can be followed by using protein chimeras in which a protein of interest is fused to a fluorescent protein and expressed from its native promoter (7). Several cell-type-specific fluorescent constructs and fluorescent protein chimeras have been described for live-imaging SAMs (7). The cell-type-specific or tissue-specific enhancer trap lines (http://www.plantsci.cam.ac.uk/Haseloff) have also been generated and they should form an excellent resource for live-imaging.

3. Methods

3.1. Preparation of Plants for Live-Imaging

Plants are germinated on MS-agar plates and allowed to grow for 10 days before transferring them into clear plastic boxes containing MS-agar (4, 7). The plants are maintained in aseptic conditions until bolting. Upon bolting, when the shoot apex emerges out of the rosette, the plants are ready for live-imaging. Prior to imaging, the MS-agar surface is overlaid with 1% agarose to minimize contamination as this would prevent the exposure of nutrient surface. The older floral buds are carefully removed by using the tweezers so as to expose the SAM. Molten agarose (1.5%) is applied onto the stem so that it makes a continuous agarose block to anchor the rosette to the base and thus stabilizes the rosette. Care should be taken to avoid covering the tip of the SAM with agarose as this would cause a drastic reduction in signal intensity of fluorescence images. The plastic boxes are filled with water to submerge the plant prior to each imaging session, which can last for 30s to 1 min depending on the scan rate. The water is then discarded and the plants are returned to normal growth conditions. This process is repeated for imaging intervals as required for individual experiments. The plants continue to grow during imaging which can last for 3–5 days. Therefore, they have to be augmented with fresh supply of agarose; the growing flower buds have to be removed so that the light path is devoid of any physical obstacles. Therefore, the process of live-imaging of SAMs is an interactive session which requires constant attention and adjustments.

3.2. Plant Performance During Live-Imaging and Validation of Imaging Data

Since the older flower buds are removed prior to imaging and the plants are imaged repeatedly, it is essential to assess the performance of plants during imaging. In some cases, dissecting the early-stage floral buds can result in desiccation and such plants are easy to recognize and are removed from experiments. The vertical growth of the plant is also a good measure of plant health and it can be verified at the end of each imaging session by recording the growth in the *Z*-axis and the plants that stop growing will have to be discontinued from imaging further. In some cases, the plants continue to grow but exhibit a gradual decrease in the SAM size and they need to be excluded from the analysis. It is also essential to test whether the imaging conditions had any adverse effect on the meristematic activity and this can be done by comparing the live-imaging data with growth patterns deduced from non-invasive methods (4). In general, the total duration of imaging varies with the imaging intervals. The plants imaged at shorter intervals such as 1–1.5 and 3 h can survive for only 40–66 h, whereas the plants imaged at longer intervals such as 6 or 12 h can survive for about 5 days. However, this survival period is applicable to experiments wherein only a single laser line is used for exciting fluorophores and the total survival period will be much shorter when multiple laser lines are used.

3.2.1. Optimization of Imaging Conditions for Live-Imaging

a. Single color live-imaging: The cell division dynamics of SAM cells can be followed by single color imaging of plants carrying 35S::YFP29-1 or 35S::H2B:mYFP. YFP can be stimulated with an argon laser at 515 nm at 25–50% of its output and by using neutral density filters at 4–7% to attenuate the laser line. The emission can be filtered by using 530–590 nm band-pass filter. For example, an ideal Z-stack can be acquired by using 1s scans of a 512 × 512 pixel frame consisting of a total of 30 optical sections sliced approximately 1.5 μm in thickness. Illuminating the specimen with appropriate amount of laser is critical to maintain healthy plants. Therefore, it is desirable to achieve required spatial resolution of images without increasing the laser power. This can be achieved by altering the amount of light collected through adjustments in PINHOLE aperture and by electronically increasing the detector sensitivity. Alternatively, several other strategies can be employed to increase the brightness of fluorescence constructs used for live-imaging (*see* next section).

b. Multi-color live-imaging: The fluorescent proteins with distinct excitation and emission wavelengths can be employed to label multiple cell types of SAMs or multiple proteins and follow them simultaneously by using multi-spectral imaging (6, 7). Spectrally distinguishable fluorescent proteins expressed from native promoter elements have been employed to follow specific gene expression patterns and cellular identity in the SAMs (6, 7). Alternatively the protein chimera in which a protein of interest is fused to a fluorescent protein and expressed from native promoter has been used to follow protein dynamics (7). The major challenge in multi-spectral imaging is to achieve a balance in signal intensities between different fluorescent proteins in order to minimize bleed-through of signals into inappropriate channels. Several approaches can be tried out to solve the problem. The brightness of fluorescent proteins should be considered in using them with appropriate promoters so that the fluorescent proteins with higher quantum yield can be used in conjunction with weakly expressed promoters. Multimerized versions of fluorescent proteins consisting of 2X or 3X tandem repeats or multimerized constitutive promoter along with a translational enhancer can provide higher signal intensity (7, 16). Alternatively, targeting of fluorescent proteins to specific intracellular compartments can increase its brightness (16).

For example, GFP and dsRED combination and GFP and YFP combinations have been imaged simultaneously by using multi-tracking option. This option allows switching between appropriate laser lines, the primary and secondary dichroic settings after every X–Y scan. Further details of multi-channel imaging of SAMs can be found in recent studies (6, 7).

3.3. Visualization and Image Analysis

A need for image processing platform has been emphasized to address several biological questions: first, to visualize 3D cellular shapes and sizes, measurement of concentrations of individual proteins of interest in a given 3D cellular volume; second, to understand the relationship between cell deformation dynamics, cell expansion patterns, and cell division orientation; third, to trace cell lineages by tracking the progeny through successive cell divisions with an aim to understand the causal link between cell division patterns and morphogenesis; and fourth, to explore the inherent variability in local cell division patterns and its influence on gene expression patterns and morphogenesis. The rules deduced from these analyses can also be applied, in the long run, in generating integrated and dynamic maps of development consisting of gene expression patterns and growth dynamics.

The following are the specific image processing (IP) and analysis tasks that are needed to analyze SAM imagery. A number of image analysis methods have been described and they can be useful in automatic analysis of image sequences. However, the IP is not a one-way street whereby existing methods can be applied directly to analyze SAM imagery. In fact, SAMs presents a number of unique challenges which necessitates development of new methods which in turn would lead to significant new developments in the image processing area. Some of the main tasks, the challenges, and possible strategies in analyzing the SAM imagery are highlighted below. The following section is slightly speculative as an effort has been made to predict possible strategies and future research directions that would be useful. Therefore, some of it should be taken as a pointer to existing work in other image processing applications and their usefulness in analyzing SAM imagery.

3.3.1. Image Registration

The time series of confocal Z-stacks requires to be aligned, and the 3D alignment can be done by using available software packages which utilize information theory to maximize mutual information across image stacks to register images at sub-pixel resolution (17). The 3D registration/alignment can also be done by using appropriate image registration module in commercial software packages such as AMIRA (Visage Imaging). The registered stacks can then be reconstructed in three dimensions, rendered and animated to play continuous movies by using the Zeiss LSM3.2 software (4). The cells in the L1 layer, located at various depths on the curved surface, are projected onto a single 3D-reconstructed view by using maximum intensity projection.

3.3.2. Segmentation, Representation, and Visualization of 3D Cell Shape

This is the most basic step that will enable rest of the image processing tasks. The main challenge is to determine a robust method for extracting the 3D space of individual cells, visualize connections between individual cells of the SAMs, and represent them by using a suitable shape descriptor.

Four-dimensional fluorescence image stacks in which cell outlines are labeled with yellow fluorescent protein (YFP) marker to track cellular dynamics can be used as an input for computationally segment cells. Cells of the SAM are isodiametrical in shape and each one of them measures about 5 μM in size. Therefore, a single cell is captured approximately three times when sliced at a thickness of 1.5 μM along their *Z*-axis. Existing image segmentation algorithms can be employed to segment individual cells from a registered 2D image stacks of SAMs. Existing level set-based methods work very well when applied to a single 2D layer (**Fig. 15.1**). The segmentation protocol can be repeated for each layer separately and then the correlation between individual layers can be established by combining segmented 2D stacks, using a region growing procedure starting from the topmost layer (18). The net result of this effort will be the identification of 3D profiles of cells across layers and visualization of the connections between individual cells in a 3D space. Different shape descriptors have been described and they can be useful for this representation (19, 20).

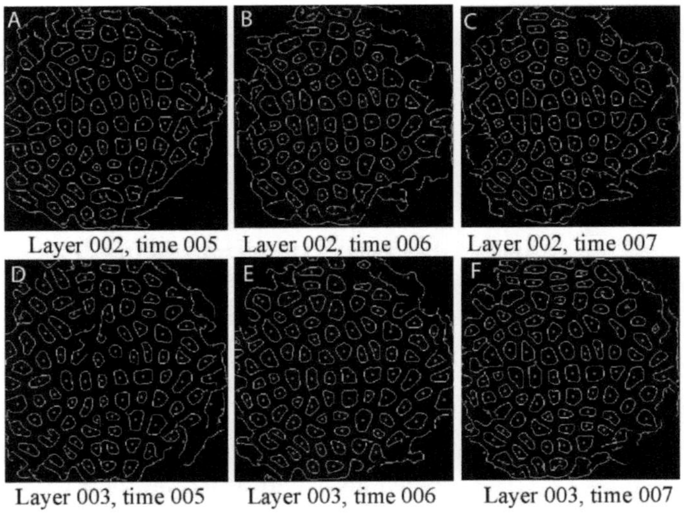

Fig. 15.1. Computational segmentation of a multi-layered SAM, labeled with ubiquitously expressed plasma membrane-localized YFP (35S::YFP29-1). The imaging data from 35S::YFP29-1 has been used to computationally segment cells. (**A–F**) Results of segmentation for two consecutive Z-sections from the SAM taken at three different time instances are shown. Consistency in the segmentation across the layers will help in tracking and identifying cell divisions even if some of the segmentation results are erroneous.

The isolation of the 3D structure of the cell within a given space will enable biologists to analyze the geometrical constrains of individual cells in a multicellular field and that would lead to quantitative understanding of relationship between individual cell shapes/sizes and the organ shape/size. Accurate estimation of cellular volumes will also be useful in measuring concentrations of individual proteins of interest within cells.

3.3.3. Three-Dimensional Tracking of Cells and Cell Divisions

The tracking of cells through successive cell divisions within the SAMs will allow reconstruction of individual lineages and this is essential to understand the dynamic interplay between growth and gene expression changes. Cells undergo continuous deformations and exhibit topology changes during their growth. Therefore, the tracking algorithm must be robust to accommodate both cellular deformations and cellular topology changes.

Tracking methods have been developed and they work well for individual 2D layer (21, 22). A 2D annealing method for tracking individual cells in a growing bacterial colony has been shown to produce reliable tracks (21). An independent study has documented a method for 3D modeling, visualization, and analysis of the cells in the reconstructed surface layer (L1 layer) of the *Arabidopsis* SAMs (22). However, new methods will have to be developed to track individual cells in 3D space and across time.

Cell divisions can be modeled by combining methods that can handle cellular deformation dynamics and changes in topology during cell division (23). Similar approaches have been used in human motion modeling, but the dynamical models for cell development may be different and therefore identification and estimation of such models can be challenging (24, 25). To identify and track changes in topology, level set-based approaches can be used. The desired output of this procedure should be to obtain correspondence between individual slices of a 3D cellular volume across time. Tracking the cells and identifying cell divisions between two time instants will enable biologists to analyze the relationship between cell deformation dynamics, cell expansion patterns, and cell division orientation.

3.3.4. Long-Term Tracking for Identifying Cell Lineages

The computational extraction of cell lineages requires identification of cells and tracking their progeny through successive cell divisions. A major challenge is to maintain consistency in the tracks over long periods of time. This is because a single mistake in tracking between two consecutive frames can lead to a completely erroneous track (lineage) later in time.

The goal here should be to developing inference strategies for identifying mistakes in the tracking result by assimilating multiple two-frame tracking results. Thereafter the tracking algorithm should be able to automatically correct the mistakes through a self-correcting procedure. This should be a significant research task from the image processing aspect. A possible approach includes representation of each 3D cell by a node on a graph with two-frame correspondences as edges of this graph. One can consider the tracks on this graph till a certain time point and then compute some long-term biological properties that would provide an estimate of the correctness of the tracks. Such long-term properties could include the morphogenic events within the SAMs such as periodic spatiotemporal appearance of bulge formation/extension of certain regions of the SAM leading to the outgrowth of differentiating organs and the appearance of regions of limited cell expansion leading to the formation of boundaries that separate differentiating organs from the SAM. This process may help in inferring whether the mistakes have been made and may also identify the locations of the mistakes. Thereafter, a process can be implemented to update the graph till a certain biologically inspired criterion is met. It has been recently shown that such an adaptive strategy is applicable in tracking objects reliably over a camera network (26). The ability to track cells and their progeny over extended periods of time will enable identification of cell lineages through successive cell divisions from large volumes of image data acquired to explore the causal link between local cell division patterns and morphogenesis.

3.3.5. Learning Dynamical Models of Cell Lineage Patterns

The experimental evidence indicates that the cell division patterns are not entirely stereotypic both within a given SAM and across different SAMs of *Arabidopsis* except for a local co-ordination in which three to four adjacent cells divide simultaneously (4). Therefore the challenge is to understand how an invariant pattern (both the gene expression pattern and organogenesis) develops from non-stereotypical cell division patterns. The computational challenge in achieving this is to learn models that describe the growth patterns of a cell lineage that contribute to organogenesis.

The tracked cell lineages can be used to learn models for describing the variations of cell growth patterns within a given SAM and between SAMs of given species. This includes variations in the cell division rates and orientations of cell division. Prior work in human activity analysis from video sequences provides some pointers to future research possibilities. For example, it is possible to learn a function space for walking or running by considering variations between different people (27). A similar concept can be used to represent the cell lineages by using a nominal pattern and learning the variations around that pattern as a function space. This function space will provide a mathematical description of all possible variations of the cell growth dynamics between different plants as well as within a plant. Once this representation is obtained, extent of similarities and dissimilarities between different cell lineages can be computed within this function space. Since this approach requires computing distances between dynamic patterns, measures like dynamic time warping (DTW) will have to be considered. These learned models can also be incorporated into the tracking algorithms later, thus making them robust to variations. This would allow biologists to explore the inherent variability in local cell division patterns and its influence on gene expression patterns and organogenesis. The rules deduced from this analysis can also be applied, in the long run, in generating integrated and dynamic maps of development consisting of representation of gene expression patterns and growth dynamics.

4. Notes

1. Alternative live-imaging setup has also been described (5). This method is suitable for imaging SAMs by using an inverted microscope. This method requires the use of napthylphthalamic acid (NAA) treatment of plants at 10^{-5}–10^{-6} M concentration since germination, so that it leads to the development of naked inflorescence which can be imaged without any interference from developing flower buds. The naked SAMs are examined using TCS-NT

confocal laser scanning microscope (Leica, Heidelberg, Germany), with an argon/krypton laser (Omnichrome, Chino, CA) mounted on inverted DM IRB microscope (Leica).

Acknowledgments

The authors thank Min Liu for providing computationally segmented images used in **Fig. 15.1**. Work in the Reddy laboratory is supported by National Science Foundation grant IOS-0718046.

References

1. Meyerowitz, E.M. (1997) Genetic control review of cell division patterns in developing plants. *Cell.* **88**, 299–308.
2. Steeves, T.A. and Sussex, I.M. (1989) Shoot apical meristem mutants of Arabidopsis thaliana. *Patterns in Plant Development.* Cambridge University Press, New York.
3. Baurle, I. and Laux, T. (2003) Apical meristems: the plant's fountain of youth. *Bioessays.* **25**, 961–970.
4. Reddy, G.V., Heisler, M.G., Ehrhardt, D.W., and Meyerowitz, E.M. (2004) Real-time lineage analysis reveals oriented cell divisions associated with morphogenesis at the shoot apex of *Arabidopsis thaliana. Development.* **131**, 4225–4237.
5. Grandjean, O., Vernoux, T., Laufs, P., Belcram, K., Mizukami, Y., and Traas, J. (2004) In vivo analysis of cell division, cell growth, and differentiation at the shoot apical meristem in Arabidopsis. *Plant Cell.* **16**, 74–87.
6. Reddy, G.V. and Meyerowitz, E.M. (2005) Stem-cell homeostasis and growth dynamics can be uncoupled in the Arabidopsis shoot apex. *Science.* **310**, 663–667.
7. Heisler, M.G., et al. (2005) Auxin transport dynamics and gene expression patterns during primordium development in the Arabidopsis inflorescence meristem. *Curr. Biol.* **15**, 1899–1911.
8. de Reuille, P.B., Bohn-Courseau, I., Godin, C., and Traas, J. (2005) A protocol to analyse cellular dynamics during plant development. *Plant J.* **44**, 1045–1053.
9. Gor, V., Elowitz. M., Bacarian, T., and Mjolsness, E. (2005) *Tracking Cell Signals in Fluorescent Images.* In: Proceedings of the 2005 IEEE Computer Society Conference on Computer Vision and Pattern Recognition, San Diego,DA.
10. Moreno, N., Bougourd, S., Haseloff, J., and Feijo, J.A. (2006) Imaging plant cells. In J. Pawley (ed.) *Handbook of Biological Confocal Microscopy,* 3rd. edition, ch. 44, Springer, New York, pp. 769–787.
11. Feijo, J.A. and Moreno, N. (2004) Two-photon microscopy applications to plant cells. *Protoplasma.* **223**, 1–32.
12. Shaner, N.C., Steinbach, P.A., and Tsien, R.Y. (2005) A guide to choosing fluorescent proteins. *Nat. Methods.* **2**, 905–909.
13. Haseloff, J. and Siemering, K.R. (2006) The uses of green fluorescent protein in plants. *Methods Biochem. Anal.* **47**, 259–284.
14. Cutler, S.R., Ehrhardt, D.W., Griffitts, J.S., and Somerville, C.R. (2000) Random GFP::cDNA fusions enable visualization of subcellular structures in cells of Arabidopsis at a high frequency. *Proc. Natl. Acad. Sci. USA.* **97**, 3718–3723.
15. Boisnard-Lorig, C., et al. (2001) Dynamic analyses of the expression of the HISTONE::YFP fusion protein in arabidopsis show that syncytial endosperm is divided in mitotic domains. *Plant Cell.* **13**, 495–509.
16. Kohler, R.H., Zipfel, W.R., Webb, W.W., and Hanson, M.R. (1997) The green fluorescent protein as a marker to visualize plant mitochondria in vivo. *Plant J.* **11**, 613–621.
17. Maes, F., Collignon, A., Vandermeulen, D., Marchal, G., and Suetens, P. (1997) Multimodality image registration by maximization of mutual information. *IEEE Trans. Med. Imaging.* **16**, 187–198.

18. Adams, R. and Bischof, L. (1994) Seeded region growing. *IEEE Trans. Pattern Anal. Mach. Intell.* **16,** 641–647.
19. Kendall, D., Barden, D., Carne, T., and Le, H. (1999) *Shape and Shape Theory.* John Wiley and Sons, New Jersey, USA.
20. Dryden, I. and Mardia, K. (1998) *Statistical Shape Analysis.* John Wiley and Sons, New Jersey, USA.
21. Gor, V., Elowitz, M., Bacarian, T., and Mjolsness, E. (2005) *Tracking Cell Signals in Fluorescent Images.* In: Proceedings of the 2005 IEEE Computer Society Conference on Computer Vision and Pattern Recognition, San Diego, CA.
22. Barbier de Reuille, P., Bohn-Courseau, I., Godin, C., and Traas, J. (2005) A protocol to analyse cellular dynamics during plant development. *Plant J.* **44**, 1045–1053.
23. Osher, S. and Paragios, N. (2003) *Geometric Level Set Methods in Imaging, Vision, and Graphics*, Springer, New York, USA.
24. Veeraraghavan, A., Roy-Chowdhury, A., and Chellappa, R. (2005) Matching shape sequences in video with applications in human motion analysis. *IEEE Trans. Pattern Anal. Mach. Intell.*, 1896–1909.
25. Roy-Chowdhury, A. (2005) A measure of deformability of shapes, with applications to human motion analysis. *IEEE Comput. Vis. Pattern Recog.*
26. Song, B. and Roy-Chowdhury, A. (2007) *Stochastic Adaptive Tracking in a Camera Network.* IEEE International Conference on Computer Vision.
27. Veeraraghavan, A., Chellappa, R., and Roy-Chowdhury, A. (2006) The function space of an activity. *IEEE Comput. Vis. Pattern Recog.*

Chapter 16

Computer Vision as a Tool to Study Plant Development

Edgar P. Spalding

Abstract

Morphological phenotypes due to mutations frequently provide key information about the biological function of the affected genes. This has long been true of the plant *Arabidopsis thaliana*, though phenotypes are known for only a minority of this model organism's approximately 25,000 genes. One common explanation for lack of phenotype in a given mutant is that a genetic redundancy masks the effect of the missing gene. Another possibility is that a phenotype escaped detection or manifests itself only in a certain unexamined condition. Addressing this potentially nettlesome alternative requires the development of more sophisticated tools for studying morphological development. Computer vision is a technical field that holds much promise in this regard. This chapter explains in general terms how computer algorithms can extract quantitative information from images of plant structures undergoing development. Automation is a central feature of a successful computer vision application as it enables more conditions and more dependencies to be characterized. This in turn expands the concept of phenotype into a point set in multidimensional condition space. New ways of measuring and thinking about phenotypes, and therefore the functions of genes, are expected to result from expanding the role of computer vision in plant biology.

Key words: Computer vision, morphometrics, plant development, image processing.

1. Introduction

Late in the nineteenth century, Gregor Mendel made his now famous use of morphological phenotypes of pea mutants to uncover the basic laws of inheritance, which led to the discovery of genes. Now in the twenty-first century, morphological phenotypes are still an important source of information, sometimes providing critical clues about gene function. It is difficult to imagine modern plant biology ever not depending on morphological phenotypes for information. Yet, despite their key role, techniques for extracting information from morphology and its development have changed little during the

Dmitry A. Belostotsky (ed.), *Plant Systems Biology, vol. 553*

DOI 10.1007/978-1-60327-563-7_16 Springerprotocols.com

lifetimes of most biologists, especially when compared to the astounding technical advances that have propelled the study of genomes. It is useful to remember that not long ago, sequencing DNA required radioactive nucleotides, large cumbersome electrophoresis gels, photographic emulsions, film development, and at least one person to decode the band patterns. Determining the mRNA level for a single gene was equally complicated, and measuring two different genes was approximately twice the work of measuring one. Now genes are sequenced, manipulated, and studied in ways that bear little resemblance to the methods used 20 years ago, thank goodness! On the other hand, morphological development, which may be considered the output side of a gene function model, is still likely to be manually measured. No revolution in the study of plant form or development of morphology has taken place in parallel with the genomics revolution. Thus, there exists today a large imbalance between the degree of sophistication with which genes and phenotypes are investigated, despite phenotype development being a deep source of information about gene function. To extract the most information about gene function from mutants and from natural variation observed in plant populations, advanced methods of studying the development of plant morphology and behavior are required. To be maximally effective, the methods should be automated and have high spatiotemporal resolution to serve purposes described below. This chapter outlines a computer vision approach designed to raise the sophistication of plant development studies.

Computer vision refers to artificial systems that obtain information from images (1). A related term is machine vision, which usually includes the engineering topics of motion-control hardware and software and carries less the connotation of using technology to replace the human sense (*2*). Wikipedia contains excellent entries that will assist a reader interested in learning more about computer vision, machine vision, and distance transforms. Regardless of the term, the goal is much the same – to extract and quantify information about the size and shapes of objects from digital images using computer algorithms and mathematics, which may be borrowed from the related fields of pattern recognition and signal processing. The images can be single frames, such as a fingerprint to be analyzed for identification purposes, or the images can come in the form of a time series or movie that covers a developmental process from which time-dependent changes are quantified.

2. Image Data Acquisition

The input data in its rawest form are pixels (electronic picture elements). The number of pixels in an image is typically on the order of 10^6. If the image is 8-bit grayscale, each pixel is represented by an intensity value between 0 (code for black) and 256 (code for white). Thus, a grayscale image is actually an *x,y* array or a grid of 10^6 numbers ranging between 0 and 256. Relatively inexpensive electronic cameras employing charge-coupled device (CCD) sensors and interfaced with a standard computer can acquire such images at programmed time intervals. When equipped with an appropriate lens, these CCD cameras are suitable for monitoring aspects of plant growth and development. Although acquiring the electronic images is relatively straightforward from a technical standpoint, the following issues require some thought and planning in order for the images to be most conducive to accurate analysis.

a. Spatial resolution on the order of a few microns is readily achievable with lenses and cameras commonly used in industrial machine vision applications. However, high resolution comes at the expense of field of view. High spatial resolution may mean that only one object can be accommodated in the field of view, and the object may grow out of the field of view too soon. Zooming out increases the field of view, but the object will be represented by fewer pixels (i.e., each pixel represents a larger real area). The experimenter must decide what balance of spatial resolution versus field of view is appropriate for the experiment. Zoom in for high detail over short periods of time. Zoom out for longer observation periods, potentially of more objects, but with lower spatial resolution.

b. Maintaining focus during the experiment is important. Depth of field is typically so shallow that sharpness of focus degrades if the object is allowed to move much along the optical axis (*z*-direction). Yet constraining the object to the plane of focus may interfere with the biology under study. One method that works well with seedlings is to culture them on the surface of a vertical agar plate. Adherence of the seedling to the medium by surface tension constrains the seedling to the 2D plane without undue interference. A much more complicated (and computationally intensive) but potentially much more informative solution would be 3D imaging of unrestrained organisms. Such techniques will not be covered in this article.

c. Consistent and high contrast between the object and the background facilitates all subsequent steps. The desired image qualities are achieved by judicious choice of optics and adequate lighting of the sample. However, light is one of the strongest environmental influences on plant development, which creates a potential conflict between image acquisition requirements and biological considerations. One proven solution is to use infrared radiation to acquire the image. Plants do not sense infrared wavelengths in the 800–900 nm range but CCD chips do (after removing the internal infrared filter found in most CCD cameras). For experiments that require the ambient light to be changed during the experiment, such as in de-etiolation studies, a long-pass filter mounted over the lens allows the infrared to create the image on the sensor but prevent changes in the visible wavelengths from affecting the image. The infrared illumination/long-pass filter technique has also been used in image-analysis studies of rhythmic phenomena.

3. Segmentation and Feature Extraction

After the process of interest is captured in a series of electronic images, the next task is to extract useful information. The best approach to take depends upon the shape of the organ or structure under study. For example, an ovate leaf like that of *Arabidopsis* might be usefully described by the combination of its area, major axis (longest straight line within the object), and minor axis (longest straight line perpendicular to the major axis) as shown in **Fig. 16.1A** and **B**. The ratio of these two axes, known as eccentricity, could be a useful single-value shape descriptor. An elongated structure such as a stem or a root may be best described by the length and shape of its midline, technically known as the medial axis (**Fig. 16.1C** and **D**). For any given descriptor, there are usually multiple image processing means to the same end, each with its own set of advantages and disadvantages. To illustrate this point, the example of finding the midline of a hypocotyl image will be considered. The grayscale image of etiolated seedlings a few millimeters in length (**Fig. 16.2A**) is readily separated from the white background by a thresholding operation that replaces every pixel having a gray level value lower than a certain threshold by 0 (black) and every pixel above that threshold by a 1 (white). The image is said to be binarized and the object of interest segmented from the background. All black pixels belong to the object of interest and all the white pixels belong to the background (**Fig. 16.2B**). One strategy for finding the midline entails eroding away the black pixels from the outside iteratively until the least set of contiguous black pixels remains. After some refinement, the coordinates of this pixel set define a midline. Another method of finding the midline of the seedling shown in the figure employs the Euclidean distance transform (3). Each pixel within the object is mapped to its nearest contour point. The distance between each point and its nearest contour point is calculated. If mapped onto the object image, the distance values form a "ridge" that runs the length of the seedling with the highest values (local maximal distances from

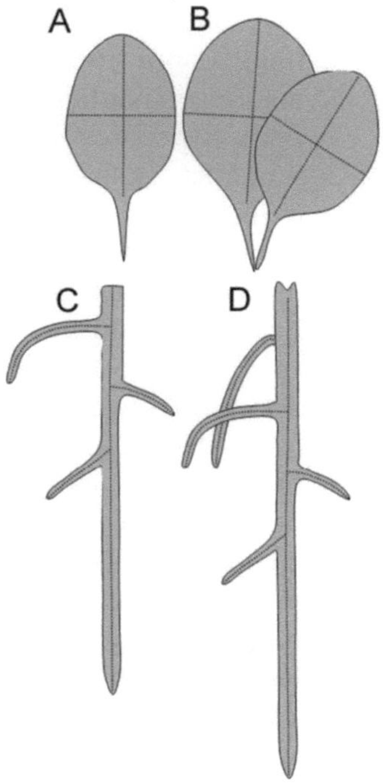

Fig. 16.1. Simple morphology descriptors. (**A**) The shape of a leaf like that of an *Arabidopsis* rosette may be adequately described by the combination of its major axis length and perpendicular minor axis length, in addition to its area. (**B**) Determining even simple shape parameters from images is complicated by real conditions such as overlap of one leaf by another. (**C**) The primary root is well described by its medial axis, or midline, but after a few days of growth, the *Arabidopsis* root system begins to branch, which complicates midline finding. (**D**) Again, real conditions such as overlapping lateral roots pose image-processing challenges that must be overcome in order to make computer vision a useful tool in plant functional genomics studies.

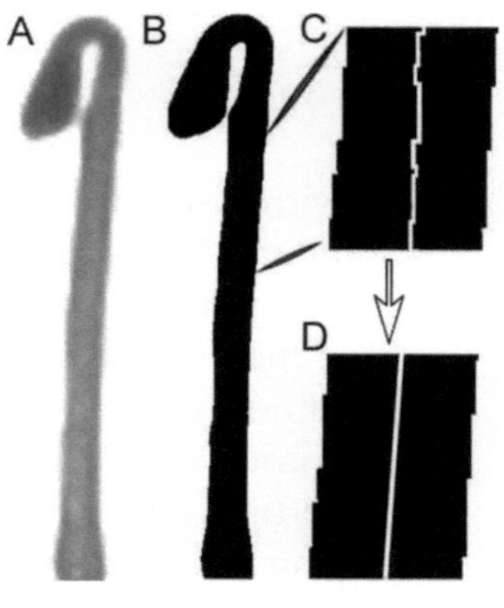

Fig. 16.2. Determining the midline. (**A**) An unprocessed grayscale image of an etiolated *Arabidopsis* seedling. (**B**) Binarized image following a thresholding operation to isolate the object of interest from the background (image segmentation). (**C**) A portion of the hypocotyl midline as determined by Euclidean distance transform or by erosion, two different skeletonization methods. (**D**) Smoothing achieves the desired result, a set of points that faithfully captures the length and shape of the object, in this case the hypocotyl. Further analysis is performed on this set of midline points to determine parameters such as length, local curvature, and angle.

the contour) defining the midline. Because the contour of the object is not smooth, the midline determined by either of the two mentioned methods will not be smooth. From the enlarged portion of the hypocotyl with midline superimposed (**Fig. 16.2C**), it can be seen that the length of a jagged midline is longer than the true midline of the structure. Also, local curvature is extremely sensitive to noise in the midline. To obtain accurate measurements of these main determinants of midline shape (length and local curvature), a noisy or jagged midline must be smoothed (**Fig. 16.2D**). Miller et al. (4) employed a local filter that processes along the distance transform peak in one-pixel-sized steps to obtain a smoothed, ordered set of midline points. This process is repeated for each frame in a time series. The result is a time series of smoothed midline point sets. Each individual midline can be analyzed separately and compared to the previous, or the entire time series of midline point sets can be treated as a surface and analyzed with differential geometry techniques to extract features such as local curvature, total length, growth rate, or angle (integrated curvature) as previously described (4), a webpage that details how to assemble an image-acquisition apparatus for studies of plant growth and development is http://phytomorph.wisc.edu/parts_list.htm). The many different approaches for extracting the relevant information are not discussed here to avoid obscuring the main point, which is that a complex biological process captured in image sequences (many hundreds of megabytes) can be distilled down to a time series of midline point sets (a few kilobytes) that is amenable to mathematical analysis and quantification.

4. An Example of Gravitropism

Gravitropism is a classic response that has been investigated with computer vision techniques. A generalized treatment will show here how a computer vision approach differs from the more common methods of investigation. As shown in **Fig. 16.3**, the initial condition is a straight root rotated horizontally to initiate the response. Typically, a measurement of root tip angle is made some time later (an endpoint measurement) by manually determining the angle formed by "before" and "after" lines drawn through the apex on an electronic image to approximate the tip angle. Computer vision provides an alternate description of the phenomenon. Analysis of a time series of images would produce a family of midlines that collectively describes this classic response of development to an environmental signal in more detail. Growth rate, tip angle, and distribution of curvature along the root axis can all be determined as a continuous function of time from the family

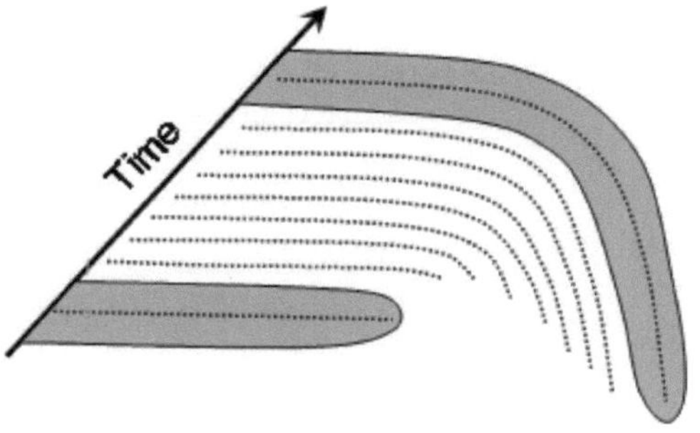

Fig. 16.3. Capturing development with a time series of midlines. The midline of a root placed on its side is initially straight. After several hours, it has reoriented by almost 90°. An endpoint measurement of tip angle provides no information about how that endpoint was reached. Analyzing a family of midlines extracted from images acquired every few minutes yields a much richer description of the gravitropic response. The impact of genetic variation on growth control, auxin response, and signal transduction mechanisms may have informative though subtle effects on the *process* of gravitropism that cannot be captured in an endpoint measurement.

of midlines depicted as dotted curves in **Fig. 16.3.** When images are taken every 2 min and when each pixel represents 5 μm of tissue, a very detailed depiction of the process emerges.

5. Expanding the Concept of Phenotype with Computer Vision

In addition to providing high spatiotemporal resolution, the computer vision approach brings automation to the quantification of development. Automation enables higher throughput because analysis is no longer the rate-limiting step and acquisition can be made parallel (either through simultaneous operation of multiple image acquisition stations or moving a camera sequentially through an array of samples with a robotic device). The combination of parallel acquisition and automated analysis expands the scope of a typical investigation of phenotype development. Rather than a snapshot of a process in a given condition, studies of development over time in an array of conditions are made possible by automation and parallel acquisition. This could lead to the discovery of novel aspects of mutant phenotypes or natural variation that are not apparent and may be difficult (or even impossible) to detect in a single-culture scenario. For example, a mutation may have no significant effect on resistance to salt or on growth during water stress. But behavior different from wild type may be observed in that mutant if it and the wild type were cultured and examined with high resolution in a grid of three salt by three water potential conditions. How the mutant responds to salt as a function of water status may be significantly different from wild type even though a standard salt or water stress test did not reveal a phenotype. The key to discovering such a condition-difference phenotype is high-resolution/high-throughput phenotyping technology. High resolution is necessary because a difference between genotypes at any single point on the condition grid may be subtle (though significant in terms of success in the wild); high throughput is necessary because even the simple example just stated requires nine different water potential/salt concentration conditions to be assayed many times to characterize each population. Machine vision technologies can deliver that combination of resolution and throughput. The result is an expanded, multidimensional view of a phenotype. If the wild-type behavior can be thought of as values (quantified by computer vision techniques) in a higher-dimension condition space, a phenotype may be thought of as an altered distribution of these values within such a space. To characterize such a phenotype, the condition space must be surveyed to an extent that the data distributions within it can be meaningfully characterized and evaluated using rigorous statistical criteria. With 25,000 genes in the *Arabidopsis* genome, many genes may be expected to play roles in adaptation, hence shaping the limits of phenotypic plasticity, or determining how a response is modulated according to changes in conditions. Their phenotypes may only be detectable when the development is assayed under a range of conditions. Automated high-resolution quantification of development is expected to be a useful approach to finding key, informative phenotypes of mutations that appear superficially like wild type. This brings the argument back to the thesis that automation and throughput make it feasible to expand and broaden the concept of phenotype.

6. Achieving Higher Throughput

In principle, there are two ways to increase throughput at a given resolution. One is to replicate the acquisition apparatus so that multiple experiments can be performed simultaneously. The technical challenges to scaling up by replicating or cloning the apparatus are minor. A single computer can readily manage acquisition from multiple cameras sampling at intervals on the order of minutes, then pass the images to other machines for analysis and storage. However, at some point scaling up by cloning is costlier than the alternative, which is to use a motion-control device to move a single camera at programmed intervals between an array of samples. On the other hand, the need to reposition a camera repeatedly within microns of a previous position over a range approximately a million-fold greater creates significant hardware and software engineering challenges. This robotics approach is simplified as the resolution requirement is relaxed whereas all critical technical issues remain constant when throughput is increased by apparatus cloning.

7. Development to Be Studied by Computer Vision

A few specific computer vision studies of wild-type and mutant plant developmental processes have been effective enough to validate the approach, but for computer vision to become generally useful and a major element of the functional genomics toolkit, applicability to many more plant processes will have to be developed. Two of the more obvious potential applications in the model plant *Arabidopsis* include quantification of root system branching and elaboration of the rosette leaf system (**Fig. 16.2.**) Quantifying them in high resolution over time is a tractable problem but both pose formidable challenges, particularly at the image segmentation phase (separation of object of interest from non-interest). Rosette development is complicated by the leaves overlapping (**Fig. 16.2B**) and root systems, while starting out fairly simple, become complicated due to crossing and bundling of lateral roots (**Fig. 16.2D**). Also, establishing plant culturing systems that enable the development to be captured in images, keeping in mind the issues that were raised in Section 2, is usually not straightforward. Hopefully, as practices and tools improve, the computer vision approach can be applied in increasingly natural (meaning increasingly complicated) situations.

8. Large-Scale Projects

The potential for high throughput in computer vision studies of plant development makes some large-scale functional genomics projects possible. For example, phenotype-space characterization of *Arabidopsis* T-DNA insertion mutants could add a large amount of useful information about function for thousands of genes. QTL analyses and other methods of learning the genetic basis of a response or process by exploiting natural genetic variation could proceed much more efficiently and with higher resolution if automated computer vision algorithms, rather than human eyes, performed the measurements. As more genomes become fully sequenced, there will be an increasing need to adapt the techniques for various species. Perhaps the biggest differences to accommodate in the near future are those associated with monocot development. For example, there is little in common morphologically between the emergent shoot of dicot and monocot seedlings, yet the growth and development of both are highly dependent on orchestration of genetic programs and responses to environmental signals, and both are critical to determining the success of an individual in the wild – or a crop in the field. Therefore, it is not difficult to imagine crop breeders incorporating computer vision techniques into the process of selecting for subtle desirable traits.

The morphometric data produced by computer vision techniques are arrays of numbers and are therefore formatted generally like results from mRNA, proteomic, and metabolomic profiling experiments. This means that large-scale morphometric data sets could be included in systems-level computational modeling studies. Advancing computer vision techniques for studying plant development to match their potential will enable the formation of quantitative models that link the genome through its molecular products to the generation of plant form and behavior.

Acknowledgments

The author thanks Nathan Miller, Tessa Durham, and Liya Wang for many discussions that generated the ideas expressed in this chapter. This project was supported by NSF grant DBI-0621702.

References

1. Shapiro, L.G.. and Stockman, G.C. (2001) Computer Vision. Prentice Hall, New York, 608 pp.
2. Davies, E.R. (2005) Machine Vision: Theory, Algorithms, Practicalities. Elsevier, Amsterdam, 934 pp.
3. Mauer, C.R., Qi, R., and Raghavan, V. (2003) A linear time algorithm for computing exact Euclidean distance transforms of binary images in arbitrary dimensions. *IEEE Trans. Pattern Anal. Machine Intell.* **25**, 265–270.
4. Miller, N.D., Parks, B.M., and Spalding, E.P. (2007) Computer-vision analysis of seedling responses to light and gravity. *Plant J.* **52**, 374–381.

Section IV

Chemical Genetic, Metabolomic, and Integrative Methods for Plant Systems Biology

Chapter 17

Metabolomics of Plant Volatiles

Anthony V. Qualley and Natalia Dudareva

Abstract

Plants communicate with their surrounding ecosystems using a diverse array of volatile metabolites that are indicative of the physiological status of the emitter. A variety of systems have been adapted to capture, analyze, identify, and quantify airborne metabolites released by plants. Metabolomic experiments typically involve four steps: sample collection, preparation, product separation, and data analysis. To date, two different types of headspace sampling, static and dynamic, are widely used for volatile metabolome investigation. For static headspace analysis, solid-phase microextraction (SPME) is used to sample volatiles while push and pull as well as closed-loop stripping methods are used for dynamic headspace sampling. After collection, volatile blends are most efficiently and routinely separated prior to analysis using gas chromatography (GC). Sample preparation is simplified because derivatization is not needed with volatile metabolites. GC coupled to detection by electron impact mass spectrometry (EI-MS) provides high chromatographic resolution, sensitivity, compound-specific detection, quantitation, and the potential to identify unknowns by characteristic and reproducible fragmentation spectra in addition to retention time. A variety of resources can be used to identify unknown compounds in a given volatile sample including >600,000 compounds with known mass spectra catalogued in searchable mass spectral libraries.

Key words: Plant volatiles, solid-phase microextraction, static headspace, dynamic headspace, closed-loop stripping, metabolomics, gas chromatography, mass spectrometry.

1. Introduction

An impressive variety of volatile compounds are biosynthesized and released into the atmosphere by plants and make up more than 1% of plant secondary metabolites. These compounds are responsible for attracting pollinators and other beneficial insects, providing a means of inter-plant communication, and directly repelling or intoxicating attacking herbivores (1, 2). Volatiles are typically small molecules with low boiling points and high vapor pressure at

Dmitry A. Belostotsky (ed.), *Plant Systems Biology, vol. 553*

DOI 10.1007/978-1-60327-563-7_17 Springerprotocols.com

ambient temperature. Unconjugated volatiles can cross membranes freely to be released from flowers, fruits, and vegetative tissues into the atmosphere and from roots into the soil. To date, 1,700 compounds were identified in the scent of flowers belonging to 90 plant families (3) in addition to 700 flavor volatiles known to be present in the aromas of fruits and vegetables (4) with many compounds found in both flowers and fruits.

Plant volatiles are represented by terpenoids, phenylpropanoid, and benzenoid compounds, amino acid derivatives, and fatty acid derivatives, which together reflect the diversity of their origins within the metabolome. Due to the established roles of volatiles in plant defense and pollination syndromes, exciting opportunities exist for manipulation of plant volatile profiles with the goal of improving crop productivity and quality. Enhancing the aroma quality of edible crops, especially vegetables and fruits, is the common goal of many breeding and biotechnological programs and is of keen public interest. As a result, much research effort has been dedicated to elucidating the biochemical pathways responsible for their biosynthesis as well as in determining their ecological significance.

A key prerequisite to understanding the function and biosynthesis of volatiles is the identification of compounds within the complex mixtures released from different plant tissues under various physiological conditions. Thus, sensitive yet unbiased methodologies are needed to provide researchers with comprehensive and accurate representations of a plant species' volatile metabolome. Metabolomics, as it has been idealistically defined, involves the isolation of all metabolites from a whole organism, organ, tissue, or cell type of interest followed by identification of individual components. However, current methodologies are limited in their ability to isolate, and even more critically to identify, many of the compounds present in a given sample (5, 6). In volatile metabolomics, the process of sample acquisition is greatly simplified. The emitting plant has already completed the first step, isolating metabolites away from tissues, by releasing compounds into the surrounding atmosphere. Researchers only need to temporarily trap these metabolites in such a way that they can be released unadulterated for separation and identification while introducing as little bias as possible. A variety of technologies have been developed over the years. In these methods, the sample of interest (a plant or its parts) is enclosed in a collection chamber and the released volatiles present in the airspace surrounding the sample (headspace) are trapped onto an adsorbent. To date two different types of sampling, static and dynamic headspace sampling, are widely used for volatile metabolome investigations. This chapter should serve as a guide to implementation of the most popular techniques currently in use worldwide by groups engaged in plant volatile analysis or the study of plant–plant and plant–insect interactions.

1.1. Static Headspace Sampling

Static headspace sampling is a passive technique for volatile collection where no air circulation is used for concentrating volatiles on a sorbent matrix. As a result, static headspace methods typically require specialized techniques that are more successful at concentrating airborne volatiles during collection and reduce or eliminate dilution of sample during desorption and sample preparation. In addition, background is drastically reduced due to the absence of a continuous airflow that can contain impurities, masking compounds released at trace amounts. In static methods, plant materials are typically sealed inside a container to retain released volatiles and the headspace is either sampled directly using a gas-tight syringe or ad/absorbed to a SPME fiber. SPME is currently the most widely used method for static headspace sampling and sees use in an unusually wide range of applications within and outside of plant biology.

1.1.1. SPME

SPME is a robust and sensitive technique for volatile headspace sampling. It is based on the adsorption of volatiles on an inert fiber from which compounds can be thermally desorbed inside a GC inlet. SPME provides detection limits in parts per billion by volume (5), a sensitivity that is achieved by a concentration of the analytes on the fiber. SPME sampling is selective because it is an equilibrium-based technique (7) and fibers are available with combinations of different coating materials to offer a strategy for avoiding trade-offs between compounds that may vary in their affinity to the fiber. Fiber coatings fall into two categories, liquid polymers of high molecular weight or solids of high porosity, and their combination has been shown to be the most effective at collecting a broader spectrum of compounds ranging in volatility (8). Compounds trapped by solid-coated fibers are adsorbed inside pores on the fiber surface whereas in the case of liquid-coated fibers they are absorbed into the matrix. The quantity of captured volatiles is governed by two equilibrium constants, the rate of volatile release from the plant tissues into the surrounding air and the partitioning of airborne volatiles to the SPME fiber (partition ratio). In the case of liquid-coated fibers such as polydimethylsiloxane (PDMS), absorption of volatile compounds occurs relatively rapidly and is governed by diffusion constants similar to those in organic solvents whereas with solid coatings the diffusion constants are so small that absorption does not occur and adsorption to the surface is the mechanism of sampling (9). With both coatings, highly volatile molecules of low molecular weights are typically trapped first as they are the most concentrated of the headspace compounds. Later, as compounds of higher molecular weight and lower volatility are captured by the fiber, they must displace the smaller more highly volatile compounds due to limitations in fiber volume (10). In addition, ad/absorption parameters can be adversely affected by humidity and temperature especially during static headspace collection due to a lack of gas exchange in sealed collection vessels.

SPME sampling requires only the exposure of the fiber to a plant headspace for volatile collection and no pump or hardware is needed for air circulation, making it ideal for use in the field. In addition, because the SPME fiber can be used independently of any other hardware, it allows for high-throughput automated analysis of multiple samples (11). Following equilibration between the fiber and volatile sample, the fiber is subjected to direct thermal desorption onto a gas chromatograph. Because SPME does not require the use of organic solvents it eliminates bias arising from differences in analytes' solubility and avoids the introduction of impurities which may be present in the solvent and interfere with sample analysis. Typically SPME is used for qualitative and semi-quantitative analysis of plant volatiles, although quantification of volatiles is generally possible, but challenging (5, 12). This method also fails to provide sufficient amounts of volatiles for structure elucidation of unknown compounds.

An alternative to SPME is direct injection of headspace, which involves removing an air sample from static headspace using a gas-tight syringe and loading it directly into the GC inlet. This simple technique requires no specialized equipment and is readily automated but suffers from a lack of sensitivity. Thus, when sampling from small amounts of tissues that give off very minute volatile emissions, SPME is the ideal choice due to its facile implementation, flexibility, and remarkable sensitivity.

1.2. Dynamic Headspace Sampling

Dynamic headspace sampling of airborne compounds offers the researcher a highly concentrated sample that can be desorbed into a solvent at volumes suitable for multiple analyses. To date, it is the most frequently used technique in all areas of plant volatile analysis (13). Unlike SPME where the entire sample is desorbed inside the GC injection liner and subsequently lost as a result of analysis, a big disadvantage of this method, dynamic headspace sampling collects a much larger quantity of compounds at higher concentrations because the continuous stream of air allows the sorbent to act as a filter trapping the volatiles. Up-scaling of a sampling could be achieved by increasing the amount of sorbent, the airflow rate, sample tissue mass, and sampling time. In addition, accurate quantitative analysis of airborne compounds is more feasible because multiple columns can be connected in tandem to estimate volatile breakthrough and to compensate for differences in affinity of sorbents to the wide range of airborne compounds emitted by plant tissues, a technique known as multiple layer adsorption. Also, push and pull headspace sampling, two examples of dynamic headspace sampling, allow researchers to avoid problems often encountered with the sealed systems used in static headspace and closed-loop stripping (see below) methods including heat, water vapor, and ethylene accumulation that can affect not only sampling efficiency but also plant physiology.

1.2.1. Push and Pull Headspace Sampling

Push and pull headspace sampling techniques utilize a unidirectional flow of air as a mobile phase to carry plant volatiles to a trapping system, i.e., a cartridge packed with adsorbing material such as Tenax, Porapak, and activated charcoal. While both push and pull systems essentially perform the same task, differences between them affect their ease of implementation and the complexity of required equipment. Pull headspace sampling utilizes a vacuum pump to draw air over plant materials and through a sorbent cartridge. The flexibility of the pull headspace sampling method offers the researcher the ability to tailor its design to the needs and conditions of a particular experiment. For example, an adsorbent trap connected to a vacuum pump could be placed next to a sample releasing volatiles without enclosure of the plant into a collecting chamber, or simple open-top chambers could be used for volatile collection. In both cases, unfiltered ambient air brings the risk of trapping impurities unrelated to the investigated volatile blend and obscuring the detection of minute amounts of volatiles during GC analysis. Thus, these two simplified sampling methods are best suited for high-level emitters. Enclosure of material in glass containers or cooking bags with an opening for incoming air allows for the concentration of volatiles in the isolated airspace and reduction of their loss through diffusion. Additionally, cleaning incoming air through a purifying filter (activated charcoal, for example) will eliminate many impurities, reducing the background and increasing the sensitivity of detection.

Push headspace sampling involves placing plant samples inside a sealed, positive-pressure system into which air is pumped and then forced to exit through a volatile trapping cartridge. Push headspace sampling provides the advantage of eliminating ambient background contaminants through the use of pressurized, purified carrier gasses as the mobile phase. Despite this advantage, push headspace sampling typically requires bottled gasses, expensive flow regulators, and unwieldy airtight sampling containers that drive up expense and difficulty of volatile sampling. In addition, it becomes difficult to sample from tissues that have not been excised. It is also far more cumbersome to use in the field where lightweight and inexpensive materials are desirable for large-scale samplings in often remote areas. Here we provide a protocol for sampling volatile headspace in the field or in the laboratory using an inexpensive, portable, and undemanding pull headspace sampling method that can be used on plant tissues in situ without the need for excision.

1.2.2. Closed-Loop Stripping Method

Closed-loop stripping, like push and pull headspace sampling, utilizes airflow over a plant specimen to carry volatiles to a trapping cartridge. Its distinguishing feature is the continuous recirculation of the stripped air back to the plant sample subsequent to volatile adsorption allowing quantitative trapping of the emitted volatiles. Typically, excised plant materials are placed inside a vacuum desiccator and a circulating pump attached at the top. A volatile trapping cartridge is housed within the intake port of the pump. Following capture of the volatiles on traps they can be eluted with organic solvents and analyzed directly by GC/MS. The greatest advantage of the closed-loop stripping method is the reduction of airborne contaminants, and thus background noise, resulting in heightened sensitivity for collection from specimens that emit minimal amounts of volatiles. As in the sealed systems used with static headspace methods, artifacts can arise from the lack of airflow through the system and comparisons should be made between closed-loop and open dynamic headspace methods (5). This system is ideal for use inside controlled climate growth chambers, greenhouses, and laboratory conditions, though with portable power supplies excised plant materials can also be sampled in the field. It may, however, be preferable to adapt a more lightweight and rugged vessel from which to sample headspace volatiles.

2. Materials

2.1. High-Throughput Static Headspace Sampling via SPME

1. Flowering *Petunia x hybrida* cv. Mitchell diploid (*see* **Note 1**).
2. Gas chromatograph (e.g., Agilent 6890) coupled to a mass spectrometer (e.g., Agilent 5975B inert MSD) and Combi-PAL autosampler with SPME option (CTC Analytics, Zwingen, Switzerland).
3. Ultra-high-purity (99.998%) helium for GC carrier gas.
4. SPME fiber assembly (50/30 μm divinylbenzene/Carboxen on polydimethylsiloxane coating [PDMS/DVB/CAR]) (Supelco, Bellefonte, PA, USA) (*see* **Note 2**).
5. Capillary column, HP-5MS (30 m × 0.25 mm, 0.25-μm film thickness; Agilent, Wilmington, DE, USA).
6. SPME inlet liner (Supelco, Bellefonte, PA, USA).
7. 2-mL glass autosampler vials with polypropylene caps and PTFE/silicone septa.

2.2. Dynamic Headspace Sampling In Situ

1. Flowering *Antirrhinum majus* plants.
2. Gas chromatograph (e.g., Agilent 6890) coupled to a mass spectrometer (e.g., Agilent 5975B inert MSD).

3. Ultra-high-purity (99.998%) helium for GC carrier gas.
4. Capillary column, HP-5MS (30 m × 0.25 mm, 0.25-μm film thickness; Agilent, Wilmington, DE, USA).
5. 24-oz clear polyethylene terephthalate (PET) beverage cups with dome top (Lollicup USA, Inc., City of Industry, CA, USA) (*see* **Note 3; Fig. 17.1F**).

Fig. 17.1. **(A–D)** Equipment used for the closed-loop stripping method. **(A)** Rotary vane pump with servo-motor attached shown together with pump adaptors, stainless steel column housing, and volatile trapping column. **(B)** Tapered glass volatile trapping column filled with Porapak-Q and sealed using borosilicate glass wool. **(C)** Stainless steel column housing and pump adaptor illustrating enclosure of glass column. **(D)** Fully assembled components of closed-loop stripping method including desiccator and plant specimen. **(E–G)** Equipment used for pull headspace sampling method. **(E)** Glass volatile trapping column filled with Porapak-Q and sealed using borosilicate glass wool. **(F)** Illustration showing enclosure of snapdragon in modified PET cup. **(G)** Implementation of pull headspace volatile sampling inside growth chamber using whole snapdragon inflorescence.

6. Porapak-Q resin, 80/100 mesh (Waters, Millford, MA, USA).
7. Glass columns, 100 × 7 × 5 mm.
8. Borosilicate glass wool to seal Porapak-Q inside glass columns.
9. Flexible PTFE tubing, internal diameter 6 mm.
10. Portable vacuum pump with flow meters.
11. 2-mL V-bottom graduated vials with PTFE-faced, silicon-lined caps.
12. 2-mL glass autosampler vials with polypropylene caps and PTFE/silicone septa.
13. 500-μL glass autosampler vial insert with polymer feet (Agilent).
14. Acetone, redistilled.
15. Dichloromethane, redistilled.
16. N_2 for drying down and concentrating eluted samples.

2.3. Closed-Loop Stripping Method

1. Flowering plants, *Petunia x hybrida* cv. Mitchell diploid.
2. Gas chromatograph (e.g., Agilent 6890) coupled to a mass spectrometer (e.g., Agilent 5975B inert MSD).
3. Ultra-high-purity (99.998%) helium for GC carrier gas.
4. Capillary column, HP-5MS (30 m × 0.25 mm, 0.25-μm film thickness; Agilent, Wilmington, DE, USA).
5. Glass vacuum desiccators fitted with PTFE stoppers, pre-drilled for pump adaptors (*see* **Note 4**).
6. Rotary vane pumps fitted with electric servo-motors (DC 12/16FK; Fürgut Germany, Tannheim, Germany) (**Fig. 17.1A and D**).
7. Custom-fabricated pump adaptors with stainless steel housing for volatile trapping cartridges (**Fig. 17.1C**).
8. Variable DC power supply with range of 6–12 V and 50–300 mA.
9. Custom-fabricated glass cartridges, 66 × 5 × 3 mm with single tapered end (last 1.5 cm) to assist adsorbent retention (**Fig. 17.1B**).
10. Porapak-Q resin, 80/100 mesh (Waters, Millford, MA, USA).
11. Borosilicate glass wool to seal Porapak-Q inside glass columns.
12. 1-mL V-bottom graduated vials with PTFE-faced, silicon-lined caps.
13. 2-mL glass autosampler vials with polypropylene caps and PTFE/silicone septa.

14. 100-μL glass autosampler vial insert with polymer feet (Agilent).
15. Acetone, redistilled.
16. Dichloromethane, redistilled.
17. N_2 for concentrating eluted samples.

3. Methods

3.1. Static SPME Headspace Sampling of Volatiles from Excised Petunia Floral Organs

3.1.1. Preparation of Tissues for SPME-GC/MS

1. Using a new razor blade, excise 10 petunia pistils from open flowers 1-day post-anthesis. Allow the cut pistils to drop directly into an autosampler vial.
2. Quickly cap the vial once the 10 pistils are inside.
3. Prepare remaining samples in the fashion described above.

3.1.2. Combi-PAL Autosampler Configuration

1. Condition SPME fiber at 280°C for 10 min (add note, initial conditioning).
2. Collect headspace from each vial for 20 min.
3. Desorb volatiles inside inlet for 2 min and repeat conditioning/sampling cycle for each subsequent sample.

3.2. Dynamic Pull Headspace Sampling of Volatiles Emitted from Antirrhinum majus Inflorescence In Situ

3.2.1. Column Preparation

1. Ball up glass wool to form a plug dense enough to retain the Porapak-Q and insert it in a glass column.
2. Weigh out 100-mg Porapak-Q and add it to the column.
3. Insert second glass wool plug to trap Porapak-Q inside column (**Fig. 17.1E**).
4. Pre-condition columns by flushing with 5-mL acetone.
5. Wash columns with 5-mL dichloromethane.
6. Dry columns in a 37°C oven overnight to remove solvent and store in clean, airtight container until use.

3.2.2. Preparation of Enclosure for Plant Material

1. Create a 5-mm hole in the bottom of the beverage cup.
2. Insert a 5-cm length of PTFE tubing through the hole so that <1 cm of the tubing protrudes to the inside of the enclosure. The tubing elasticity should provide a fit tight enough so that the tubing and enclosure produce a near airtight seal.
3. Connect a packed glass column via the PTFE tubing attached to the enclosure.
4. Cut the dome top of the cup longitudinally to allow passage of an inflorescence stem through the dome lid (**Fig. 17.1F**).

3.2.3. Sample Collection

1. Slide enclosure over snapdragon inflorescence taking caution not to damage the tissues.
2. Stabilize the enclosure using any convenient method.
3. Put a dome around selected inflorescence and use tape to seal the cut. Snap dome onto the enclosure.
4. Connect column to the vacuum pump via PTFE tubing and begin sampling at a rate of approximately 2 L/min (**Fig. 17.1G**).
5. When sampling is completed, switch off the vacuum pump and remove the column making sure to place it in an individual airtight container.

3.2.4. Sample Preparation

1. Elute samples from the columns using 2 mL of redistilled dichloromethane. Collect eluate in a 2-mL V-bottom graduated vial and cap with PTFE-faced, silicon-lined cap.
2. Dry samples down to final volume of 0.5 mL and transfer it to an autosampler vial insert (*see* **Note 5**).
3. Add internal standard.
4. Analyze by GC/MS (*see* **Section 3.4**).

3.3. Dynamic Closed-Loop Stripping Headspace Sampling of Volatiles from Excised Petunia Flowers

3.3.1. Column Preparation

1. Assemble the column as described in 3.2.1 using 35-mg Porapak-Q (**Fig. 17.1B**).
2. Pre-condition columns by flushing with 1-mL acetone.
3. Wash columns with 2-mL dichloromethane.
4. Dry columns in 37°C oven overnight to remove solvent and store in clean, airtight container until use.

3.3.2. Sample Collection

1. Excise three petunia flowers at the base of the pedicel with razor blade. Immediately place only the cut end in a beaker containing 5% sucrose in water.
2. Place the beaker with cut flowers into the glass desiccator and close the lid.
3. Insert pre-drilled PTFE stopper in the desiccator lid.
4. Place a clean, dry column inside steel housing and attach it in pump intake port. Be sure that the airflow is pulled first through the column into the pump and recirculated to the desiccator.
5. Slide pump adaptors through pre-drilled holes of the PTFE stopper in desiccator lid (**Fig. 17.1D**).
6. Adjust voltage to provide airflow of 2-L/min and begin sampling (*see* **Note 6**).
7. When sampling is completed, disconnect the power supply and remove the columns making sure to place them in individual airtight containers.

3.3.3. Sample Preparation

1. Elute samples from the columns using 1 mL of redistilled dichloromethane. Capture eluate in a 1-mL V-bottom graduated vial and cap with PTFE-faced, silicon-lined cap until all samples are eluted.
2. Dry samples down to final volume of 0.1 mL and transfer to an autosampler vial insert.
3. Add internal standard.
4. Analyze by GC/MS (*see* **Section 3.4**).

3.4. GC/MS Parameters

a. Inlet temperature is set to 280°C.
b. GC interface temperature set to 280°C.
c. MS source set to 250°C.
d. Quadrupole set to 150°C.
e. Mobile phase flow rate is 1.0 mL/min.
f. GC temperature gradient programmed as follows: Initial temperature of 30°C held for 2 min followed by gradient of 5°C/min to 260°C, hold for 6 min (*see* **Note** 7).

3.5. Data Analysis

After collection of volatiles from plant samples and their separation on GC/MS, the data will be presented as a total ion chromatogram of individual constituents in the analyzed blend. Peaks within this chromatogram contain two types of information that can be used for compound identification: retention time and mass spectrum consisting of a characteristic ion fragmentation pattern. Identification of a compound based on only one of these parameters is risky and often leads to its misidentification. To date, several comprehensive mass spectral libraries are commercially available (Wiley, NIST MS Database) and can guide researchers in their choice of authentic standards for comparison. These standards should be run along with a sample and their retention times and mass spectra should match that of the analyte of interest if identification is correct. It is also recommended to match retention time between authentic standard and analyte of interest on a column of different polarity. **Figure 17.2** represents the analysis of volatile compounds collected from snapdragon flowers using the closed-loop stripping method (**Fig. 17.2B**) and complementary authentic standards (**Fig. 17.2A**) run on GC/MS under the same conditions. All identified compounds were matched to standards using retention time and mass spectrum as it is shown for 3,5-dimethoxytoluene (see inserts in **Fig. 17.2A and B**). Since the biological activity and function of particular volatile blend often depends on its enantiomeric composition, determination of compound chirality is often required. To achieve this goal a variety of capillary columns are available for separation of enantiomers using GC, and their applications, advantages, and disadvantages are thoroughly discussed in recent reviews (14, 15). **Figure 17.3** represents an enantiomeric

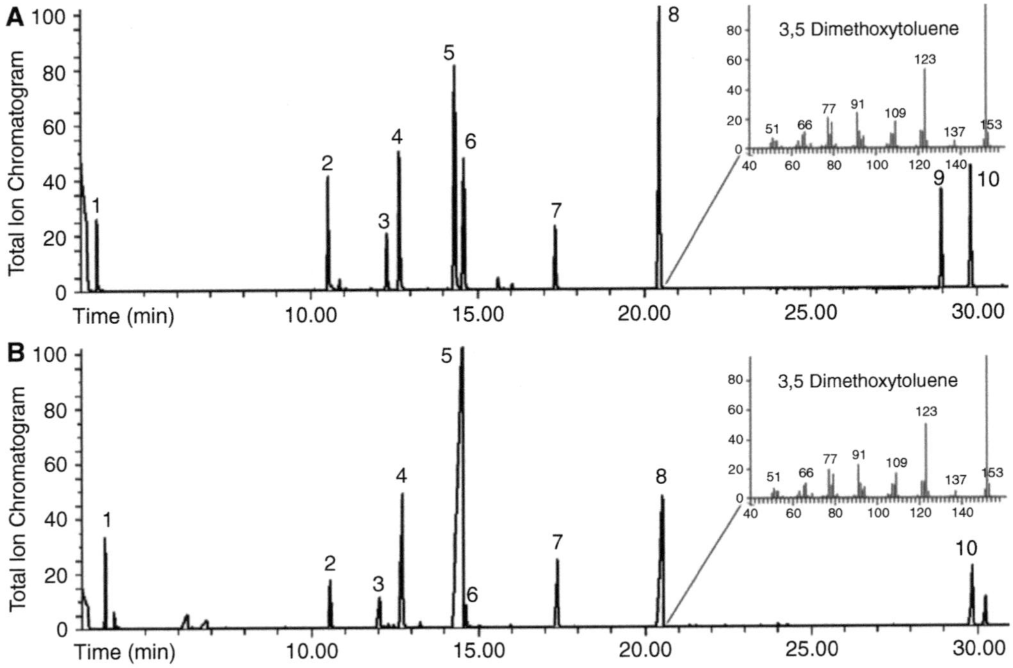

Fig. 17.2. Total ion chromatograms of floral scent authentic standards representative of snapdragon cv. Maryland True Pink (MTP) floral scent blend with insert depicting mass spectrum from 3,5-dimethoxytoluene (**A**) and volatile profile of emitted MTP snapdragon floral scent with insert depicting mass spectrum from 3,5-dimethoxytoluene (**B**). Numbered peaks (1–10) represent toluene (1, internal standard); ß-myrcene (2); *trans*-ocimene (3); *cis*-ocimene (4); methyl benzoate (5); *S*-linalool (6); naphthalene (7, internal standard); 3,5-dimethoxytoluene (8); *cis*-nerolidol (9); *trans*-nerolidol (10).

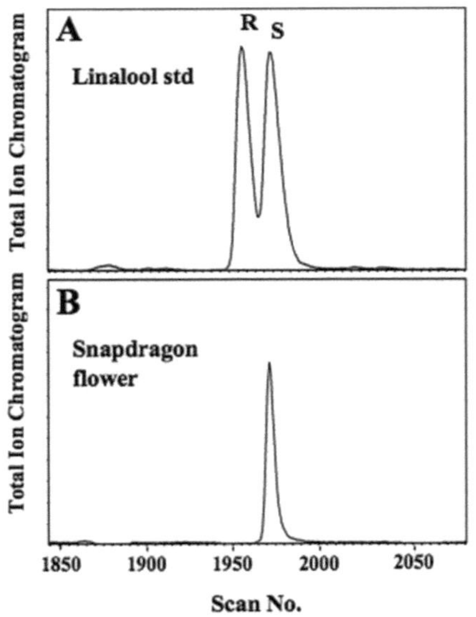

Fig. 17.3. Total ion chromatograms depicting separation of linalool enantiomers. (**A**) Enantiomeric separation using racemic mixture of *R*- and *S*-linalool on ß-cyclodextrin enantioselective GC column. (**B**). The floral scent of snapdragon flowers contains only *S*-linalool, as demonstrated by the separation of floral scent on the same column.

separation of racemic mixture of linalool authentic standard in comparison with that emitted by snapdragon flowers. In this case, scent was collected by SPME and volatiles were analyzed using GC-MS equipped with 2,3-di-*O*-methyl-6-*O*-*tert*-butyl dimethylsilyl beta cyclodextrin doped into 14% cyanopropylphenyl/86% dimethyl polysiloxane (Rt^{TM}-βDEXsm) column (30 m × 0.25 mm ID) (Restek, Bellefonte, PA) under conditions provided by the manufacturer.

4. Notes

1. Sample acquisition in metabolomic experiments requires consistency. Because plant volatile emissions are linked to the physiological status of the emitter, special care must be taken to control not only the plant-growing environment but also all other possible variables concerning the growth of the plant to limit unwanted fluctuations in metabolism that might affect collected data. This includes time of day, photoperiod, temperature, humidity, water conditions, etc. Whenever possible, growth chambers must be used for plant cultivation and volatile collection. Pests and pathogens must be excluded at all costs (unless the subject of the experiment) and conditions that may alter plant metabolism in other ways must be avoided, i.e., environmental stresses. For this method, make sure that the flowers are at the same developmental stage and that they come from a large set of plants growing under near-identical conditions. Also, it is best to minimize the time period that samples spend on the autosampler tray prior to SPME headspace collection.
2. SPME fibers are available in with wide range of coatings that allow sampling of volatiles, semi-volatiles, polar analytes, and flavor and odor compounds. Because the goal of a metabolomic analysis is to sample as many metabolites as possible, the use of PDMS/DVB/CAR fibers is recommended to increase the number of analytes that the fiber is capable of trapping. Depending on the application, the sensitivity of trapping could be optimized through the use of a more specific fiber type to increase affinity to a particular volatile compound. Although SPME fibers can be re-used many times before they should be discarded (~100 desorptions), the number of uses should be controlled in order to avoid sampling errors introduced by losses in fiber sensitivity.
3. When sampling from plants in situ, options become increasingly limited by the weight of the sampling apparatus, restricting the use of glass enclosures. Researchers have ingeniously modified PET beverage cups found often in the food service industry for

volatile sampling and polyacetate cooking bags have also been employed (5, 16). The setup is flexible and could be improvised as well as optimized for the working conditions and available resources in each particular case. These supplies are readily available and usually inexpensive; however, it is necessary to account for background contaminants originating from the enclosure materials or their adsorption of plant volatiles. Controls should include samplings without plant materials to identify any compounds leaching from plastics and those present in the ambient air. Also, a comparative quantitative analysis of volatiles collected using different types of open-top enclosures will account for losses due to unintended adsorption by enclosure material. It is advisable to avoid recycling enclosures to minimize carryover of analytes between samples.

4. The main disadvantage of this technique is the accumulation of water vapor, heat, and ethylene inside the sealed container (5) that can adversely alter plant metabolism and affect sorbent efficiency, changing the spectrum of both emitted and collected volatiles. Thus, it may be necessary to limit sampling times as a result to eliminate introduction of artifact. It may also be beneficial to compare volatiles captured using closed-loop stripping with those collected in more vented systems, such as push or pull headspace to identify artifacts of the sampling method.
5. It may be necessary to decrease the final sample volume to increase concentration of trace volatiles.
6. Since volatile emissions from many plant species vary with respect to the time of day, collection strategies should consider volatile sampling over a 24-h period to prevent unintentional exclusion of volatile blend components. This can be done by using one column with a 24-h collection period or by changing columns after specified intervals.
7. Initial temperature of 30°C is a recommended temperature for GC/MS analysis of volatile compounds dissolved in dichloromethane due to its low boiling point (39°C). GC/MS analysis of volatiles collected with SPME does not require such low temperature and can be started at 40°C, thus drastically reducing GC cycling time.

Acknowledgments

This work was supported by grants from the National Science Foundation (grant no. MCB-0615700), the USDA Cooperative State Research, Education, and Extension Service (grant no. 2005-35318-16207), and the Fred Gloeckner Foundation.

References

1. Pichersky, E., Noel, J.P., and Dudareva, N. (2006) Biosynthesis of plant volatiles: nature's diversity and ingenuity. *Science.* **311**, 808–811.
2. Dudareva, N., Negre, F., Nagegowda, D.A., and Orlova, I. (2006) Plant volatiles: recent advances and future perspectives. *Crit. Rev. Plant Sci.* **25**, 417–440.
3. Knudsen, J.T., Eriksson, R., Gershenzon, J., and Stahl, B. (2006) Diversity and distribution of floral scent. *Bot. Rev.* **72**, 1–120.
4. Acree, T. and Arn, H. (2004) Flavornet and human odor space. http://www.flavornet.org/index.html. Accessed January 10, 2008.
5. Tholl, D., Boland, W., Hansel, A., Loreto, F., Rose, U.S.R., and Schnitzler, J.P. (2006) Practical approaches to plant volatile analysis. *Plant J.* **45**, 540–560.
6. Tholl, D. and Rose, U.S.R. (2006) Detection and identification of floral scent compounds. In: Dudareva, N., Pichersky, E., editors. *Biology of Floral Scent.* Boca Raton, FL, USA: CRC/Taylor & Francis;. pp. 3–25.
7. Arthur, C. and Pawliszyn, J. (1990) Solid phase microextraction with thermal desorption using fused silica optical fibers. *Anal. Chem.* **62**, 2145–2148.
8. Bicchi, C., Drigo, S., and Rubiolo, P. (2000) Influence of fibre coating in headspace solid-phase microextraction–gas chromatographic analysis of aromatic and medicinal plants. *J. Chromatogr. A.* **892**, 469–485.
9. Górecki, T., Yu, X., and Pawliszyn, J. (1999) Theory of analyte extraction by selected porous polymer SPME fibres. *Analyst.* **124**, 643–649.
10. Barták, P., Bednár, P., Cáp, L., Ondráková, L., and Stránsky, Z. (2003) SPME – A valuable tool for investigation of flower scent. *J. Sep. Sci.* **26**, 715–721.
11. Aharoni, A., Giri, A.P., Deuerlein, S., Griepink, F., de Kogel ,W.J., Verstappen, F.W.A., Verhoeven, H.A., Jongsma, M.A., Schwab, W., and Bouwmeester, H.J. (2003) Terpenoid metabolism in wild-type and transgenic Arabidopsis plants. *Plant Cell.* **15**, 2866–2884.
12. Vas, G. and Vékey, K. (2004) Solid-phase microextraction: a powerful sample preparation tool prior to mass spectrometric analysis. *J. Mass Spectrom.* **39**, 233–254.
13. D'Alessandro, M. and Turlings, T.C.J. (2006) Advances and challenges in the identification of volatiles that mediate interactions among plants and arthropods. *Analyst.* **131**, 24–32.
14. Shurig, V. (2001) Separation of enantiomers by gas chromatography. *J. Chromatogr. A.* **906**, 275–299.
15. Shurig, V. (2002) Chiral separations using gas chromatography. *Trends. Anal. Chem.* **21**, 647–661.
16. Raguso, R.A. and Pellmyr, O. (1998) Dynamic headspace analysis of floral volatiles: a comparison of methods. *Oikos.* **81**, 238–254.

Chapter 18

Chemical Genomics Approaches in Plant Biology

Lorena Norambuena, Natasha V. Raikhel, and Glenn R. Hicks

Abstract

Chemical genomics (i.e., genomics-scale chemical genetics) approaches are based on the ability of low-molecular-mass molecules to modify biological processes. Such molecules are used to affect the activity of a protein or a pathway in a manner that is tunable and reversible. A major advantage of this approach compared to classical plant genetics is the fact that chemical genomics can address loss-of-function lethality and redundancy. Bioactive chemicals resulting from forward or reverse chemical screens can be useful in understanding and dissecting complex biological processes due to the essentially limitless variation in structure and activities inherent in chemical space. An important aspect of utilizing small molecules effectively is to characterize bioactive chemicals in detail including an understanding of structure activity relationships (SARs) and the identification of active and inactive analogs. Bioactive chemicals can be useful as reagents to probe biological pathways directly. However, the identification of cognate targets and their pathways is also informative and can be achieved by screens for genetic resistance or hypersensitivity in *Arabidopsis thaliana* or other organisms in which the results can be translated to plants. Here, we describe approaches to screen for bioactive chemicals that affect biological processes in Arabidopsis. We will also discuss considerations for the characterization of bioactive compounds and genetic screens for target identification. This should provide those who are considering this approach some practical knowledge of how to design and establish a chemical genomics screen.

Key words: Chemical genomics, screening, target identification, structure–activity relationship (SAR), Sortin, endomembrane, vacuole.

1. Introduction

To fully understand the biology of a biological pathway, it is crucial to manipulate it and study the function of its corresponding genes. In Arabidopsis, although T-DNA inactivation mutants have become a valuable tool for understanding gene function, the availability of viable and informative knockout lines is limited because

Dmitry A. Belostotsky (ed.), *Plant Systems Biology, vol. 553*

DOI 10.1007/978-1-60327-563-7_18 Springerprotocols.com

the loss of function for a substantial number of genes is lethal. Another, and perhaps more formidable, challenge is that knockouts may display no phenotype due to functional gene redundancy which may complicate analysis and interpretation. Chemical genomics approaches are based on the ability of low-molecular-mass molecules to modify the activity of a protein or a pathway overcoming the limitations of mutational approaches. Such compounds can trigger tunable and reversible plant responses which are difficult or impossible to achieve using conventional genetics in plants. In forward chemical genomic screens, thousands of compounds are tested for their ability to alter a specific pathway resulting in a phenotype. Ultimately, bioactive compounds can be useful to understand and dissect molecular and biochemical processes. The power of bioactive molecules in plant biology has been amply illustrated by the use of specific chemical inhibitors of biological processes. Examples include chemicals such as brefeldin A (1, 2), latrinculin B (3), and auxin transport inhibitors (e.g., NPA) (4). Recently, several new bioactive compounds in Arabidopsis have been found via screening (5–8). Subsequent identification of chemical targets can be achieved by biochemical or genetic approaches. Biochemical identification of targets can be difficult because success depends on the type of chemical–target interaction, the abundance of the target site and the binding affinity of a bioactive chemical for its target. Arabidopsis genetic screens for resistance can be time consuming; however, they have resulted in the identification of the corresponding cognate targets for several novel chemicals in the past few years (9–12).

In principle, a chemical screen can be performed in any plant system. However, the advantage of using Arabidopsis is that the available genetics and genomic tools, which are substantial, can be used to identify genes involved in a target pathway. The more critical aspect for success is to have a simple, reliable, and robust phenotypic assay that can be done in a high-throughput manner. The use of robots for performing a screening assay can improve reliability and speed as well. The methods described in this chapter outline (i) a chemical screen based upon the root length of Arabidopsis seedlings, (ii) characterization of one hypothetical bioactive compound, *Chemical A*, and (iii) a genetic screen to identify hypersensitive and resistant mutants to *Chemical A* and, ultimately, a potential cognate target.

2. Materials

2.1. Medium

The culture medium for making plates is 0.5 × Murashige and Skoog (MS) medium (PlantMedia, Dublin, OH) pH 5.6 containing 2% sucrose and 0.3% GELRITE (RPI, Illinois, IL).

2.2. Plant Material

1. For the chemical screen discussed, wild-type Arabidopsis (ecotype Columbia-0) will be used. The screen for genetic resistance or hypersensitivity will be performed using a collection of EMS-treated Arabidopsis (ecotype Columbia-0) seeds (M2 population) prepared as described in Lightner and Caspar (13).
2. Seeds are sterilized and stratified in darkness for 48 h at 4°C prior to plating. They are germinated and grown in an incubator at 22°C under standard conditions of humidity and photoperiod appropriate for growing Arabidopsis.

2.3. Chemical Library Sources

Chemical collections can be purchased from numerous commercial sources. Usually the criteria for choosing a library are the number and structures of compounds that are to be screened for activity. Depending on their size, chemical libraries come in 96- or 384-well format plates. The chemical library used in this example screen is the DIVERSet library (ChemBridge, San Diego, CA) which comes in 96-well format plates. To prepare master plates with stock solutions, DMSO (Fischer Scientific) is used as solvent. The advantage of purchasing libraries from commercial sources over attempting synthesis is that those compounds that trigger interesting phenotypes can usually be re-ordered individually for further characterization.

2.4. Chemical Treatments

The DIVERSet library (ChemBridge, San Diego, CA) contains 10,000 small organic molecules in a 96-well format plate. In a 96-well format it is possible to pipet chemicals by hand using multi-channel pipetters. However, this is much more difficult for 384-well format library plates. In such cases, access to liquid-handling robots is extremely valuable, if not a necessity. For example, our laboratory has access to several robots including a relatively simple Precision 2000 pipetting robot (Bio-Tek Instruments, Winooski, VT) as well as a more sophisticated BioMek (Beckman Instruments) that is capable of liquid transfer by either pipette tips or pin tools for small volume transfers (for instance, 0.2 μl).

For labs that wish to pipet chemicals, the following protocol should prove useful.

1. For master plates, dissolve 0.1 mg of each compound in 20 μl of 100% DMSO and store at –20°C (*see* **Note 1**).
2. Prepare working solution plates by diluting each master plate fivefold with water in polypropylene 96-well plates (Corning, # 3355). The final concentration of the compound solutions will range between 2 and 4 mM (depending upon the mass of the compounds) in 20% DMSO.
3. For direct screening of Arabidopsis seedlings, re-array the working solution library from 96-well format to a 24-well format (*see* **Note 2**).

4. For chemical screen:
   4.1 Add 10 μl of each chemical from working solution plates to each well of a 24-well plate (Corning, #3526).
   4.2 To each well, add 390 μl of culture media (at 50°C) and mix it by shaking the plate gently. Allow the agar to solidify for 30–40 min. The final concentration of compound is 25 μg/ml in 0.5% DMSO (*see* **Note 1**).
5. Screens for genetic resistance and hypersensitivity:
   The screening is performed using large format Petri plates (23.5 cm × 23.5 cm). Pour 120 ml of 0.5 × MS agar medium containing *Chemical A* at the concentration for resistance (C_R) or hypersensitivity (C_{Hy}) screening (*see* **Section 3.4**).

3. Methods

3.1. Practical Considerations Before Starting a Chemical Screen

3.1.1. Define the Phenotype You Would Like to Score for

As mentioned above, the phenotypic assay is crucial for the design of the chemical screen. To work in a high-throughput manner, the assay has to be as quick and straightforward as possible. Depending on the assay, developmental and physiological phenotypes can be analyzed directly. Phenotypic screens for responses to a stimulus have been successful for screening thousands of chemicals and have resulted in novel bioactive chemicals (6, 7). On the other hand, effects of the compounds may be monitored at the subcellular level, via monitoring of marker proteins fused to GFP and targeted to different compartments or cell domains. Organelle or membrane-localized GFP markers can also permit the visualization of compartment morphology. For example, we have utilized the tonoplast marker GFP-δTIP to great effect in examining the morphology of vacuoles (14).

Although this can be done using a conventional confocal microscope, such screens can be greatly aided by the use of a high-throughput confocal microscope such as the Pathway HT (Atto Biosciences). For such laborious assays, an alternative primary screen can be designed using an easily scored phenotype (if available) that is associated with the phenotype of interest. For example, inhibition of root growth is a phenotype associated with many different bioactive molecules at high concentrations. Such generalized primary screens can then be followed up by more specific secondary screens for subcellular phenotypes of interest. This two-step approach to screening can save considerable time and effort (7). Other developmental phenotypes such as size or the presence/absence of organs (cotyledons, tricoms, root, etc.) can be useful for prescreening a large chemical library.

As a final alternative, the primary phenotypic screen can be done in a simpler system such as yeast. This approach works well for processes that are conserved evolutionarily such as endomembrane trafficking, transcription, or translation. Chemicals of interest can then be tested in Arabidopsis (5).

3.1.2. Choice of Chemical Library: Advantages of Diverse, Tagged, and Focus Libraries

Chemical companies have collected a large variety of compounds to create diverse libraries. In principle, the advantage of such large collections is that they are fairly structurally unbiased, although bias is impossible to eliminate fully. An advantage of using a relatively unbiased collection is that the chances of finding a novel and interesting chemical are increased. However, without any notion of what is bioactive from a structural viewpoint, a large number of compounds must be screened in order to find a hit.

If the purpose of doing a chemical screen is to identify biological targets, the use of a tagged chemical library can be more convenient, at least in principle. Such libraries incorporate into their structures a tag such as a reactive amine. Once a bioactive tagged chemical is identified, a biochemical approach can be taken to isolate the target using an affinity matrix to which the chemical is immobilized. These types of libraries are built on scaffolds that are suitable for tagging and, thus, are more biased than untagged libraries. Although reported, such libraries are not commercially available yet (15). Nonetheless, there are already some reports of successful target identification using this strategy in non-plant models (16, 17). It is also possible to add a linker attachment tag to many chemicals of interest. However, loss of activity poses a significant risk with this strategy.

In cases where more in-depth structural variants of a particular compound are desired, so-called focus libraries can be synthesized based on a known chemical. Usually such libraries are a systematic variation of a well-characterized chemical and require the expertise of an experienced synthetic chemist who is interested in collaborating with biologists.

3.2. A Chemical Screen

In this hypothetical example, a primary screen is performed using a 10,000 compound library in order to identify chemicals that inhibit root growth. Once the primary hits are identified, a secondary screen could be done using a specific assay of interest (not described here).

1. Place five to eight stratified Arabidopsis seeds in a line along the center of each well of a 24-well plate (when placed vertically) containing media supplemented with chemicals.
2. Incubate plates vertically in light at 22°C for 5 days.
3. Score the plates and record the size of seedlings by imaging the plates using a flat-bed scanner (for example, model 2450; Epson, Long Beach, CA).

4. A chemical is considered a primary hit when the majority of seedling roots in the well are shorter than the control. Root and hypocotyl lengths can also be quantified from images using software such as NIH image or Scion Image (Scion Corp, Frederick, MD) which can be downloaded free of charge (http://www.scioncorp.com/).
5. Primary hits have to be confirmed. Perform the assay under the same conditions as the primary screen at concentrations of 0, 0.25, 5, and 10 μg/ml in a 24-well plate format.
6. A chemical is considered as a confirmed positive if (i) root growth inhibition is present in the re-test and (ii) the effect of the chemical appears to be dose dependent.

3.3. Characterizing Bioactive Chemicals: Specificity and Analogs

Once bioactive dose-dependent chemicals are identified several aspects must be tested in order to demonstrate that they are valuable probes for a particular phenotype or pathway. Aspects such as inducibility, reversibility, and specificity will be discussed.

1. Inducibility is the ability of a compound to yield a phenotype in a relatively short period of time, on the order of hours or days. For inducibility, seeds are sown in the absence of chemical for 5 days in light at 22°C. Then, seedlings are transferred to a plate with *Chemical A* for 3 days to test for inhibition of root growth due to a bioactive chemical.
2. Reversibility is the loss of phenotype over time due to chemical metabolism, modification, exclusion, sequestration, or other form of metabolic clearing. For reversibility, this assay is carried out under the same conditions as the primary screen for bioactive dose-dependent chemicals. Seedlings are then transferred to a plate without the chemical, to test for recovery of wild-type phenotype; such recovery would indicate that the chemical effect is reversible.
3. Specificity refers to the ability of a chemical to affect a specific biological process. One of the important questions in working with bioactive chemicals is whether a resulting phenotype is due to a chemical or to a more generalized effect, for example, on growth and development. In other words, does the interaction of a chemical with a specific cognate target result in the observed phenotype? A useful strategy for examining specificity is to test active and inactive structural analogs. Having similar molecules that are inactive would show that the chemical effect is specific in terms of chemical structure. Along these lines, studies of structure–activity relationships (SARs) can be conducted in order to gain insights into the specific moieties and domains important for bioactivity. These structural analogs can be searched using databases such as ChemMine (18) which is a publically available database that has associated with it a number of highly useful tools

for searches. For example, structural similarity searches can be done among millions of chemicals, many of which are available commercially. The database also permits substructure searches and can produce graphical representations of relatedness. ChemMine centralizes compound structure and activity information from a growing number of public providers and vendors of chemical screening libraries. Thus, in most cases, structural analogs and substructures can be identified and obtained with surprising ease.

3.4. Target Identification: Screens for Genetic Resistance and Hypersensitivity

Once a bioactive compound is identified, its use can be combined with genetic screens to identify genes related to the chemical's target pathway. Screens for genetic resistance and hypersensitivity can be done for this purpose. Based upon the dose–response behavior of one of our hits, *Chemical A*, the design of a genetic screen would be straightforward in this case where the chemical of interest results in inhibition of root growth (**Fig. 18.1A**). The minimum concentration to observe the phenotype reliably will correspond to the chemical concentration for a resistance screen (C_R) (**Fig. 18.1B**). In contrast, a hypersensitive screen should be done at the maximum chemical concentration that does not result in phenotype in wild type (C_{Hy}) (**Fig. 18.1C**).

3.4.1. Screen for Genetic Resistance

1. Place stratified Arabidopsis EMS mutant seeds at a density of 150–200 seeds per row with 5 rows per Petri plate (**Section 2.4**) on medium supplemented with C_{hy} of *Chemical A* (up to 1000 seeds per plate).
2. Incubate plates in a vertical position for 7 days in light at 22°C.
3. Seedlings with roots longer than 0.3 cm are putative resistant mutants (**Fig. 18.1B**).
4. To recover, transfer putative resistant seedlings to medium without *Chemical A*. Incubate for three more days.
5. Transplant *Chemical A*-resistant seedlings to soil and collect M3 generation seed.
6. Re-test the resistance of M3 generation seedlings. To do this, plate 10–20 seeds on the presence or absence of C_R of *Chemical A* and rescore.

3.4.2. Screen for Genetic Hypersensitivity

1. Place stratified Arabidopsis EMS mutant seeds at a density of 150–200 seeds per row with 5 rows per Petri plates (**Section 2.4**) containing media supplemented with C_{Hy} of *Chemical A* (up to 1000 seeds per plate).
2. Incubate plates in a vertical position for 7 days in light at 22°C (**Fig. 18.1C**).
3. Transfer seedlings with roots less than 0.3 cm to medium without *Chemical A*. Incubate the transferred seedlings for five more days.

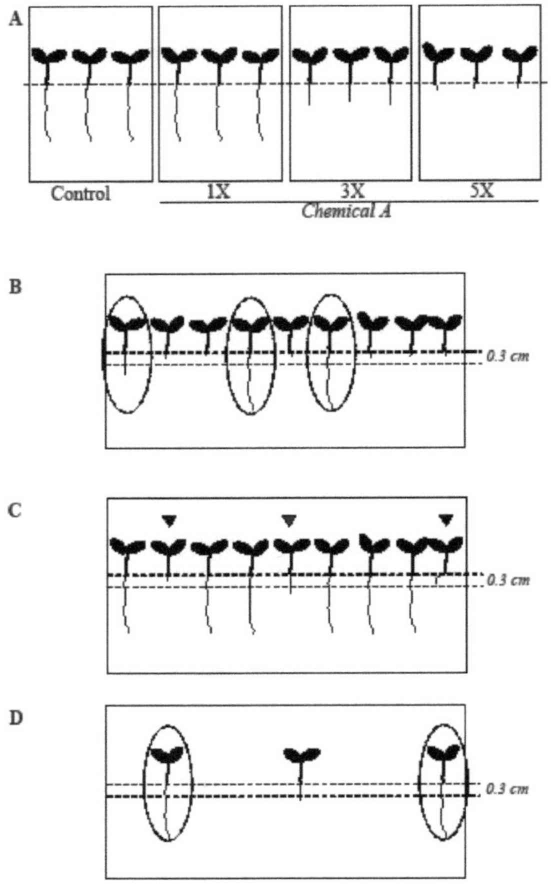

Fig. 18.1. (**A**) *Chemical A* dose response. The phenotype is tested at several concentrations of *Chemical A*. C_{Hy} (1X) and C_R (*5X*) are defined for genetic screens. (**B**) Genetic screen for resistance. Seedlings (circled) display genetic resistance to C_R *of Chemical A*. (**C**) Genetic screen for hypersensitivity. Seedlings (arrowed) display sensitivity to *Chemical A* at a concentration that has no effect upon seedling growth of wild-type plants (C_{Hy}). A convenient cut-off length for seedling roots after 7 days is 0.3 cm. (**D**) These seedlings are transferred to a non-chemical plate for 5 days. Seedlings that resume root growth in the absence of chemical are considered drug-dependant mutants (circled).

4. Seedlings that resume root growth in the absence of *Chemical A* display a phenotype that is drug dependent (**Fig. 18.1D**). Seedlings that do not resume growth in the absence of *Chemical A* may be developmental mutants whose phenotype does not depend upon the presence of the chemical (*see* **Note 3**).
5. Transplant seedlings displaying a *Chemical A* drug-dependent phenotype to soil and collect M3 generation seed.
6. Re-test the hypersensitivity of M3 generation seedlings as well as their *Chemical A* drug-dependent phenotype. For this, plate 10–20 seeds in the presence or absence of C_{Hy}.

7. Determine the mean root lengths (n = 10–20 per mutant) using Scion Image (Scion Corp, Frederick, MD). By quantifying from scanned images, normalized ratios of relative hypersensitivity can be generated.

4. Notes

1. Typically, the chemicals should be screened at as high a concentration as feasible without the concentration of DMSO exceeding about 1%. At such percentages, DMSO will inhibit growth complicating the results. Typically the final screening concentrations of chemicals in a primary screen can range from 50 to 100 μM. Although this may seem like a high concentration, remember that these compounds have been optimized only to be "drug-like" in that their properties that should permit them to be membrane permeable in mammalian cells. It is more difficult to assess other properties such as transport through the vascular system, stability in terms of inactivation, breakdown, sequestration, or other detoxifying mechanisms.
2. The re-array of the working solution library to a 24-well format can be done by using either a handheld multi-channel pipetter or a Bio-Tek Precision 2000 liquid-handling robot or similar fluid-handling robot. If more screens are planned, generate working solution plates in the 24-well format.
3. Resistant and hypersensitive mutants to *Chemical A* can be classified based on the strength of phenotype: strong, moderate, or weak phenotypes for instance. Using the example of root length, this can be quantified from scanned images and "strong mutants" can even be defined (for example, those with roots greater than 1 cm in length in the presence of *Chemical A*). Such definitions are extremely useful in prioritizing mutants for further studies or mapping.

Acknowledgments

This work was supported by a grant from the National Science Foundation (MCB-0515963 and NSF2010-MCB-0520325).

References

1. Dinter, A. and Berger, E.G. (1998) Golgi-disturbing agents. *Histochem. Cell Biol.* **109**, 571–590.
2. Nebenfuhr, A., Ritzenthaler, C., and Robinson, D.G. (2002) Brefeldin A: Deciphering an enigmatic inhibitor

of secretion. *Plant Physiol.* **130**, 1102–1108.

3. Friml, J., Wisniewska, J., Benkova, E., Mendgen, K., and Palme, K. (2002) Lateral relocation of auxin efflux regulator PIN3 mediates tropism in Arabidopsis. *Nature.* **415**, 806–809.
4. Geldner, N., Friml, J., Stierhof, Y.D., Jurgens, G., and Palme, K. (2001) Auxin transport inhibitors block PIN1 cycling and vesicle trafficking. *Nature.* **413**, 425–428.
5. Zouhar, J., Hicks, G.R., and Raikhel, N.V. (2004) Sorting inhibitors (Sortins): Chemical compounds to study vacuolar sorting in Arabidopsis. *Proc. Natl. Acad. Sci. USA.* **101**, 9497–9501.
6. Armstrong, J.I., Yuan, S., Dale, J.M., Tanner, V.N., and Theologis, A. (2004) Identification of inhibitors of auxin transcriptional activation by means of chemical genetics in Arabidopsis. *Proc. Natl. Acad. Sci. USA.* **101**, 14978–14983.
7. Surpin, M., Rojas-Pierce, M., Carter, C., Hicks, G.R., Vasquez, J., and Raikhel, N.V. (2005) The power of chemical genomics to study the link between endomembrane system components and the gravitropic response. *Proc. Natl. Acad. Sci. USA.* **102**, 4902–4907.
8. DeBolt, S., Gutierrez, R., Ehrhardt, D.W., Melo, C.V., Ross, L., Cutler, S.R., Somerville, C., and Bonetta, D. (2007) Morlin, an inhibitor of cortical microtubule dynamics and cellulose synthase movement. *Proc. Natl. Acad. Sci. USA.* **104**, 5854–5859.
9. Zhao, Y., Dai, X., Blackwell, H.E., Schreiber, S.L., and Chory, J. (2003) SIR1, an upstream component in auxin signaling identified by chemical genetics. *Science.* **301**, 1107–1110.
10. Dai, X., Hayashi, K., Nozaki, H., Cheng, Y., and Zhao, Y. (2005) Genetic and chemical analyses of the action mechanisms of sirtinol in Arabidopsis. *Proc. Natl. Acad. Sci. USA.* **102**, 3129–3134.
11. Walsh, T.A., Bauer, T., Neal, R., Merlo, A.O., Schmitzer, P.R., Hicks, G.R., Honma, M., Matsumura, W., Wolff, K., and Davies, J.P. (2007) Chemical genetic identification of glutamine phosphoribosylpyrophosphate amidotransferase as the target for a novel bleaching herbicide in Arabidopsis. *Plant Physiol.* **144**, 1292–1304.
12. Rojas-Pierce, M., Titapiwatanakun, B., Sohn, E.-J., Fang, F., Larive, C., Blakeslee, J., Cheng, Y., Cuttler, S., Peer, W., Murphy, A., and Raikhel, N.V. (2007) Arabidopsis P-Glycoprotein19 participates in the inhibition of gravitropism by Gravacin. *Chem. Biol.* **14**, 1366–1376.
13. Lightner, J. and Caspar, T. (1998) Seed mutagenesis of Arabidopsis. *Methods Mol. Biol.* **82**, 91–102.
14. Avila, E.L., Zouhar, J., Agee, A.E., Carter, D.G., Chary, S.N., and Raikhel, N.V. (2003) Tools to study plant organelle iogenesis. Point mutation lines with disrupted vacuoles and high-speed confocal screening of green fluorescent protein-tagged organelles. *Plant Physiol.* **133**, 1673–1676.
15. Ahn, Y.-H. and Chang, Y.-T. (2007) Tagged small molecule library approach for facilitated chemical genetics. *Acc. Chem. Res.* **40**, 1025–1033.
16. Khersonsky, S.M., Jung, D.W., Kang, T.W., Walsh, D.P., Moon, H.S., Jo, H., Jacobson, E.M., Shetty, V., Neubert, T.A., and Chang, Y.T. (2003) Facilitated forward chemical genetics using a tagged triazine library and zebrafish embryo screening. *J. Am. Chem. Soc.* **125**, 11804–11805.
17. Min, J., Kyung Kim, Y., Cipriani, P.G., Kang, M., Khersonsky, S.M., Walsh, D.P., Lee, J.-Y., Niessen, S., Yates, J.R., Gunsalus, K., Piano, F., and Chang, Y.-T. (2007) Forward chemical genetic approach identifies new role for GAPDH in insulin signaling. *Nat. Chem. Biol.* **3**, 55–59.
18. Girke, T., Cheng, L.-C., and Raikhel, N. (2005) ChemMine. A compound mining database for chemical genomics. *Plant Physiol.* **138**, 573–577.

Chapter 19

Comparison of Quantitative Metabolite Imaging Tools and Carbon-13 Techniques for Fluxomics

Totte Niittylae, Bhavna Chaudhuri, Uwe Sauer, and Wolf B. Frommer

Abstract

The recent development of analytic technologies allows fast analysis of metabolism in real time. Fluxomics aims to define the genes involved in regulation of flux through a metabolic or signaling pathway. Flux through a metabolic or signaling pathway is determined by the activity of its individual components; regulation can occur at many levels, including transcriptional, posttranslational, and allosteric levels. Currently two technologies are used to monitor fluxes. The first is pulse labeling of the organism with a tracer such as C13, followed by mass spectrometric analysis of the partitioning of label into different compounds. The second approach is based on the use of flux sensors, proteins that respond with a conformational change to ligand binding. Fluorescence resonance energy transfer (FRET) detects the conformational change and serves as a proxy for ligand concentration. Both methods provide high time resolution. In contrast to mass spectrometry assays, FRET nanosensors monitor only a single compound, but the advantage of FRET nanosensors is that they yield data with cellular and subcellular resolution.

Key words: Flux, FRET, nanosensor, carbon-13.

1. Introduction

Metabolic fluxes underlie all biological activity, ultimately manifesting phenotype and functioning of an organism. Metabolic flux is highly dynamic and is controlled through signaling networks to acclimate appropriate cellular responses to environmental challenges. Fluxomics aims at quantifying and modeling these fluxes in the entire metabolic network of an organism, a feat that has not yet been attained, as well as the factors affecting all fluxes. Flux is measured either by determining the flow of a label (typically a radiotracer) in metabolic networks or by measuring changes in

Dmitry A. Belostotsky (ed.), *Plant Systems Biology, vol. 553*

DOI 10.1007/978-1-60327-563-7_19 Springerprotocols.com

substrate and product concentrations. At present, none of the existing experimental techniques is capable of comprehensively resolving all of the metabolic fluxes of entire metabolic networks in any organism; however for apparent reasons, most progress has been made in single cell microbes. This review provides a comparison of quantitative metabolite imaging and carbon-13-based approaches for flux analysis. The particular focus is on the methods that were developed recently for metabolite imaging and how they can be used to measure the rate changes of metabolite concentrations with subcellular resolution, and how the information gained from the use of quantitative imaging can be applied to estimate net flux. For a more detailed review of FRET-based analysis of in vivo metabolite levels, cf. Okumoto et al. (1).

2. Isotope-Based Flux Analysis

Isotope tracers have proven very effective for determining pathway structure in the past (e.g., the dark reactions in photosynthesis (2, 3)) and are currently being applied to obtain comprehensive flux analysis with the aim of producing system-wide flux maps of metabolic networks. The increased sensitivity of mass spectrometry (MS) and nuclear magnetic resonance (NMR) techniques obtained over the past years, and the development of powerful tools for data analysis, begins to make system-wide flux analysis possible in microorganisms as well as plants. ^{13}C-based flux analysis has been pioneered in microorganisms such as bacteria and yeast (4–6) but is increasingly used in plants. For recent reviews of isotope flux measurements in plants confer the special issue on fluxomics in *Phytochemistry* (7) and recent reviews in other journals (8, 9).

Isotope flux measurements can be classified into two categories: steady-state analysis which measures the distribution of a label after the system has attained an isotopic and metabolic steady state, i.e., the point at which the labeling of each metabolite in a network is constant. This method is most powerful when applied in microbes, because they can easily be cultivated under such steady-state conditions. Steady-state labeling has been also been used to create flux maps of central carbon metabolism in plants (10, 11) and has helped, for example, to establish a previously unknown role for Rubisco as CO_2 scavenger during oil synthesis in *Brassica napus* seeds (12). The second isotope flux measurement approach is dynamic, using time-course analysis of label distribution to calculate flux. In plants, dynamic analysis has been mainly used to characterize secondary metabolite pathways.

Notable examples include the characterization of 38 fluxes involved in the production of benzenoid compounds in *Petunia* petals (13) and the regulation of phenylpropanoid biosynthesis in potato tubers (14).

The major advantage of ^{13}C flux measurements is that it allows the determination of net fluxes in a network and, in some cases, provides the individual forward and backward fluxes of bidirectional steps using the information embedded in the isotopomer distribution (15). For this purpose, isotope-based flux analysis requires mathematical models that represent the possible isotopic states of the metabolic network. The distribution of fluxes is then estimated as a best fit of intracellular fluxes to the actually measured isotope distributions and physiological fluxes in and out of the cell. The main challenge in flux analysis of plants (and other eukaryotes) using isotopes comes from the complexity of the metabolic networks arising from different cell types and the subcellular compartmentalization of metabolism.

Another challenge is that analysis of isotope experiments relies on the current structural understanding of the networks involved: in plants these are only known accurately for a few pathways in primary metabolism. Even for primary metabolism the subcellular compartmentalization of the pathways is not always clear and is still being revised as apparent from, for example, the recent discovery of a plastidic maltose transporter and maltose metabolizing cytosolic glucosyltransferase, both of which are essential for starch degradation in leaves (16–18). Another example of the limited understanding of metabolic compartmentalization in plants is the debate on sucrose transport in and out of vacuoles, which contributes to carbon storage in leaves, in stems of sugarcane, and in roots of sugar beet (19, 20). Only recently one of the sucrose transporters SUT4 was localized to the tonoplast membrane (21), although it remains unclear how exactly SUT4 contributes to vacuolar sucrose accumulation.

Our current understanding of the compartmental distribution of metabolites relies mostly on the destructive analysis of whole organs. Compartmentalization of metabolic reactions and metabolite flux within and between cells can only be understood if the cellular and subcellular flux of the metabolites can be established by non-destructive dynamic monitoring techniques. Therefore the application of methods for the non-destructive determination of metabolite fluxes in subcellular compartments and different cell types is of major importance. A comparison of extractable in vitro enzyme activities and steady-state in vivo fluxes in *B. napus* embryos showed no clear correlation between the two (22), emphasizing the necessity for developing non-invasive in vivo analysis techniques with cellular and subcellular resolution.

3. Imaging-Based Flux Analysis

As an alternative to the isotope-based flux methods, metabolite imaging-based flux analysis, which measures dynamic changes in metabolite concentration, provides both cellular and subcellular resolution. The development of Förster resonance energy transfer (FRET)–based nanosensors was the first step toward in vivo flux measurements (23). Genetically encoded FRET sensors enable both the analysis of steady-state concentration of metabolites and dynamic changes in response to perturbations in living tissue with high temporal and, most importantly, subcellular resolution. FRET sensors report conformational changes of proteins (recognition elements) as a change in the rate of energy transfer between two coupled fluorophores (reporter elements) (24). Thus when the recognition element changes conformation in response to analyte binding, a change in the FRET efficiency reports a change in analyte levels. Importantly, such FRET sensors report changes in steady-state levels over time, e.g., glucose nanosensors, after addition of glucose to a cell or an intact organ, provide information on the sum of the rate of influx and the rate of metabolism. The principle of inferring flux information from these metabolite nanosensors is thus based on the analysis of dynamic responses of metabolite concentrations when the composition of the external medium is manipulated. Apparently the sensor reports only a single metabolite or the change in any of the flux components that affect the steady state.

The concept for genetically encoded FRET sensors was originally developed 10 years ago for measuring calcium by Persechini's and Tsien' s groups (25, 26). In short, a calmodulin was fused between two fluorescent proteins (e.g., cyan and yellow variants of the green fluorescent protein, GFP). When the cyan FP in the fusion protein is excited with 435 nm light, a fraction of the energy will be transferred to the yellow FP provided the yellow FP is in Förster distance (50% energy transfer at ~5 nm distance). When Ca^{2+} binds to calmodulin, the domain undergoes a conformational change which results in a change in FRET and thus into a change in the ratio of emission of the two fluorescent proteins. Miyawaki et al. (25) used an additional actuator, a calmodulin-binding domain to increase the conformational rearrangement of the binding moiety. FRET nanosensors are essentially ratiometric dyes that provide for quantitative measurements and, since they are DNA encoded, analyses can be performed in any type of transiently or stably transformed cells. Moreover, the addition of targeting sequences allows targeting of the fusion proteins to specific cellular compartments. Subsequent imaging of compartment-specific fluxes then does not require high-resolution microscopy due to the specific localization of the sensors. Based on this concept, a variety of nanosensors have been developed for small molecules (phosphate, carbohydrates, and amino acids) using bacterial periplasmic binding proteins or transcriptional regulators as the backbone (27–35) (**Table 19.1**). All published FRET sensors developed by the Frommer lab can be obtained from Addgene (http://carnegiedpb.stanford.edu/research/frommer/nanosensors/index.html) for a nominal fee.

Table 19.1
FRET sensors for ion and metabolite analysis

Analyte	Recognition element	Sensor construct	Reference
Glucose/ galactose	Bacterial periplasmic binding protein	FLIPglu	(27, 41, 60)
Maltose	Bacterial periplasmic binding protein		(28, 43)
Sucrose	Bacterial periplasmic binding protein		(64)
Ribose	Bacterial periplasmic binding protein		(34)
Arabinose	Bacterial periplasmic binding protein		(43)
Glutamate	Bacterial periplasmic binding protein		(27, 35)
Tryptophan	Bacterial repressor protein		(33)
Arginine	Bacterial periplasmic binding protein		(65)
Calcium	Calmodulin, troponin C		(52, 62)
Phosphate	Bacterial periplasmic binding protein		(32)
Other FRET sensors	Various		For review, cf., e.g., (47, 49)

Therefore the combination of in vivo metabolite imaging techniques and mass spectrometry-based fluxomics is likely to be required to understand the dynamics of metabolic systems. For a comparison of the two approaches, cf. **Table 19.2**.

Table 19.2
Side-by-side comparison of ^{13}C- and nanosensor-based fluxome analyses

Application	^{13}C-fluxomics	Nanosensor-based fluxomics
Pathway coverage	High	Limited to single node
Sensitivity	> 0.1 mmol analyte per hour and g of cells (dry weight)	Depends on the K_d and dynamic range of the nanosensor: e.g., ultrahigh affinity sensors such as FLIPglu170nΔ13 can provide nM analyte per second per cell
Information content (cells, tissues, or cell populations)	Population average	Population average, single cells, or subpopulations of cells
Compartment specific	Currently not available	Targeting allows to specifically analyze cytosolic and organellar levels and flux across intracellular membranes
Temporal resolution	0.8 s intervals (39)	Full frame acquisition is possible in 200–1000 ms intervals (cf. (51, 60); small regions can be imaged at 10–30 Hz (53)
Suitability for dynamic analysis	Good	Excellent
Sensitivity to changes in other parameters	N/A	Sensor conformation may be affected by pH, ionic strength, posttranslational modification, binding to other proteins, or proteolysis (can be controlled for by using set of affinity mutants; effect of pH can be determined in vitro and corrections can be applied if in vivo pH is monitored)
Invasiveness	High	Relatively low; however the additional ligand buffer may affect the cell's physiology (can be evaluated by comparing wt and transformant using ^{13}C-fluxomics)
Suitability for high throughput	Hundreds of mutants (19)	Thousands of mutants, complete genomes possible

4. Comparison of FRET Sensor and Carbon-13-Based Fluxomics

When integrating ^{13}C-data, extracellular fluxes, and biosynthetic requirements with computer models, ^{13}C-based fluxomics can reach high pathway coverage for those parts of the metabolism where a particular tracer molecule is converted and suitable analytes are available to track the resulting isotope patterns of these conversions. The flux distribution is typically identified by iterative fitting of fluxes to the measured data, whereby the difference between observed and simulated isotope spectra is minimized (36). Essentially, this is a parameter fitting procedure where the relation between unknown fluxes and measured data is described by mathematical models of varying complexity. Most published data sets were obtained from (quasi) steady-state growth in glucose media, while other substrates remained largely unexplored although they are principally amenable to the current methods. Its strength, in particular with respect to FRET sensors, is the ability to resolve fluxes through several competing or diverging pathways such as in the ubiquitous central metabolism. If one accepts the limitation that cells are cultured in a stable steady state, appropriate isotope experiments typically resolve the distribution of flux between competing pathways with an accuracy of about 5% (5, 36). While standard ^{13}C-flux methods are based on relatively tedious experiments and data analysis, a simplified method based on a direct and local interpretation of selected labeling patterns – so-called flux ratio analysis – enables high-throughput monitoring of intracellular flux distributions (37, 38).

A major limitation of most current methods is that they rely on the detection of isotope patterns in amino acids bound in cell proteins, which requires that these amino acids be actually synthesized from a labeled source molecule, thus precluding the analysis of non-growing cells or cells cultivated in complex media. A second major disadvantage that also relates to pattern detection in proteinogenic amino acids, is the restriction of current ^{13}C-methods to steady-state conditions. Analyzing metabolism under biologically relevant dynamic conditions requires different methods. One of these is the detection of isotope patterns in free intracellular intermediates, where an isotopic steady state can be attained within minutes to a few hours, enabling dynamic analyses at this time scale. To achieve higher dynamic time resolution, alternative methods that measure during the period of isotopic instationarity are currently under development (39, 40). The down side is that these methods will be even more tedious than the above ^{13}C-flux methods. With the exception of certain in vivo NMR experiments with a relatively low resolution and sensitivity, essentially all ^{13}C-methods are destructive.

FRET sensors typically analyze a single metabolite at a time and they cannot detect flux changes unless there is a change in the concentration of the metabolite. Multiplexing is possible by either targeting sensors to different compartments and analyzing the cellular regions separately or using sensors with separated spectral properties. Even when there is an observable rate of concentration change of a metabolic intermediate the way in which the change relates to flux has to be studied on a case-by-case basis. The FRET change reflects the sum of total flux change, which consists of all possible components affecting the influx and efflux of the metabolite. In vivo glucose measurements in *Arabidopsis* roots using glucose FRET sensors illustrate the point (**Fig. 19.1**) and (41). The rate of glucose concentration change in the cytosol can be calculated from the slope of the FRET change. This rate reflects the influx (import/uptake and synthesis) and efflux (export, subcellular transport, and metabolism) of glucose in the cytosol of *Arabidopsis* root cells, provided the perfusion system is not limiting. Additional experiments are required to establish how much each of these components contributes to the measured rate. This typically involves manipulation of the system with genetic or chemical (specific inhibitors) tools and/or the use of isotopes.

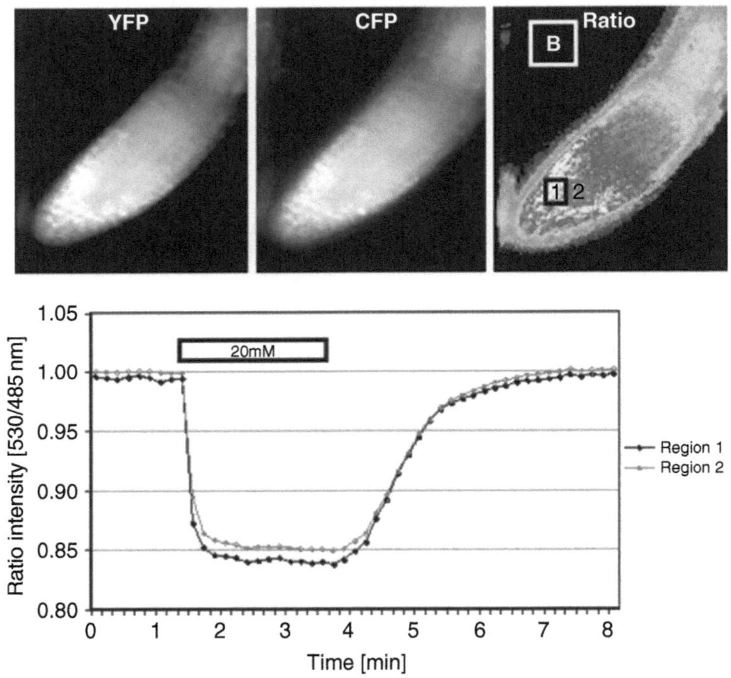

Fig. 19.1. Glucose-induced FRET changes in the cytosol of intact *Arabidopsis* roots. The FRET sensor FLIPglu-600μΔ13 with an affinity for glucose of 600 μM in stably transformed *rdr6-11 Arabidopsis* plants (41) responds to perfusion with 20 mM glucose. *Top panel*: Images of the root tip for the YFP and the CFP channels as well as the ratiometric image with pseudocolor converted to grayscale are shown (at time 0). The regions from which the quantitative data are calculated are shown on the ratiometric image as black (1) and gray (2) boxes. Data from the white box (B) was used for the background correction. Quantitative data were derived by pixel-by-pixel integration of the regions in the ratiometric image. *Bottom panel*: The graph shows the ratio of eYFP intensity divided by eCFP intensity (normalized to the staring ratio) for the two regions (gray trace corresponds to gray box (2) after background subtraction and normalization, black trace (1) corresponds to black box) at different time points (here 10 s intervals) measured over 8 min. The bars on top of the trace give the concentration and the duration of the glucose perfusion. The response is fully reversible. Note that accumulation and elimination phases show different rate constants. Also note the low noise as apparent from the smooth trace.

The advantage of FRET sensors is their applicability for in vivo determination of cellular, tissue-specific, and subcellular metabolite concentration changes (29, 30, 41), measurement of steady-state concentration of metabolites (35, 41), and screening of signaling networks affecting metabolite concentrations in vivo (Haerizadeh and Frommer, unpublished). Apparently, even imaging at low magnification can provide cellular resolution (41). Since the sensors are genetically encoded, they can be targeted to subcellular compartments as demonstrated for the glucose sensor, which by fusion to a nuclear targeting sequence was successfully employed to measure nuclear glucose flux (29), or by fusion to an ER targeting and retention sequence could be used to monitor glucose flux across the ER membrane (31). Exocytosis of glutamate was monitored by targeting and anchoring the glutamate sensor to the cell surface (35). Apparently, extracellular analysis can simply by performed by adding purified sensor to the cells or tissues of interest (42).

These attributes qualify FRET sensors uniquely for studies of how and when the concentration or net flux of a metabolite varies across an organ, tissue, or cell. Another great advantage of FRET sensor technology is their applicability to large-scale screens of chemicals or mutant collections. In the case of single cells, fluorescence microplate readers may be used instead of imaging to analyze FRET responses of a large number of samples in a short time (43).

The use of FRET sensors already provided new insights into metabolic processes. FRET sensors with different affinities for glucose were used to show that the cytsolic glucose concentration in soil-grown roots can drop to <100 nM in the absence of external glucose supply in *Arabidopsis* roots (41). This estimate is much lower than the previous estimation of cytosolic glucose concentrations in heterotrophic tissues (potato tuber) measured using non-aqueous fractionation (NAF) to provide subcellular resolution (44). Concentration estimations using disruptive extraction and analysis methods rely on estimations of cellular compartment volumes. Farré et al. estimated the volumes from electron microscopy pictures of cellular cross sections (44). The sensors were also used to carefully characterize the protonophore-insensitive accumulation of glucose and sucrose in root tips of *Arabidopsis* (45). FRET sensors measure steady-state levels and detect concentration change directly in vivo and are therefore superior tools for the analysis of factors affecting metabolite concentrations in the cell of interest. They may even allow more accurate estimation of subcellular compartment volumes when combined with NAF analysis of total metabolite amounts in subcellular compartments. FRET sensors also provide a tool to test for the potential metabolite oscillations, as used to analyze cytosolic calcium waves (46).

A detailed comparison of the specific advantages and drawbacks of FRET sensor and ^{13}C-flux technologies is presented in **Table 19.2.**

5. In Vivo FRET Imaging in Arabidopsis – A How to Guide

FRET can be measured either in a fluorimeter or by imaging. Many excellent overviews over the use of FRET in biology have been published (47–50). Quantitative analysis of FRET data derived from imaging approaches has been used most extensively to determine changes in calcium in neurobiology. Several excellent how-to-guides have been published (51–53). While written for applications in the animal field, the technical approach is highly similar for plants as are the challenges, e.g., how to carry out analyses in live organs. The reader is thus referred to these reviews for details in the methodology. FRET sensors for calcium and fluorescent indicators for pH have also been used by a small number of plant labs (46, 54–57). Thus here, mainly aspects relating to metabolite imaging will be covered.

5.1. Expression of FRET Sensor Constructs in Plants

Glucose FRET sensors have successfully been used to monitor glucose levels in intact roots and in leaf slices of *Arabidopsis* (41). Stable transgenic *Arabidopsis* lines for the FRET sensors of interest are created using standard transformation protocols. Most calcium imaging studies have been carried out in guard cells (46). Apparently, FRET sensors are subject to gene silencing in *Arabidopsis* (41). This has not precluded the analysis in guard cells since these cells, at least when mature, are protected from gene silencing (58). Thus to be able to monitor FRET sensors in other tissues, gene silencing has to be overcome. This can be achieved either by the use of gene silencing mutants (41) or by analyzing young seedlings at stages before silencing has reduced fluorescence below levels necessary for obtaining high-quality FRET images. Alternatively, it may be possible to use cell-specific or regulated promoters to circumvent gene silencing.

For all of the metabolite sensors developed so far a series of affinity mutants are available (e.g., for FLIPglu (41), FLIPE (35), and FLIPPi (32)). It is recommended to use several affinity mutants to exclude artifacts due to changes in other parameters that may either affect the fluorophores or the recognition element. If the FRET change I due to c change in analyte levels, the response curves should shift according to the affinity of the sensors used (cf. (41)). If affinity mutants of the sensor, which typically differ only in a single amino acid, show identical responses, additional controls such as analysis for pH shifts may be necessary. pH shifts can be monitored using fluorescent indicator proteins expressed in control plants (56).

FRET is analyzed by determining the relative fluorescence intensity of the two fluorophores, typically YFP and CFP. The fluorescence intensity is measured with either a fluorimeter or a fluorescence microscope.

5.2. Instrumentation for Imaging-Based FRET Metabolite Analysis in Plants

The FRET sensors for metabolites described here contain a recognition element fused to two spectral variants of GFP. FRET between CFP and YFP can be measured using a variety of methods such as fluorescence lifetime imaging (FLIM), fluorescence recovery after photobleaching (FRAP), anisotropy decay or simply by rationing the relative fluorescence intensities of FRET donor (CFP) and FRET acceptor (YFP). Due to the fixed molar ratio of the two fluorophores, the simplest method, i.e., ratiometric analysis of emission intensities, is sufficient for most applications. The signal-to-noise ratio of the described metabolite FRET sensors is sufficient to use "poor human's FRET", i.e., simple recording of the emission intensities at two wavelength. More sophisticated approaches may be recommended that correct for bleed-through (direct excitation of the acceptor at excitation wavelength) or for changes in sensor levels or proteolysis of the sensor (by normalization to acceptor amount obtained by recording acceptor emission at the acceptor's excitation wavelength) (59). It is important to note that a ratio change cannot necessarily be attributed to a change of FRET, e.g., during photobleaching or due to interference of other parameters; the two GFP variants may differ in their sensitivity to photobleaching or other parameter changes or changes in the focal plane may mimic a FRET change. Inspection of the raw data (individual fluorescence emission intensities and correction as described above) will help identify potential artifacts.

Due to the low intrinsic noise of metabolic signals such as glucose (**Fig. 19.1** and (60)), the relatively slow rate changes compared to calcium spikes together with the ability to express the sensors to high levels in stably transformed plants allow the use of simpler acquisition systems. Since the sensors can be targeted genetically to subcellular compartments, epifluorescence imaging is sufficient for most cases. Since spatial resolution is not relevant, essentially a single or few pixels per cell are sufficient, thus allowing pixel binning to enhance the signal-to-noise ratio. It is also possible to record FRET using a confocal microscope, e.g., to observe spatial differences inside a cell.

For ratiometric FRET analysis the following instruments are required: a microscope stand with fluorescence optics, a fluorescence excitation light source, appropriate filters, a filter switching device or image splitter and a digital camera for acquisition of emission, a perfusion system to be able to change the analyte levels in the perfusion medium, and software for driving the instruments. A complete and workable epifluorescence FRET imaging system suitable for metabolite imaging can be assembled for below $50,000. Apparently, if a suitable microscope and camera are available, a FRET imaging system can be assembled at minimal cost. Factors that determine the cost include quality of the stand, number of objectives, sensitivity of the camera, the use of free or commercial software, and the versatility of the devices such as fast multi-wavelength acquisition and computer-controlled perfusion. Notwithstanding, systems are available that enable spectral imaging to obtain full spectra of donor and acceptor as well as background fluorescence. Such data can then be used for spectral unmixing to obtain reliable FRET data even in cells with significant fluorescence background (61).

Epifluorescence microscopes are well suited for whole tissue analysis as well as single cell analysis. Apparently, fluorescence intensity drops when tissues deeper inside an organ are analyzed. However, when analyses are performed in roots expressing the sensor in all cell types as described by Deuschle et al. (41), it is not possible to determine cellular responses reliably. The use of specific promoters active only in certain cell layers provides an alternative to the use of confocal microscopes in order to obtain tissue layer or subcellular resolution. Confocal microscopes have to be used with caution as changes in focal plane due to swelling or shrinking of the cells as a consequence of changes in the composition of the perfusion medium may lead to artifacts. Tracking of the z-stacks may be required to verify that the same z-section is analyzed. The apparent advantage of confocal microscopes is that they reduce the background fluorescence originating from tissues outside the region of interest.

Several parameters affect the signal-to-noise ratio and thus the quality of the data as well as the detection range, e.g., fluorescence intensity over background and signal change of the sensor. Therefore a lot of effort has been invested in improvements in sensor responses (27, 60, 62). On the other hand, typically, the higher the emission intensity, the higher the signal to noise. Due to the occurrence of photobleaching, apparently too high excitation light may be damaging. At low magnification, the amount of excitation light from a normal fluorescence light source is limiting, thus lower angle/high-magnification lenses, high-intensity light sources (Hg band at 435 nm of mercury lamps, high-power Xenon lamps, high-power LED lights or lasers), high-transmission filters (such as high-throughput modified magnetron sputter-coated filter sets), and high-sensitivity cameras with on-chip multiplication gain have proven advantageous. If high excitation leads to photobleaching, one may reduce excitation intensity by neutral density filters, reducing the frequency of acquisition while increasing integration times for acquisition or camera gain.

To monitor responses of FRET glucose sensors (containing eCFP as donor and Venus or eYFP as acceptor) we mount roots of intact seedlings in a perfusion chamber (e.g., P1 Warner Instruments, USA) and on a stage adapter (41, 60). A wide spectrum of open and closed perfusion chambers suitable for different applications is available from different companies. Ratio imaging is performed on an inverted fluorescence microscope (DM IRE2, Leica) with a QuantEM digital camera (Roper) and a 20 × oil objective (HC PL APO 20 × /0.7IMM CORR, Leica, Germany). Essentially, any high-quality inverted microscope can be used for this purpose. Dual emission intensities are simultaneously recorded using a DualView with a dual CFP/YFP-ET filter set (high-transmission modified magnetron sputter-coated filter sets ET470/24m (470 indicates emission wavelength, /24 indicates bandwidth); ET535/3, Chroma, USA) and Slidebook software (Intelligent Imaging Innovations, Inc., USA). The Dualview (or similar image splitter from other companies) enables simultaneous recording of both emission wavelength without mechanical filter switching. For most metabolic imaging studies, filter wheels that automatically switch between the two emission wavelengths are equally suitable. Software for FRET image acquisition is available from a variety of commercial vendors, as scripts from individual labs, or can be implemented using the free software package ImageJ (rsb.info.nih.gov/ij/). The use of EM-gain cameras may be advantageous when analyzing low-fluorescence samples or when using low magnification, but in general is not crucial. Excitation (filter ET430/24x, Chroma) is provided by a Lambda DG4 light source (Sutter Instruments; http://www.sutter.com), which enables rapid switching between several excitation wavelengths, a feature used when normalization to YFP emission is intended. Simpler light sources are available from a variety of vendors. Images are acquired within the linear detection range of the camera and depend on the expression level. Exposure times used for measuring glucose flux vary typically between 300 and 600 ms, with software binning 2 and at an EM gain of 300. Typical values for acquisition with anon-EM gain camera such as the Coolsnap HQ (Roper) have been described (41). Fluorescence intensities for eCFP and eYFP are typically in the range of 2000–8000 and 6000–14,000, respectively. YFP, CFP, and ratio images of an *Arabidopsis* root tip are shown in **Fig. 19.1.** The software allows to select regions for analysis that can be freely chosen, e.g., to determine the YFP/CFP ratio of individual cells or of groups of cells (grey and black squares). Regions outside the tissue are analyzed for background subtraction (large white square). Image stacks derived from time laps analysis are used to obtained traces of the ratio over time (lower panel **Fig. 19.1.**) Typically the software provides an option for real-time monitoring of images for the individual channels (**Fig. 19.1**) and traces of the intensities for each fluorophore as well as traces of the ratio for the individual regions that were selected. Acquired image stacks can be analyzed by selecting different regions and the quantitative data can be transferred to data analysis programs for more detailed analysis (e.g., ASCI or spreadsheet export function). Some software also provide options to implement corrections, e.g., bleed-through correction obtained from cells expressing CFP and YFP alone as well as automatic background subtraction or normalization to YFP excitation/emission.

5.3. Perfusion Chamber

One of the difficulties of observing live organisms under the microscope, especially in the context of quantitative imaging, is the necessity to exclude movement under perfusion and during time lapse, while ensuring free exchange of the perfusion medium. To prevent movement, roots were mounted on coverslips using medical adhesive (stock no. 7730, Hollister). Alternatively, tracking software may be used to register images in stacks (e.g., stackreg in ImageJ).

Control of perfusion buffer composition, temperature, flow rate, and chamber volume is of paramount importance to ensure reproducible experiments. If rates will be recorded rather than just steady state, it is also important to be able to change the perfusion media surrounding the specimen at velocities that are not limiting. Minimizing the chamber volume and efficient peristaltic pump or pressurized gravity-operated systems ensures that FRET response is not limited by substrate supply. Precise event marking and knowledge of the dead volume of the perfusion system (time until new buffer reaches cells) are important for correlating the response to the change in perfusion. Computer control over the perfusion system and TTL-linked acquisition to the valves of the perfusion system increase the data quality. Root perfusions as shown in **Fig. 19.1** are performed with full nutrient medium containing typical macro- and micronutrients buffered with 20 mM MES to pH 5.8 at 3 ml/min containing the molecule of interest. Apparently accessibility of the perfusion medium to the tissue is essential. Therefore roots are apparently ideal objects. For analysis of other organs such as hypocotyls, leaf, or stem, access to the perfusion medium needs to be ensured, e.g., by removing the cuticle, by using cuticle mutants, or by using organ slices (41).

5.4. Analysis of FRET Data

Baseline shifts of the FRET response can be corrected using second- or third-order polynomial fits of the ratio measured in the absence of treatment. The obtained function describes the baseline aberration (photobleaching) as a function of time during perfusion. To correct for this effect, the difference between the ratio at the beginning of the experiment r(0) and the baseline aberration f(t) is calculated at each time point of the measurement and added to the value of the measured ratio at the respective time point r(t): $r_{corr}(t) = r(t) + [r(0) - f(t)]$ (63).

Example or flux calculation from FRET slope data: the cytosolic glucose concentration can be calculated using the equation: $[gluc]_{cytosol} = K_d \times (r - 1)/(R_{max} - r)$. R_{max} is the maximum Δratio, which can be determined by measurement of the ratio at 95% saturation, K_d is the in vitro glucose binding affinity of the sensor, and r is the Δratio at each glucose concentration. The in vivo apparent $K_{0.5}$ of each nanosensor can be determined by fitting to the Michaelis–Menten equation; $r = [gluc] \times R_{max}/(K_{0.5} + [gluc])$; [gluc] is extracellular glucose concentration; and r is the initial ratio change rate after glucose loading (Δratio/s). This calculation relies on the assumption that the sensor has the same K_d in vivo as in vitro. To determine the initial flux rate in vivo, the initial accumulation rate is calculated by using time-ratio plot 2–20 s after glucose loading (provided sensor dynamics and perfusion are not limiting.

6. ^{13}C-fluxomics – How-to-Guides

Recent how-to-guides to ^{13}C-fluxomics can be found for local flux ratio analysis by Nanchen et al. (4) and for ^{13}C-flux balancing by Ratcliffe and Shachar-Hill (8).

Acknowledgments

This work was supported by grants to WBF from NIH (NIDDK, 1RO1DK079109-01) and DOE (DE-FG02-04ER15542). TN was supported in part by an HFSP fellowship.

References

1. Okumoto, S., Takanaga, H., and Frommer, W.B. (2008) Tansley review: quantitative imaging for discovery and assembly of the metabo-regulome. *New Phytol.* **180**, 271–295.
2. Aronoff, S., Benson, A., Hassid, W.Z., and Calvin, M. (1947) Distribution of C^{14} in photosynthesizing barley seedlings. *Science* **105**, 664–665.
3. Benson, A. and Calvin, M. (1947) The dark reductions of photosynthesis. *Science* **105**, 648–649.
4. Nanchen, A., Fuhrer, T., and Sauer, U. (2007) Determination of metabolic flux ratios from ^{13}C-experiments and gas chromatography-mass spectrometry data: protocol and principles. *Methods Mol. Biol.* **358**, 177–197.
5. Sauer, U. (2006) Metabolic networks in motion: 13C-based flux analysis. *Mol. Syst. Biol.* **2**, 62.
6. Schuetz, R., Kuepfer, L., and Sauer, U. (2007) Systematic evaluation of objective functions for predicting intracellular fluxes in Escherichia coli. *Mol. Syst. Biol.* **3**, 119.
7. Kruger, N.J. and Ratcliffe, R.G. (2007) Dynamic metabolic networks: going with the flow. *Phytochemistry* **68**, 2136–2138.
8. Ratcliffe, R.G. and Shachar-Hill, Y. (2006) Measuring multiple fluxes through plant metabolic networks. *Plant J.* **45**, 490–511.
9. Wiechert, W., Schweissgut, O., Takanaga, H., and Frommer, W.B. (2007) Fluxomics: mass spectrometry versus quantitative imaging. *Curr. Opin. Plant Biol.* **10**, 323–330.
10. Schwender, J., Ohlrogge, J., and Shachar-Hill, Y. (2004) Understanding flux in plant metabolic networks. *Curr. Opin. Plant Biol.* **7**, 309–317.
11. Schwender, J., Ohlrogge, J.B., and Shachar-Hill, Y. (2003) A flux model of glycolysis and the oxidative pentosephosphate pathway in developing Brassica napus embryos. *J. Biol. Chem.* **278**, 29442–29453.
12. Schwender, J., Goffman, F., Ohlrogge, J.B., and Shachar-Hill, Y. (2004) Rubisco without the Calvin cycle improves the carbon efficiency of developing green seeds. *Nature* **432**, 779–782.
13. Boatright, J., Negre, F., Chen, X., Kish, C.M., Wood, B., Peel, G., Orlova, I., Gang, D., Rhodes, D., and Dudareva, N. (2004) Understanding in vivo benzenoid metabolism in petunia petal tissue. *Plant Physiol.* **135**, 1993–2011.

14. Matsuda, F., Morino, K., Ano, R., Kuzawa, M., Wakasa, K., and Miyagawa, H. (2005) Metabolic flux analysis of the phenylpropanoid pathway in elicitor-treated potato tuber tissue. *Plant Cell Physiol.* **46**, 454–466.
15. Wiechert, W. and de Graaf, A.A. (1996) In vivo stationary flux analysis by 13C labeling experiments. *Adv. Biochem. Eng. Biotechnol.* **54**, 109–154.
16. Chia, T., Thorneycroft, D., Chapple, A., Messerli, G., Chen, J., Zeeman, S. C., Smith, S.M., and Smith, A.M. (2004) A cytosolic glucosyltransferase is required for conversion of starch to sucrose in Arabidopsis leaves at night. *Plant J.* **37**, 853–863.
17. Niittyla, T., Messerli, G., Trevisan, M., Chen, J., Smith, A.M., and Zeeman, S.C. (2004) A previously unknown maltose transporter essential for starch degradation in leaves. *Science* **303**, 87–89.
18. Zeeman, S.C., Smith, S.M., and Smith, A.M. (2007) The diurnal metabolism of leaf starch. *Biochem. J.* **401**, 13–28.
19. Neuhaus, H.E. (2007) Transport of primary metabolites across the plant vacuolar membrane. *FEBS Lett.* **581**, 2223–2226.
20. Uys, L., Botha, F.C., Hofmeyr, J.H., and Rohwer, J.M. (2007) Kinetic model of sucrose accumulation in maturing sugarcane culm tissue. *Phytochemistry* **68**, 2375–2392.
21. Endler, A., Meyer, S., Schelbert, S., Schneider, T., Weschke, W., Peters, S.W., Keller, F., Baginsky, S., Martinoia, E., and Schmidt, U.G. (2006) Identification of a vacuolar sucrose transporter in barley and Arabidopsis mesophyll cells by a tonoplast proteomic approach. *Plant Physiol.* **141**, 196–207.
22. Junker, B.H., Lonien, J., Heady, L.E., Rogers, A., and Schwender, J. (2007) Parallel determination of enzyme activities and in vivo fluxes in Brassica napus embryos grown on organic or inorganic nitrogen source. *Phytochemistry* **68**, 2232–2242.
23. Förster, T. (1948) Zwischenmolekulare Energiewanderung und Fluoreszenz. *Ann. Physik* **6**, 55.
24. Looger, L.L., Lalonde, S., and Frommer, W.B. (2005) Genetically encoded FRET sensors for visualizing metabolites with subcellular resolution in living cells. *Plant Physiol.* **138**, 555–557.
25. Miyawaki, A., Llopis, J., Heim, R., McCaffery, J.M., Adams, J.A., Ikura, M., and Tsien, R.Y. (1997) Fluorescent indicators for Ca2+ based on green fluorescent proteins and calmodulin. *Nature* **388**, 882–887.
26. Romoser, V.A., Hinkle, P.M., and Persechini, A. (1997) Detection in living cells of Ca2+-dependent changes in the fluorescence emission of an indicator composed of two green fluorescent protein variants linked by a calmodulin-binding sequence. A new class of fluorescent indicators. *J. Biol. Chem.* **272**, 13270–13274.
27. Deuschle, K., Okumoto, S., Fehr, M., Looger, L.L., Kozhukh, L., and Frommer, W.B. (2005) Construction and optimization of a family of genetically encoded metabolite sensors by semirational protein engineering. *Protein Sci.* **14**, 2304–2314.
28. Fehr, M., Frommer, W.B., and Lalonde, S. (2002) Visualization of maltose uptake in living yeast cells by fluorescent nanosensors. *Proc. Natl. Acad. Sci. USA* **99**, 9846–9851.
29. Fehr, M., Lalonde, S., Ehrhardt, D.W., and Frommer, W.B. (2004) Live imaging of glucose homeostasis in nuclei of COS-7 cells. *J. Fluoresc.* **14**, 603–609.
30. Fehr, M., Lalonde, S., Lager, I., Wolff, M.W., and Frommer, W.B. (2003) In vivo imaging of the dynamics of glucose uptake in the cytosol of COS-7 cells by fluorescent nanosensors. *J. Biol. Chem.* **278**, 19127–19133.
31. Fehr, M., Takanaga, H., Ehrhardt, D.W., and Frommer, W.B. (2005) Evidence for high-capacity bidirectional glucose transport across the endoplasmic reticulum membrane by genetically encoded fluorescence resonance energy transfer nanosensors. *Mol. Cell Biol.* **25**, 11102–11112.
32. Gu, H., Lalonde, S., Okumoto, S., Looger, L.L., Scharff-Poulsen, A.M., Grossman, A.R., Kossmann, J., Jakobsen, I., and Frommer, W.B. (2006) A novel analytical method for in vivo phosphate tracking. *FEBS Lett.* **580**, 5885–5893.
33. Kaper, T., Looger, L.L., Takanaga, H., Platten, M., Steinman, L., and Frommer, W.B. (2007) Nanosensor detection of an immunoregulatory tryptophan influx/kynurenine efflux cycle. *PLoS Biol.* **5**, e257.
34. Lager, I., Fehr, M., Frommer, W.B., and Lalonde, S. (2003) Development of a fluorescent nanosensor for ribose. *FEBS Lett.* **553**, 85–89.
35. Okumoto, S., Looger, L.L., Micheva, K.D., Reimer, R.J., Smith, S.J., and Frommer, W.B. (2005) Detection of glutamate release from neurons by genetically encoded surface-displayed FRET nanosensors. *Proc. Natl. Acad. Sci. USA* **102**, 8740–8745.
36. Wiechert, W. (2001) ^{13}C metabolic flux analysis. *Metab. Eng.* **3**, 195–206.

37. Blank, L.M., Kuepfer, L., and Sauer, U. (2005) Large-scale ^{13}C-flux analysis reveals mechanistic principles of metabolic network robustness to null mutations in yeast. *Genome Biol.* **6**, R49.

38. Fischer, E. and Sauer, U. (2005) Large-scale in vivo flux analysis shows rigidity and suboptimal performance of *Bacillus subtilis* metabolism. *Nat. Genet.* **37**, 636–640.

39. Nöh, K., Gronke, K., Luo, B., Takors, R., Oldiges, M., and Wiechert, W. (2007) Metabolic flux analysis at ultra short time scale: isotopically non-stationary ^{13}C labeling experiments. *J. Biotechnol.* **129**, 249–267.

40. Nöh, K., Wahl, A., and Wiechert, W. (2006) Computational tools for isotopically instationary ^{13}C labeling experiments under metabolic steady state conditions. *Metab. Eng.* **8**, 554–577.

41. Deuschle, K., Chaudhuri, B., Okumoto, S., Lager, I., Lalonde, S., and Frommer, W.B. (2006) Rapid metabolism of glucose detected with FRET glucose nanosensors in epidermal cells and intact roots of Arabidopsis RNA-silencing mutants. *Plant Cell* **18**, 2314–2325.

42. Dulla, C., Tani, H., Okumoto, S., Frommer, W.B., Reimer, R.J., and Huguenard, J.R. (2008) Imaging of glutamate in brain slices using FRET sensors. *J. Neurosci. Methods* **168**, 306–319.

43. Kaper, T., Lager, I., Looger, L.L., Chermak, D., and Frommer, W.B. (2008) FRET sensors for quantitative monitoring of pentose and disaccharide accumulation in bacteria. *Biotechnol. Biofuels.* **1**,11.

44. Farré, E.M., Tiessen, A., Roessner, U., Geigenberger, P., Trethewey, R.N., and Willmitzer, L. (2001) Analysis of the compartmentation of glycolytic intermediates, nucleotides, sugars, organic acids, amino acids, and sugar alcohols in potato tubers using a nonaqueous fractionation method. *Plant Physiol.* **127**, 685–700.

45. Chaudhuri, B., Hörmann, F., Lalonde, S., Brady, S.D.O., Benfey, P., and Frommer, W.B. (2008) Protonophore- and pH-insensitive glucose and sucrose accumulation detected by FRET nanosensors in Arabidopsis root tips *Plant J.* **56**, 948–962.

46. Allen, G.J., Chu, S.P., Harrington, C.L., Schumacher, K., Hoffmann, T., Tang, Y.Y., Grill, E., and Schroeder, J.I. (2001) A defined range of guard cell calcium oscillation parameters encodes stomatal movements. *Nature* **411**, 1053–1057.

47. Lalonde, S., Ehrhardt, D.W., and Frommer, W.B. (2005) Shining light on signaling and metabolic networks by genetically encoded biosensors. *Curr. Opin. Plant Biol.* **8**, 574–581.

48. Miyawaki, A. (2003) Visualization of the spatial and temporal dynamics of intracellular signaling. *Dev. Cell* **4**, 295–305.

49. Tsien, R.Y. (2006) Breeding and building molecules to spy on cells and tumors. *Keio J. Med.* **55**, 127–140.

50. Vogel, S.S., Thaler, C., and Koushik, S.V. (2006) Fanciful FRET. *Sci.* STKE 2006, re2.

51. Fiala, A. and Spall, T. (2003) In vivo calcium imaging of brain activity in Drosophila by transgenic cameleon expression. *Sci.* STKE 2003, PL6.

52. Palmer, A.E. and Tsien, R.Y. (2006) Measuring calcium signaling using genetically targetable fluorescent indicators. *Nat. Protoc.* **1**, 1057–1065.

53. Roe, M.W., Fiekers, J.F., Philipson, L.H., and Bindokas, V.P. (2006) Visualizing calcium signaling in cells by digitized widefield and confocal fluorescent microscopy. *Methods Mol. Biol.* **319**, 37–66.

54. Iwano, M., Shiba, H., Miwa, T., Che, F.S., Takayama, S., Nagai, T., Miyawaki, A., and Isogai, A. (2004) Ca2+ dynamics in a pollen grain and papilla cell during pollination of Arabidopsis. *Plant Physiol.* **136**, 3562–3571.

55. Monshausen, G.B., Bibikova, T.N., Messerli, M.A., Shi, C., and Gilroy, S. (2007) Oscillations in extracellular pH and reactive oxygen species modulate tip growth of Arabidopsis root hairs. *Proc. Natl. Acad. Sci. USA* **104**, 20996–21001.

56. Schulte, A., Lorenzen, I., Bottcher, M., and Plieth, C. (2006) A novel fluorescent pH probe for expression in plants. *Plant Methods* **2**, 7.

57. Young, J.J., Mehta, S., Israelsson, M., Godoski, J., Grill, E., and Schroeder, J.I. (2006) CO(2) signaling in guard cells: calcium sensitivity response modulation, a Ca(2+)-independent phase, and CO(2) insensitivity of the gca2 mutant. *Proc. Natl. Acad. Sci. USA* **103**, 7506–7511.

58. Oparka, K.J. and Roberts, A.G. (2001) Plasmodesmata. A not so open-and-shut case. *Plant Physiol.* **125**, 123–126.

59. Zal, T. and Gascoigne, N.R. (2004) Photobleaching-corrected FRET efficiency imaging of live cells. *Biophys. J.* **86**, 3923–3939.

60. Takanaga, H., Chaudhuri, B., and Frommer, W.B. (2008) GLUT1 and GLUT9 as

major contributors to glucose influx in HepG2 cells identified by a high sensitivity intramolecular FRET glucose sensor. *Biochim. Biophys. Acta.* 1778, 1091–1099.

61. Zimmermann, T., Rietdorf, J., Girod, A., Georget, V., and Pepperkok, R. (2002) Spectral imaging and linear un-mixing enables improved FRET efficiency with a novel GFP2-YFP FRET pair. *FEBS Lett.* **531**, 245–249.
62. Garaschuk, O., Griesbeck, O., and Konnerth, A. (2007) Troponin C-based biosensors: a new family of genetically encoded indicators for in vivo calcium imaging in the nervous system. *Cell Calcium* **42**, 351–361.
63. Takanaga, H., Chaudhuri, B., and Frommer, W.B. (2008) GLUT1 and GLUT9 as major contributors to glucose influx in HepG2 cells identified by a high sensitivity intramolecular FRET glucose sensor. *Biochim. Biophys. Acta* **1778**, 1091–1099.
64. Lager, I., Looger, L.L., Hilpert, M., Lalonde, S., and Frommer, W.B. (2006) Conversion of a putative Agrobacterium sugar-binding protein into a FRET sensor with high selectivity for sucrose. *J. Biol. Chem.* **281**, 30875–30883.
65. Bogner, M. and Ludewig, U. (2007) Visualization of arginine influx into plant cells using a specific FRET-sensor. *J. Fluoresc.* **17**, 350–360.

Chapter 20

Democratization and Integration of Genomic Profiling Tools

Michael R. Sussman, Edward L. Huttlin, and Dana J. Wohlbach

Abstract

Systems biology is a comprehensive means of creating a complete understanding of how all components of an organism work together to maintain and procreate life. By quantitatively profiling one at a time, the effect of thousands and millions of genetic and environmental perturbations on the cell, systems biologists are attempting to recreate and measure the effect of the many different states that have been explored during the 3 billion years in which life has evolved. A key aspect of this work is the development of innovative new approaches to quantify changes in the transcriptome, proteome, and metabolome. In this chapter we provide a review and evaluation of several genomic profiling techniques used in plant systems biology as well as make recommendations for future progress in their use and integration.

Key words: Transcriptomics, proteomics, metabolomics.

1. Introduction

Integration of data derived from transcriptome, proteome, and metabolome studies is one of the critical objectives of systems biology. However, before this goal can be achieved, the methods of data acquisition and analysis for each of these individual techniques must be optimized and standardized. Whereas transcriptome profiling via microarray analysis is fairly well established, in terms of both experimental and statistical methods, comparative advancement in proteome and metabolome studies is still ongoing. One of the key considerations in developing systems biology approaches is optimizing the input of money and time with the output of usable data. Typically, technologies for profiling RNA, proteins, and metabolites start out as specialized and expensive tools that are only available to large labs with ample

Dmitry A. Belostotsky (ed.), *Plant Systems Biology, vol. 553*

DOI 10.1007/978-1-60327-563-7_20 Springerprotocols.com

resources. This is a regrettable situation, and recent advances in several fields offer the promise of "democratizing" some of these technologies. Once these techniques become easy, affordable, and fast, the availability to all researchers regardless of funding level can skyrocket, and the data stream increases. Finally, as more systems-wide data are generated, it becomes necessary to carefully consider how these data can be integrated into the other "phenotyping" tools that a typical systems biologist would use. In this chapter, we describe some of the methods emerging in the fields of transcriptomics, proteomics, and metabolomics and then speculate on ways of integrating data from these types of studies.

2. The *Arabidopsis* Transcriptome and the AtMegaCluster

Transcriptomics, the study of the expression levels of all the RNAs in a cell, is probably the most ubiquitous of all the available systems biology approaches. Although we will not devote space in this chapter to detailing the methods involved in a microarray study, there are several excellent reviews available on the subject (e.g., 1, 2), as well as other chapters in this book. The community of *Arabidopsis* researchers benefits from a well-curated set of gene expression data in the AtGenExpress microarray database (3–6). This database is a repository for data derived from more than 1,300 microarrays representing an exhaustive variety of experimental conditions, including environmental perturbations, pathogen interactions, hormone and chemical treatment, as well as different genetic modifications, ecotypes, tissue types, and *Arabidopsis* developmental stages. Additionally, several different tools for easy visualization and data analysis are available, such as the Genevestigator (7–9), which facilitates co-expression analysis. These tools, and others like it, employ meta-analysis techniques to summarize gene expression information, providing the researcher with different methods of interpreting and analyzing data.

Wohlbach et al. (10) recently demonstrated the utility of such an approach to derive biologically relevant clusters of genes out of a hierarchical clustering by creating what they termed the AtMegaCluster (**Fig. 20.1**). The AtMegaCluster combines over 1,700 publicly available microarray experiments into a large database to facilitate gene co-expression analysis. Because these gene expression experiments had been performed at multiple different labs, Wohlbach et al. (10) obtained raw data and used the robust multi-array average (RMA) method (11–14), which corrects arrays for background, normalizes

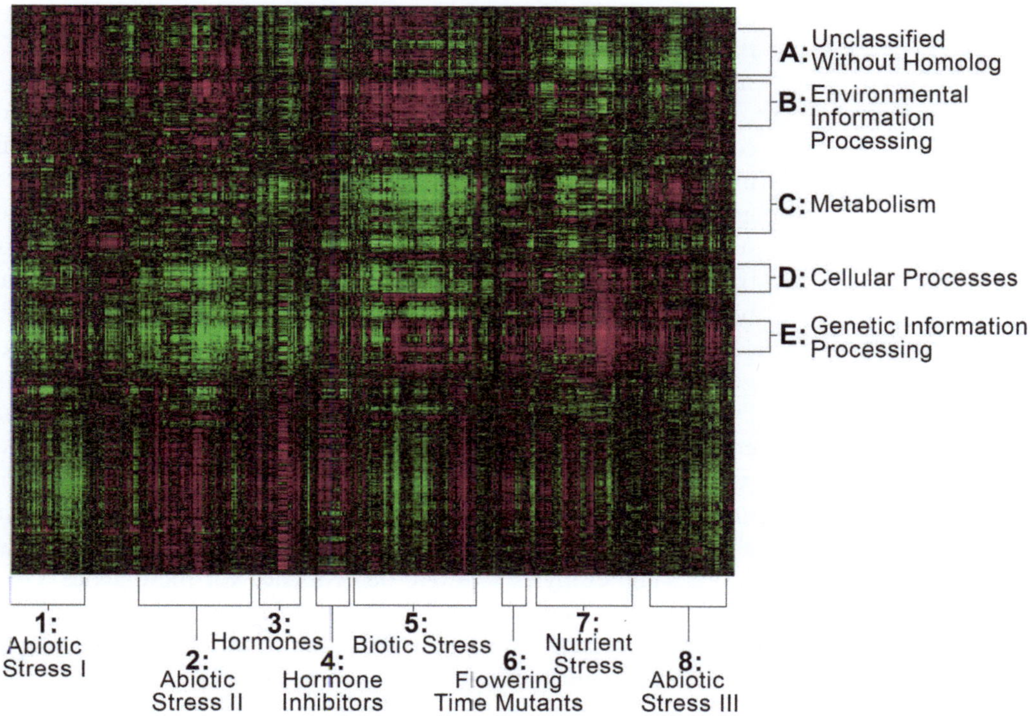

Fig. 20.1. The AtMegaCluster displays hierarchical clustering of *Arabidopsis thaliana* microarray experiments and genes. Experiments, represented on the horizontal axis, were grouped into eight clusters according to the fold change values of genes and have been named according to the classification of the majority of experiments in that cluster. Genes, represented on the vertical axis, were grouped into five clusters according to their fold change values and have been named according to the functional category of the majority of genes in that cluster. Induced fold changes are in magenta; repressed fold changes are in green. From **(10)**.

arrays based on the normal distribution, and uses a linear model to estimate log scale expression values, to preprocess all the microarray data in one set. The necessity of this kind of normalization illustrates one of the primary difficulties manifest in systems biology approaches: the difficulty of combining data obtained from different labs.

A useful outcome of this exercise was the observation that *AtHK1*, the gene encoding a plasma membrane histidine kinase that appears to act as a major osmosensor, is co-transcriptionally regulated together with the genes encoding many *Arabidopsis* response regulators (ARRs). The ARRs represent the third protein in the two-component signaling pathway that connects the AtHK1 sensory protein at the plasma membrane with a gene expression response in the nucleus. The similarity in expression profiles between *AtHK1* and the ARRs was not obvious in any one experiment, but could ONLY be revealed via the AtMegaCluster analysis. Also interesting was the observation that co-transcriptionally regulated with *AtHK1* is a gene encoding a protein with no sequence homology to any known protein, potentially representing a previously unknown member of the AtHK1 signaling pathway.

As illustrated by the AtMegaCluster, an obvious application of co-expression analysis is to identify novel members of known signaling pathways. Indeed, as over half of the genes in the *Arabidopsis* genome remain unclassified, a major goal of functional genomics studies is to assign putative functional classification to genes on the basis of sequence or expression similarities. Co-expression studies can be valuable when a gene of unknown function clusters next to a gene of known function, because genes in biological pathways tend to group together. For example, the AtMegaCluster was able to identify five distinct clusters of genes with distinct patterns of functional enrichment (**Fig. 20.1**). Gene cluster A contained 2,039 genes, of which approximately 55% were unclassified and another 15% were unclassified with no known homolog in *Arabidopsis*. All other clusters also contained approximately 50% unclassified genes; however cluster A was significantly (p = 4.64e-08) enriched for these genes. Gene cluster B contained 2,023 genes significantly (p= 6.07e-26) enriched for environmental information processing functions, such as signal transduction and ligand–receptor interaction. Gene cluster C contained 2,370 genes with an enrichment for metabolism function, including energy metabolism (p = 3.37e-14), lipid metabolism (p= 1.83e-07), and amino acid metabolism (p = 2.06e-08). Gene cluster D contained 1,133 genes with significant (p = 2.62e-13) enrichment for cellular processes such as cell communication, cell growth, and cell death. Finally, gene cluster E contained 1,357 genes functionally enriched for genetic information processing (p= 4.00e-191), including transcription, translation, and post-translational processes such as protein folding, sorting and degradation, as well as nucleotide metabolism (p= 8.53e-23).

Interestingly, the gene cluster A of the AtMegaCluster revealed that for a large number of *Arabidopsis* genes with no known homolog, co-expression analysis fails to group genes with unknown function near genes with known function. Many of these unknown genes might comprise undiscovered functional gene families whose expression patterns are unique. Therefore, integrating data from additional systems biology techniques, including proteome and metabolome profiling, may be necessary to elucidate functions of these currently unknown genes.

3. Proteome Profiling

While the central goal of transcriptomics is to monitor expression of each gene in the genome, proteomics is concerned with monitoring changes among proteins in a biological system. Two primary analytical challenges arise in proteomics experiments: identification of proteins in complex mixtures and quantitative comparison of each protein's abundance under different biological conditions. Each of these challenges is addressed by mass spectrometry in the "shotgun" proteomics strategy that has become the standard for the field. Though a complete introduction to proteomics is beyond the scope of this chapter, see Domon and Aebersold (15) for a recent review. Typically, proteins are digested using trypsin and individual peptides are identified based on their sequence derived from MS/MS fragmentation patterns (16, 17). Quantification of each peptide usually employs one of several isotopic labeling strategies including ICAT (18), ITRAQ (19), enzymatic labeling with ^{18}O (20), SILAC (21), and metabolic labeling with ^{15}N (22). For a complete discussion of plant quantitative proteomics, see a recent review by Thelen and Peck (23). Much like transcriptomics, the final output of a quantitative proteomics experiment is generally a list of observed proteins with ratios reflecting the relative abundance of each protein across each biological sample.

Though proteomics and transcriptomics share similar experimental goals, some basic differences in underlying technology influence their performance. One fundamental difference is coverage: while a typical microarray experiment will report expression levels for tens of thousands of genes, the most comprehensive quantitative proteomics experiments to date have been limited to around 5,000 proteins (24). Though higher numbers of protein identifications have been achieved, with over 13,000 unique proteins identified across multiple *Arabidopsis* tissues in one report (25), this experiment was not quantitative in nature and required an extremely large number of LC-MS analyses. Though this remarkable survey is a landmark for plant proteomics, such performance cannot presently be expected on a routine basis. Typical proteomics analyses tend to favor highly abundant proteins such as metabolic enzymes and heat shock proteins, while low-abundance proteins such as transcription factors are less frequently detected. While some of this difference can be attributed to the fact that proteomics is a less mature field whose technology is still rapidly evolving, several fundamental aspects are responsible as well. First, whereas DNA microarrays allow the simultaneous measurement of thousands of genes, proteomics measurements are inherently serial in nature: each peptide from each protein must be analyzed individually. Second, while PCR can be used to amplify small nucleic acid samples prior to analysis, no analogous technique is available for proteins. This makes low-abundance species more difficult to detect. Third, the chemical diversity of proteins is greater than the chemical diversity of mRNA molecules, complicating development of comprehensive isolation and detection methods. Given the differences in performance for each of these techniques, the challenge is to employ both proteomic and transcriptomic technologies in a way that capitalizes on the relative strengths of each to attain the most complete description of the biological system possible.

Perhaps the most obvious experimental approach for proteomics is to survey changes in protein abundance in an untargeted manner analogous to a typical microarray experiment. When microarray and proteomics results are compared under the same conditions, a moderate correlation is observed between the two (26). This is not surprising, as the abundance of a single form of any particular protein depends on its synthesis and degradation rates, as well as the rates of any post-translational modifications. While proteomics measurements reflect all of these biological processes, microarrays only indirectly measure the effects of mRNA synthesis. Though these kinds of proteomics experiments may provide interesting results, especially when applied to systems that are post-transcriptionally regulated, perhaps the greatest potential for novel biological insight comes from other types of proteomics studies. Several alternative experimental strategies for proteomics are described below. Though focused on different aspects of protein systems biology, each requires direct characterization of proteins to reveal properties that cannot generally be inferred from DNA microarrays or other kinds of large-scale biological data. Each of these approaches could provide insight into important areas of plant systems biology and likely offer the greatest potential return on one's investment of experimental resources.

One particularly useful application of proteomics technology is to characterize the make-up and dynamics of protein complexes. This may be done by using antibodies to immunoprecipitate the protein of interest along with other interacting proteins, either under native conditions or after chemical crosslinking (27). These samples are then digested and all proteins are identified via mass spectrometry. With appropriate controls to distinguish specific and nonspecific interactions, patterns of protein–protein interactions and protein complexes can be revealed (28). By combining this technique with a quantitation strategy such as SILAC or another metabolic labeling approach, one can directly compare the make-up of specific protein complexes under different biological conditions. This approach has been used to characterize protein–protein interactions within EGF signaling (29), as well as selected yeast and *Drosophila* complexes (30).

Recently protein degradation has been recognized as a critical aspect of many biological systems. Its importance in plant biology is especially clear, thanks to the recent demonstration that auxin exerts many of its effects through modulated degradation of specific proteins (3, 32). Protein turnover can be directly monitored on a proteomic scale by following the incorporation of stable isotopes such as ^{13}C and ^{15}N into each protein after their introduction into a living organism through their diet or media (33). Though this experimental approach has been most frequently applied in single-celled organisms such as yeast (34) and bacteria (35), it has also been applied in relatively complex multicellular organisms including the chicken (36). Though these kinds of proteomic turnover experiments have not yet been reported in plants, our laboratory and others have demonstrated metabolic labeling of cultured plant cells (37–41) or intact plants (42–45) for traditional quantitative proteomics experiments. By demonstrating that complete labeling of plants with ^{13}C, ^{15}N, or various isotopically labeled amino acids may be readily achieved, this work indicates that similar protein turnover studies are technically feasible in plants as well. There is a strong foundation for future studies that promise to shed light on important biological processes in plants that are otherwise invisible to systems biology.

Because peptides are typically sequenced through MS/MS fragmentation, one can often directly observe modified residues based on fragmentation patterns. Thus, modified forms of specific peptides can be identified and compared under diverse biological conditions. This technology potentially allows the direct observation of signaling cascades and other important aspects of biological regulation and cellular communication. Though some are more amenable to mass spectrometric analysis than others, a wide variety of post-translational modifications have been characterized in plants, including ubiquitination (46), glycosylation (47), and phosphorylation (48). Due to its biological importance, we now briefly consider phosphorylation as an illustration of some issues that can arise from proteomic study of post-translational modifications. In spite of unique analytical challenges, recent technological advances should enable researchers to study patterns of phosphorylation at a truly global scale.

One primary challenge for studying phosphorylation of proteins is low analyte abundance. Only a small fraction of most phosphorylation sites are modified at any particular time. Furthermore, addition of a phosphate generally inhibits ionization of phosphopeptides in the mass spectrometer, reducing their apparent signal. As a result, detection of phosphopeptides among other unmodified species is difficult. Fortunately, a number of strategies for selective isolation of phosphopeptides have been introduced. Though they are less useful to plant biologists, good antibodies are available that can be used to immunoprecipitate peptides containing phosphotyrosine residues (49). Peptides containing phosphoserine and phosphothreonine residues are isolated using their affinity for certain transition metals, via immobilized metal affinity chromatography (IMAC) (50–53) or titanium oxide chromatography (54). In some cases, anion or cation exchange chromatography is also used as a preliminary step to separate phosphopeptides from other tryptic peptides based on differences in their ionic charge in solution (52, 55, 56). By allowing the isolation of enriched phosphopeptide populations, these techniques greatly enhance our detection ability.

Another challenge for phosphoproteomics is that phosphopeptides are difficult to sequence via MS/MS using traditional methods. Typically peptides are fragmented via a process called collision-induced dissociation, in which each peptide collides with inert gaseous molecules, picking up energy with each collision until enough energy has been absorbed to break a chemical bond. While this approach allows fragmentation all along the peptide backbone for unmodified peptides and often gives rich MS/MS spectra, phosphopeptides almost exclusively fragment by losing phosphate. The resulting MS/MS spectra often contain too few fragments to allow identification of the phosphopeptide's sequence. However, two alternative methods have been introduced that are much more effective for sequencing phosphopeptides: electron capture dissociation (57, 58) and electron transfer dissociation (59). Both of these approaches induce fragmentation through donation of an electron to each peptide molecule, causing destabilization and rapid dissociation. Loss of phosphate is not favored via this mechanism, so MS/MS spectra from phosphopeptides are much richer and more easily sequenced. Now that both ETD and ECD are available on commercial instruments, excellent tools for phosphoproteomics are available for use by many researchers.

A final consideration for characterization of post-translational modifications is determining the relative abundances of modified and unmodified forms. One general approach that may be used for virtually any modification is to introduce isotopically labeled synthetic forms of both modified and unmodified peptides into the sample at known amounts and use the signal from these internal standards to determine the quantities of modified and unmodified peptides in the original sample. Typically the modified and unmodified peptides are quantified on triple-quadrupole mass spectrometers using a technique called multiple reaction monitoring (MRM) that provides especially accurate quantification. This technique was originally described in a proteomic context for peptides by Gerber and co-workers (60), though it is based on a standard method of small molecule quantitation in drug analysis that has been in use for decades. In the case of phosphopeptides, our laboratory has also introduced an alternative strategy for determining relative stoichiometries of phosphorylation sites using a combination of isotopic labeling and phosphatase treatment (61). It is clear that in plant systems biology, mass spectrometry-based proteomics is an emerging area that offers great promise in revealing important mechanisms that alter growth and development. Together with quantitative measurements of mRNA and metabolome abundance, we can achieve a more comprehensive picture of how the plant's suite of proteins respond to, and direct, changes in the transcriptome and metabolome.

4. "Sequencing" the Metabolome

Metabolomics is the study of small molecules in an organism and is usually operationally defined to include those species with molecular weights less than approximately 1,000 Da. Like transcriptomics and proteomics, the field of metabolomics is concerned with describing global changes within biological systems. Though relatively new compared to other approaches for systems biology, metabolomics is already an important tool for systems biologists. By focusing on small molecules and metabolites in the cell, one can observe a wide variety of essential cellular processes in a way that is largely orthogonal to proteomic and transcriptomic approaches. Thus metabolomics illuminates aspects of the biological system that would otherwise be invisible and in conjunction with other systems biology techniques provides a relatively complete characterization of biological systems.

Though monitoring the small molecule portion of biological systems is an essential aspect of systems biology, the complexity of the metabolome presents a variety of practical challenges that must be addressed experimentally. Perhaps the greatest of these challenges is the remarkable chemical diversity of the metabolome. Although they contain instructions for synthesis of a dizzying array of biological molecules, at a chemical level, all mRNA molecules are fairly similar: all are polymers of the same four basic chemical building blocks with similar overall structures. Though proteins are more variable, being made of 20 different amino acids with more widely varying chemical properties, they still share many fundamental chemical properties. This is especially true of peptides after tryptic digestion. As described above, these chemical similarities of peptides and mRNA molecules can be exploited to develop general sample preparation and analysis procedures that will be appropriate for most of the target biomolecules. In contrast, the metabolome includes all kinds of chemical compounds, including amino acids, sugars, lipids, and nucleotides. These compounds range from very hydrophobic to hydrophilic; some are volatile; some are chemically unstable. This range of chemical properties makes studying the metabolome in its entirety the most challenging of all phenotypes.

The second fundamental problem in metabolomics is determining the range of compounds that could exist in any particular organism. This is significantly different from the proteome and the transcriptome, the sizes of which are bounded to a first approximation based on gene sequences that are found in the genome (of course, neglecting post-translational modifications and splice variants, respectively). Current estimates indicate that metabolome size varies considerably from species to species, from yeast which may have around 600 metabolites (62) to humans, for whom estimates range from around 2,500 metabolites to over 25,000 (63). It has been estimated that 90,000–200,000 different metabolites may be found across the plant kingdom, though any individual species would only contain a small fraction of that number (64). Each metabolite must be individually identified de novo. Of course, since most of the drugs and nutrients on which human life depends are made in plants, there is a large motivation to tackle this challenging field.

One direct consequence of chemical diversity in the metabolome is that no single set of sample preparation conditions will allow simultaneous observation of all compounds. Different and often incompatible extraction and isolation procedures must be used for different classes of compounds, requiring multiple rounds of sample preparation on multiple replicate samples for comprehensive metabolome characterization. In practice, researchers often choose to focus their analysis on a particular subset of the metabolome with similar chemical properties, such as fatty acids or flavonoids. Multiple laboratories have evaluated the performance of a range of different extraction protocols for different classes of metabolites (65–67).

Another consequence of chemical diversity is that no single analytical platform is suitable for characterization of all metabolites. As a result, multiple methods must be used to identify and quantify different metabolic classes. Each method has its own advantages and disadvantages. NMR spectrometry is one mainstay of metabolomics research. Though limited by poor sensitivity, 1-D and 2-D NMR experiments provide tremendous structural information for identifying unknown metabolites. Typically up to a few dozen of the most abundant molecules in a complex mixture may be identified. Additionally, NMR can be used for quantification of selected metabolites. Usually this has been done via 1-D ^{1}H-NMR, though researchers at UW-Madison have recently demonstrated an alternative technique that allows rapid and accurate quantification of dozens of metabolites in complex mixtures via 2-D ^{1}H-^{13}C NMR experiments (68).

Another mainstay of metabolomics is mass spectrometry, coupled to various chromatographic techniques (69). While these techniques usually provide somewhat less structural information on each compound than NMR, they demonstrate vastly superior sensitivity. GC-MS has been used to identify and quantify unknown small molecules for decades. This approach is especially appropriate for volatile compounds and the combination of precise separation via GC and fragmentation by electron impact ionization provides considerable information for identifying unknowns. However, many biological compounds require derivatization prior to GC-MS analysis, making this approach less desirable for some classes of compounds. LC-MS techniques are also widely used for metabolomics analysis, employing multiple types of chromatography. Both hydrophilic interaction chromatography (HILIC) and standard reversed phase chromatography can be used. While reversed phase chromatography is good for many classes of molecules including amino acids, HILIC is especially useful for analysis of especially water-soluble analytes such as sugars (70). Furthermore, the mass spectrometers can be operated in either positive or negative ion mode to accommodate a wider variety of chemical compounds (71). LC-MS experiments generally provide intact mass measurements for each metabolite and possible MS/MS fragmentation spectra as well. Though resources for identification of metabolites via LC-MS are currently less developed, a variety of tools are available for quantification of unknown metabolites via LC-MS. One particularly useful tool is XCMS, a freely distributed program for aligning multiple LC-MS analyses and comparing abundances of selected compounds (72).

Given the range of technology that is available, several distinct experimental strategies guide metabolomics experiments. Since it is impractical to survey the metabolome in a comprehensive way, many researchers choose instead to take a more targeted approach, focusing specifically on selected classes of molecules such as fatty acids or flavonoids, or alternatively selecting specific important metabolites for targeted analyses. Though by definition these targeted approaches do not provide a comprehensive picture of all of metabolism, by carefully choosing particular metabolites based on a specific experimental question one may design an experiment that will reveal the status of key metabolic pathways. This can be one of the most efficient means of deriving biological information from metabolomics experiments.

In contrast to the targeted metabolomics approach, other researchers are attempting a comprehensive characterization of all metabolites in a number of organisms. Much like the various genome projects, whose goals were to identify the sequences of all genes making up each organism, the goal of these metabolomics projects is to use a variety of analytical techniques to create a database of metabolites that have been identified in each organism, along with a variety of experimental data from each molecule that can be used in subsequent experiments to confirm its observation. Several of these kinds of database efforts have been undertaken, though the organisms each targets and the types of experimental evidence and information they contain vary. Some of the earliest and most comprehensive of these databases were created for GC-MS. These libraries generally contain GC retention times for each compound as well as representative electron impact mass spectra displaying fragmentation patterns for each molecule. A variety of commercial libraries are available, as well as libraries from NIST (73) and an open-access database that is specifically focused on GC-MS data for metabolomics called the Golm Metabolome Database (74). Researchers at Scripps Research Institute have also created a separate database called METLIN that contains LC-MS data as well as high-resolution Fourier transform mass spectral data and MS/MS spectra for a variety of metabolites (75).

While the previous database efforts have focused on mass spectrometry data, multiple efforts have been established that contain mass spectra and NMR spectra as well as other kinds of physico-chemical data. Canadian researchers have recently established the Human Metabolome Database, which contains a wide assortment of data on human metabolites and metabolic pathways (76). This database includes data from over 2,500 compounds that was mined from the primary literature, other smaller database efforts, and experimental data including 1-D and 2-D NMR spectra as well as MS and MS/MS spectra. Separately, the Madison Metabolomics Consortium has established another database containing NMR and MS data for metabolites from a variety of different species (77). This database contains a variety of information on over 20,000 different compounds, including data from NMR experiments, as well as mass spectral data and information regarding chromatographic behavior of each compound on a variety of different column formats. This database is fully accessible via the Internet. The long-term goal of all of these metabolomics database projects is to compile sufficient information to allow identification of metabolites during future untargeted survey experiments based on each compound's chromatographic properties as well as a variety of mass spectral and NMR data.

While the aforementioned metabolite database projects are making significant progress, at the present time it is generally not possible to identify many metabolites of interest from general survey experiments using only these data. As a result, researchers have developed a third strategy for metabolome characterization. In these experiments, mass spectrometry-based survey experiments are performed to identify "features" and compare their intensities across multiple biological samples. Each feature is defined by a unique mass and retention time under specific LC conditions. Following statistical analysis to identify features whose abundances appear to vary with respect to the biological variable of interest, a subset of features are identified through an iterative process of comparison with standards.

The process of identifying unknown compounds can be a painstaking one. After identifying a feature of interest, the first step is to identify possible molecular formulas that would match the observed mass. When the observed mass is known to sufficient accuracy, there are often only limited numbers of combinations of atoms that could account for the observed mass. Based on the observed mass as well as some basic rules for eliminating formulas that are chemically impossible (78), it is often possible for features of small to medium mass to be assigned to particular molecular formulas. We have found that we can use metabolic labeling of selected organisms such as *Arabidopsis* to aid in formula assignment. After characterizing natural abundance plants as well as ^{15}N-labeled and ^{13}C-labeled plants, we can compare the masses of each unknown feature under different labeling conditions to determine the numbers of nitrogen and carbon atoms in each compound (79). Once these elemental counts are fixed, there are relatively few formulas that will fit the observed masses at a reasonable (3 ppm) mass error. This approach can allow assignment of unique formulas to features up to masses well above 1,000 Da, greatly aiding in formula assignment for metabolites of all sizes.

Once formulas have been identified that are consistent with the feature in question, it must then be determined which compound was actually observed. This is done by purchasing and analyzing purified standards for each compound in question and comparing retention times, MS/MS fragmentation spectra, and other available data to confirm the identity of the feature in the original sample. In practice, this can be a tedious process, as there are often many possible compounds to consider. Furthermore, not all metabolites may be purchased, meaning that some standards must be synthesized or may simply be unavailable for comparison. Even in cases where only a single molecular formula matches the observed mass, one must remember that multiple isomers may exist that could account for the observed signal. This is especially problematic for sugars. In these cases, confirmation of a compound's identity via analysis of standards is essential.

Though metabolomics is currently the newest and least mature of the systems biology approaches, the rapid development of databases and collection of data on many standard metabolites should streamline the process of metabolite identification in the years to come. With cooperation, researchers' painstaking efforts to identify unknown compounds via characterization of standards can greatly aid the development of comprehensive metabolome databases. Though metabolomics already plays an important role in systems biology, its significance will only grow as tools for its systematic study mature.

5. Integrating Different Types of Profiling Data

Systematic integration of complex types of profiling data is an obvious next step in a systems biology approach and one that will facilitate further data mining in order to fully describe genome-wide cellular dynamics and biochemical regulation. In the future, a complete understanding of these "systems-wide phenotypes" will allow us to group phenotypes with the types of perturbations (e.g., environmental, genetic) that create them. Then, utilizing this type of data will allow us to answer questions like "what different conditions result in the activation of a particular metabolic pathway?" An alternate way of framing this approach is to consider how many different ways the system of a plant can be altered before we have described the complete complement of plant phenotypes.

To address this question in a purely hypothetical manner, consider that there are approximately 28,000 genes contributing to a given phenotype in *Arabidopsis.* Imagine that the phenotype we wish to examine is the complete gene expression profile of a plant. If we assume that all the genes act completely independently, and that there are three possible states for the expression of a given gene (increased, decreased, or no change), the number of different possible patterns of gene expression is $3^{28,000}$, a number so high, it exceeds the number of atoms in the universe. On the other hand, if ALL of the 28,000 genes act in concert, there is only one phenotype. Obviously, the answer lies somewhere in between, but where?

We can create plants with 28,000 gene knockouts, and then make every possible combination of multiple knockouts, which is a huge number. Furthermore, we can apply various different environmental or chemical challenges (nutrient stress, hormone application, light changes, temperature changes, etc.), and then we can start making multiple stacks of mutations and environmental changes and so again, the number of possible perturbations explodes. In any case, the reductionist approach that we use in the lab seems most amenable to discovering what genetic and environmental changes cause the biggest changes in phenotype. That is, phenotype discovery should begin with conditions that cause large morphological and developmental changes, and should then be refined to examine more subtle changes, such as those that can be seen only at the molecular level (e.g., on a DNA microarray or through a metabolome analysis).

The study of plant systems biology will begin to approach its fullest potential only when all these types of studies are integrated. Although some studies have attempted to integrate analysis of metabolites, proteins, and RNA (80–82), this aspect of the field is largely in its infancy, due to several issues, including data quality and the availability of statistical methods and algorithms that can address the complexity of such data to extract usable information. Below, we describe some of these important considerations and then highlight some methods for integrating data from various high-throughput studies.

5.1. Data Quality

The quality of data and the validity of conclusions drawn from systems biology data are dependent upon the care with which the initial experiments were designed. Although this is certainly true with any experimental method, the importance of experimental design is perhaps no more important than in an approach that seeks to integrate many different types of data. In any systems biology study where complex methods of sample processing and data acquisition are utilized, both technical replicates, which represent multiple measurements taken from a single biological sample, and biological replicates, which represent measurements taken from multiple biological samples, can be crucial. Technical replicates allow the researcher to assess variability inherent only in the measurements, while biological replicates allow the researcher to estimate the effect of variability between biological samples. Therefore, when the aim of a study is to draw conclusions about populations, rather than individuals, biological replicates are a necessity. In statistical terms, the power of a study is defined as the probability of rejecting a false null hypothesis. Although many different methods of determining power exist (see also the chapter by Gadbury et al. in this volume), in general, including more replicates in a study increases the statistical power for that study. Determining the optimal number of replicates to provide adequate statistical power, while still considering the availability of resources including money and time, is a challenging question but one that must be addressed in a systems biology approach.

In particular, when combining data from a variety of studies, reducing extraneous factors that might confound analysis is extremely important. Furthermore, the use of complex technologies such as transcriptome, proteome, or metabolome analysis might necessitate collaborations between different labs with complementary expertise in different techniques, and so careful experimental design and oversight is of fundamental import. For example, biological sample collection should ideally be performed in large batch using well-established methods and then divided so that the same biological sample is processed using these different methods. Additionally, replicate samples should be analyzed by the same lab and by the same researcher, ideally on the same day, so as to minimize sample handling differences.

5.2. Data Analysis

Whether the researcher is interested in integrating multiple gene expression studies from multiple labs, as in the case of the AtMegaCluster, or in integrating genomic, proteomic, and metabolomic studies for a single biological question, data preprocessing to reduce systematic noise and variation is essential. For microarray analysis, this might include data normalization, data transformation, filtering, or background subtraction, and analogous methods exist for proteome and metabolome studies; especially when combining multiple studies for large-scale analysis, this step is extremely important to allow variation inherent in the data to be reduced before further analysis.

The final step in data analysis on a systems biology scale is to integrate information from different types of studies and classify the objects under study (e.g., genes from a microarray study or small molecules from a metabolome study) into different categories. Popular methods include data clustering via hierarchical methods or principal components analysis (PCA) (for review see (1)), and some groups studying plant systems biology have begun integrating different types of -omics data using these approaches (83–85). As the field of systems biology advances, we will surely see many advances on this front.

6. Future Challenges

Although systems biology analysis strategies like those described above can provide new models and hypotheses as part of a discovery engine, it is important to recognize that conclusions made from such approaches are predictions that should be rigorously tested in follow-up experiments. It is possible for statistical analyses utilized in systems-wide approaches to identify connections or correlations that are not biologically meaningful. Hence these kinds of results should be interpreted with caution and not be used for conclusions until they are corroborated via independent means.

This requirement of validating conclusions made from systems biology approaches with smaller-scale molecular biology approaches brings to light an additional consideration for any plant systems biologist. It would be extremely useful and perhaps necessary for the plant samples (tissue, extracts, etc.) used in large-scale analyses to be archived so that additional experiments could be performed in future investigations. These experiments might include studies to validate conclusions or even new methods of phenotyping that could be applied and combined with other data obtained using the same tissue sample. While there is a good deal of anecdotal evidence about the "lab effect" in high-throughput biology, we are not aware of any studies that have systematically analyzed how easy it is to replicate a particular expression profile in different labs, at different times. However, at some point in the near future, the ability to do this will be critical since the true power of a systems biology approach is that individually, a particular measurement is low in information content, but when coupled with many other measurements from labs around the world, it can provide the foundation of an encyclopedia that describes precisely how the transcriptome, proteome, and metabolome all work together in the plant.

In this chapter we have sought to provide the reader with a glimpse of the promise and challenges in genomic profiling tools used for systems biology analysis in plants. It is clear that technologies are continually changing, and these often result in less expensive methods with faster data streams. Although methods of genome-wide mRNA measurements are fairly well established, the remaining challenge is to accelerate proteomic and metabolomic measurements to improve sample throughput and data comprehensiveness. Systematic integration of RNA and DNA measurements with changes that are occurring in proteins, small molecules, and growth will be a further step in understanding cellular dynamics and biochemical regulation. Finally, progress in systems biology will be greatly facilitated by the creation of a tissue archive, so that scientists will not have to re-grow plants each time a new systems biology technique is developed. One can only hope that the students and postdoctoral associates that we are training now will not only have enough insight to devise the right experiments but also have enough foresight to plan ahead so that each generation of biologists will not have to rediscover the wheel.

References

1. Quakenbush, J. (2001) Computational analysis of microarray data. *Nat. Rev. Genet.* **2**, 418–427.
2. Stears, R.L., Martinsky, T., and Schena, M. (2003) Trends in microarray analysis. *Nat. Med.* **9**, 140–145.
3. The Arabidopsis Functional Genomics Network (AFGN). Web site: http://www.uni-tuebingen.de/plantphys/AFGN/atgenex.htm.
4. Schmid, M., Davison, T.S., Henz, S.R., Pape, U.J., Demar, M., Vingron, M., Schölkopf, B., Weigel, D., and Lohmann, J.U. (2005) A gene expression map of *Arabidopsis thaliana* development. *Nat. Genet.* **37**, 501–506.
5. Kilian, J., Whitehead, D., Horak, J., Wanke, D., Weinl, S., Batistic, O., D'Angelo, C., Bornberg-Bauer, E., Kudla, J., and Harter, K. (2007) The AtGenExpress global stress expression data set: protocols, evaluation and model data analysis of UV-B light, drought and cold stress responses. *Plant J.* **50**, 347–363.
6. Goda, H., Sasaki, E., Akiyama, K., Maruyama-Nakashita, A., Nakabayashi, K., Li, W., Ogawa, M., Yamauchi, Y., Preston, J., Aoki, K., Kiba, T., Takatsuto, S., Fujioka, S., Asami, T., Nakano, T., Kato, H., Mizuno, T., Sakakibara, H., Yamaguchi, S., Nambara, E., Kamiya, Y., Takahashi, H., Hirai, M.Y., Sakurai, T., Shinozaki, K., Saito, K., Yoshida, S., and Shimada, Y. (2008) The AtGenExpress hormone- and chemical-treatment data set: experimental design, data evaluation, model data analysis, and data access. *Plant J.* E-pub (ahead of print).
7. Swiss Federal Institute of Technology Zurich. Genevestigator. Web site: https://www.genevestigator.ethz.ch/.
8. Zimmermann, P., Hirsch-Hoffmann, M., Hennig, L., and Gruissem, W. (2004) GENEVESTIGATOR: *Arabidopsis* microarray database and analysis toolbox. *Plant Phys.* **136**, 2621–2632.
9. Zimmermann, P., Hennig, L., and Gruissem, W. (2005) Gene expression analysis and network discovery using Genevestigator. *Trends Plant Sci.* **9**, 407–409.
10. Wohlbach, D.J., Quirino, B.F., and Sussman, M.R. (2008) Analysis of the *Arabidopsis* histidine kinase ATHK1 reveals a connection between vegetative osmotic stress sensing and seed maturation. *Plant Cell.* E-pub (ahead of print).
11. Bolstad, B.M., Irizarry, R.A., Astrand, M., and Speed, T.P. (2003) A comparison of normalization methods for high density oligonucleotide array data based on variance and bias. *Bioinformatics.* **19**, 185–193.
12. Irizarry, R.A., Bolstad, B.M., Collin, F., Cope, L.M., Hobbs, B., and Speed, T.P.

(2003) Summaries of Affymetrix GeneChip probe level data. *Nucleic Acids Res.* **31**, e15.

13. Irizarry, R.A., Hobbs, B., Collin, F., Beazer-Barclay, Y.D., Antonellis, K.J., Scherf, U., and Speed, T.P. (2003) Exploration, normalization, and summaries of high density oligonucleotide array probe level data. *Biostatistics.* **4**, 249–264.
14. Bioconductor. Web site: http://bioconductor.org/.
15. Domon, B. and Aebersold, R. (2006) Mass spectrometry and protein analysis. *Science.* **312**(5771), 212–217.
16. Sadygov, R.G., Cociorva, D., and Yates, J.R. III (2004) Large-scale database searching using tandem mass spectra: looking up the answer in the back of the book. *Nat. Methods.* **1**(3), 195–202.
17. Steen, H. and Mann, M. (2004) The ABC's (and XYZ's) of peptide sequencing. *Nat. Rev. Mol. Cell Biol.* **5**(9), 699–711.
18. Gygi, S.P., Rist, B., Gerber, S.A., Turecek, F., Gelb, M.H., and Aebersold, R. (1999) Quantitative analysis of complex protein mixtures using isotope-coded affinity tags. *Nat. Biotechnol.* **17**(10), 994–999.
19. Ross, P.L., Huang, Y.N., Marchese, J.N., Williamson, B., Parker, K., Hattan, S., Khainovski, N., Pillai, S., Dey, S., Daniels, S., Purkayastha, S., Juhasz, P., Martin, S., Bartlet-Jones, M., He, F., Jacobson, A., and Pappin, D.J. (2004) Multiplexed protein quantitation in *Saccharomyces cerevisiae*using amine-reactive isobaric tagging reagents. *Mol. Cell Proteomics.* **3**, 1154–1169.
20. Yao, X., Freas, A., Ramirez, J., Demirev, P.A., and Fenslau, C. (2001) Proteolytic ^{18}O labeling for comparative proteomics: model studies with two serotypes of adenovirus. *Anal. Chem.* **73**, 2836–2842.
21. Ong, S.-E., Blagoev, B., Kratchmarova, I., Kristensen, D.B., Steen, H., Pandey, A., and Mann, M. (2002) Stable isotope labeling by amino acids in cell culture, SILAC, as a simple and accurate approach to expression proteomics. *Mol. Cell Proteomics.* **1**, 376–386.
22. Krijgsveld, J., Ketting, R.F., Mahmoudi, T., Johansen, J., Artal-Sanz, M., Verrijzer, C.P., Plasterk, R.H.A., and Heck, A.J.R. (2003) Metabolic labeling of *C. elegans*and *D. melanogaster*for quantitative proteomics. *Nat. Biotechnol.* **21**, 927–931.
23. Thelen, J.J. and Peck, S.C. (2007) Quantitative proteomics in plants: choices in abundance. *Plant Cell.* **19**(11), 3339–3346.
24. Graumann, J., Hubner, N.C., Kim, J.B., Ko, K., Moser, M., Kumar, C., Cox, J., Scholer, H., and Mann, M. (2008) Stable isotope labeling by amino acids in cell culture (SILAC) and proteome quantitation of mouse embryonic stem cells to a depth of 5,111 proteins. *Mol. Cell Proteomics.* 7(4), 672–683.
25. Baerenfaller, K., Grossmann, J., Grobei, M.A., Hull, R., Hirsch-Hoffmann, M., Yalovsky, S., Zimmermann, P., Grossniklaus, U., Gruissem, W., and Baginsky, S. (2008) Genome scale proteomics reveals *Arabidopsis thaliana*gene models and proteome dynamics. *Science.* **320**, 938–941.
26. Gygi, S.P., Rochon, Y., Franza, B.R., and Aebersold, R. (1999) Correlation between protein and mRNA abundance in yeast. *Mol. Cell Biol.* **19**(3), 1720–1730.
27. Gingras, A.C., Gstaiger, M., Raught, B., and Aebersold, R. (2007) Analysis of protein complexes using mass spectrometry. *Nat. Rev. Mol. Cell Biol.* **8**(8), 645–654.
28. Cravatt, B.F., Simon, G.M., and Yates, J.R. III (2007) The biological impact of mass-spectrometry-based proteomics. *Nature.* **450**(7172), 991–1000.
29. Blagoev, B., Kratchmarova, I., Ong, S.E., Nielsen, M., Foster, L.J., and Mann, M. (2003) A proteomics strategy to elucidate functional protein–protein interactions applied to EGF signalling. *Nat. Biotechnol.* **21**(3), 315–318.
30. Pflieger, D., Junger, M.A., Muller, M., Rinner, O., Lee, H., Gehrig, P.M., Gstaiger, M., and Aebersold, R. (2008) Quantitative proteomic analysis of protein complexes: concurrent identification of interactors and their state of phosphorylation. *Mol. Cell Proteomics.* 7(2), 326–346.
31. Dharmasiri, N., Dharmasiri, S., and Estelle, M. (2005) The F-box protein TIR1 is an auxin receptor. *Nature.* **435**(7041), 441–445.
32. Kepinski, S. and Leyser, O. (2005) The Arabidopsis F-box protein TIR1 is an auxin receptor. *Nature.* **435**(7041), 446–451.
33. Doherty, M.K. and Beynon, R.J. (2006) Protein turnover on the scale of the proteome. *Expert Rev. Proteomics.* **3**(1), 97–110.
34. Pratt, J.M., Petty, J., Riba-Garcia, I., Robertson, D.H., Gaskell, S.J., Oliver, S.G., and Beynon, R.J. (2002) Dynamics of protein turnover, a missing dimension in proteomics. *Mol. Cell Proteomics.* **1**(8), 579–591.
35. Rao, P.K., Roxas, B.A.P., and Li, Q. (2008) Determination of global protein turnover in

stressed mycobacterium cells using hybrid-linear ion trap-fourier transform mass spectrometry. *Anal. Chem.* **80**, 396–406.

36. Doherty, M.K., Whitehead, C., McCormack, H., Gaskell, S.J., and Beynon, R.J. (2005) Proteome dynamics in complex organisms: using stable isotopes to monitor individual protein turnover rates. *Proteomics.* **5**(2), 522–533.
37. Gruhler, A., Schulze, W.X., Matthiesen, R., Mann, M., and Jensen, O.N. (2005) Stable isotope labeling of Arabidopsis thaliana cells and quantitative proteomics by mass spectrometry. *Mol. Cell Proteomics.* **4**(11), 952–964.
38. Kim, J.K., Harada, K., Bamba, T., Fukusaki, E.-I., and Bobayashi, A. (2005) Stable isotope dilution-based accurate comparative quantification of nitrogen-containing metabolites in *Arabidopsis thaliana*T87 cells using in vivo ^{15}N-isotope enrichment. *Biosci. Biotechnol. Biochem.* **69**(7), 1331–1340.
39. Harada, K., Fukusaki, E., Bamba, T., Sato, F., and Kobayashi, A. (2006) In vivo ^{15}N-enrichment of metabolites in suspension cultured cells and its application to metabolomics. *Biotechnol. Prog.* **22**(4), 1003–1011.
40. Engelsberger, W.R., Erban, A., Kopka, J., and Schulze, W.X. (2006) Metabolic labeling of plant cell cultures with $K^{15}NO_3$as a tool for quantitative analysis of proteins and metabolites. *Plant Methods.* **2**, 14–25.
41. Lanquar, V., Kuhn, L., Lelievre, F., Khafif, M., Espagne, C., Bruley, C., Barbier-Brygoo, H., Garin, J., and Thomine, S. (2007) ^{15}N-metabolic labeling for comparative plasma membrane proteomics in Arabidopsis cells. *Proteomics.* 7(5), 750–754.
42. Ippel, J.H., Pouvreau, L., Kroef, T., Gruppen, H., Versteeg, G., van den Putten, P., Struik, P.C., and van Mierlo, C.P. (2004) In vivo uniform ^{15}N-isotope labelling of plants: using the greenhouse for structural proteomics. *Proteomics.* **4**(1), 226–234.
43. Nelson, C.J., Huttlin, E.L., Hegeman, A.D., Harms, A.C., and Sussman, M.R. (2007) Implications of ^{15}N-metabolic labeling for automated peptide identification in *Arabidopsis thaliana. Proteomics.* 7(8), 1279–1292.
44. Huttlin, E.L., Hegeman, A.D., Harms, A.C., and Sussman, M.R. (2007) Comparison of full versus partial metabolic labeling for quantitative proteomics analysis in Arabidopsis thaliana. *Mol. Cell Proteomics.* **6**(5), 860–881.
45. Hebeler, R., Oekjeklaus, S., Reidegeld, K.A., Eisenacher, M., Staphan, C., Sitek, B., Stuhler, K., Meyer, H.E., Sturre, M.J., Dijkwel, P.P., and Warscheid, B. (2008) Study of early leaf senescence in *Arabidopsis thaliana*by quantitative proteomics using reciprocal ^{14}N/^{15}N labeling and difference gel electrophoresis. *Mol. CellProteomics.* 7(1), 108–120.
46. Maor, R., Jones, A., Nuhse, T.S., Studholme, D.H., Peck, S.C., and Shirasu, K. (2007) Multidimensional protein identification technology (MudPIT) analysis of ubiquitinated proteins in plants. *Mol. Cell Proteomics.* **6**(4), 601–610.
47. Fitchette, A.C., Dinh, O.T., Faye, L., and Bardor, M. (2007) Plant proteomics and glycosylation. *Methods Mol. Biol.* **355**, 317–342.
48. Peck, S.C. (2006) Phosphoproteomics in Arabidopsis: moving from empirical to predictive science. *J. Exp. Bot.* **57**(7), 1523–1527.
49. Ding, S.J., Qian, W.J., and Smith, R.D. (2007) Quantitative proteomics approaches for studying phosphotyrosine signaling. *Expert Rev. Proteomics.* **4**(1), 13–23.
50. Vener, A.V., Harms, A., Sussman, M.R., and Vierstra, R.D. (2001) Mass spectrometric resolution of reversible protein phosphorylation in photosynthetic membranes of *Arabidopsis thaliana. J. Biol. Chem.* **276**(10), 6959–6966.
51. Ficarro, S.B., McCleland, M.L., Stukenberg, P.T., Burke, D.J., Ross, M.M., Shabanowitz, J., Hunt, D.F., and White, F.M. (2002) Phosphoproteome analysis by mass spectrometry and its application to *Saccharomyces cerevisiae. Nat. Biotechnol.* **20**, 301–305.
52. Nuhse, T.S., Stensballe, A., Jensen, O.N., and Peck, S.C. (2003) Large-scale analysis of in vivo phosphorylated membrane proteins by immobilized metal ion affinity chromatography and mass spectrometry. *Mol. Cell Proteomics.* **2**(12), 1261–1270.
53. Nuhse, T., Yu, K., and Salomon, A. (2007) Isolation of phosphopeptides by immobilized metal ion affinity chromatography. *Curr. Protoc. Mol. Biol.* **18**, 18.13.
54. Pinkse, M.W., Uitto, P.M., Hilhorst, M.J., Ooms, B., and Heck, A.J. (2004) Selective isolation at the femtomole level of phosphopeptides from proteolytic digests using 2D-nano-LC-ESI_MS/MS and titanium oxide precolumns. *Anal. Chem.* **96**(14), 3935–3943.
55. Beausoleil, S.A., Hedrychowski, M., Schwartz, D., Elias, J.E., Villen, J., Li, J.,

Cohn, M.A., Cantley, L.C., and Gygi, S.P. (2004) Large-scale characterization of HeLa cell nuclear phosphoproteins. *Proc. Nat. Acad. Sci. USA.* **101**(33), 12130–12135.

56. Gruhler, A., Olsen, J.V., Mohammed, S., Mortensen, P., Faergeman, N.J., Mann, M., and Jensen, O.N. (2005) Quantitative phosphoproteomics applied to the yeast pheromone signaling pathway. *Mol. Cell Proteomics.* **4**(3), 310–327.
57. Kelleher, N.L., Zubarev, R.A., Bush, K., Furie, B., Furie, B.C., McLafferty, F.W., and Walsh, C.T. (1999) Localization of labile posttranslational modifications by electron capture dissociation: the case of gamma-carboxyglutamic acid. *Anal. Chem.* **71**(19), 4250–4253.
58. Zubarev, R.A., Horn, D.M., Fridriksson, E.K., Kelleher, N.L., Kruger, N.A., Lewis, M.A., Carpenter, B.K., and McLafferty, F.W. (2000) Electron capture dissociation for structural characterization of multiply charged protein cations. *Anal. Chem.***72**(3), 563–573.
59. Syka, J.E., Coon, J.J., Schroeder, M.J., Shabanowitz, J., and Hunt, D.F. (2004) Peptide and protein sequence analysis by electron transfer dissociation mass spectrometry. *Proc. Natl. Acad. Sci. USA.* **101**(26), 9528–9533.
60. Gerber, S.A., Rush, J., Stemman, O., Kirschner, M.W., and Gygi, S.P. (2003) Absolute quantification of proteins and phosphoproteins from cell lysates by tandem MS. *Proc. Nat. Acad. Sci. USA.* **100**(12), 6940–6945.
61. Hegeman, A.D., Harms, A.C., Sussman, M.R., Bunner, A.E., and Harper, J.F. (2004) An isotope labeling strategy for quantifying the degree of phosphorylation at multiple sites in proteins. *J. Am. Soc. Mass Spectrom.* **15**(5), 647–653.
62. Dunn, W.B., Bailey, N.J.C., and Johnson, H.E. (2005) Measuring the metabolome: current analytical technologies. *Analyst.* **130**, 606–625.
63. Marshall, E. (2007) Metabolic Research: Canadian group claims "unique" database. *Science.* **315**, 583–584.
64. Fiehn, O., Kloska, S., and Altmann, T. (2001) Integrated studies on plant biology using multiparallel techniques. *Curr. Opin. Biotechnol.* **12**(1), 82–86.
65. Kimball, E. and Rabinowitz, J.D. (2006) Identifying decomposition products in extracts of cellular metabolites. *Anal. Biochem.* **358**(2), 273–280.
66. Want, E.J., O'Maille, G., Smith, C.A., Brandon, T.R., Uritboonthai, W., Qin, C., Trauger, S.A., and Siuzdak, G. (2006) Solvent-dependent metabolite distribution, clustering, and protein extraction for serum profiling with mass spectrometry. *Anal. Chem.* **78**(3), 743–752.
67. Rabinowitz, J.D. and Kimball, E. (2007) Acidic acetonitrile for cellular metabolome extraction from Escherichia coli. *Anal. Chem.* **79**(16), 6167–6173.
68. Lewis, I.A., Schommer, S.C., Hodis, B., Robb, K.A., Tonelli, M., Westler, W.M., Sussman, M.R., and Markley, J.L. (2007) Method for determining molar concentrations of metabolites in complex solutions from two-dimensional ^{1}H-^{13}C NMR spectra. *Anal. Chem.* **79**(24), 9385–9390.
69. Want, E.J., Nordstrom, A., Morita, H., and Suizdak, G. (2007) From exogenous to endogenous: the inevitable imprint of mass spectrometry in metabolomics. *J. Proteome Res.* **6**(2), 459–468.
70. Bajad, S.U., Lu, W., Kimball, E.H., Yuan, J., Peterson, C., and Rabinowitz, J.D. (2006) Separation and quantitation of water soluble cellular metabolites by hydrophilic interaction chromatography-tandem mass spectrometry. *J. Chromatogr.* **1125**(1), 76–88.
71. Nordstrom, A., Want, E., Northen, T., Lehtio, J., and Siuzdak, G. (2008) Multiple ionization mass spectrometry strategy used to reveal the complexity of metabolomics. *Anal. Chem.* **80**(2), 421–429.
72. Smith, C.A., Want, E.J., O'Maille, G., Abagyan, R., and Suizdak, G.(2006) XCMS: processing mass spectrometry data for metabolite profiling using nonlinear peak alignment, matching, and identification. *Anal. Chem.* **78**(3), 779–787.
73. NIST Website: http://www.nist.gov/srd/nist1.htm.
74. Kopka, J., Schauer, N., Krueger, S., Birkemeyer, C., Usadel, B., Bergmuller, E., Dormann, P., Weckwerth, W., Gibon, Y., Stitt, M., Willmitzer, L., Fernie, A.R., and Steinhauser, D. (2005) GMD@CSB.DB: the Golm metabolome database. *Bioinformatics.* **21**(8), 1635–1638.
75. Smith, C.A., O'Maille, G., Want, E.J., Qin, C., Trauger, S.A., Brandon, T.R., Custodio, D.E., Abagyan, R., and Siuzdak, G. (2005) METLIN: a metabolite mass spectra database. *Ther. Drug Monit.* **27**(6), 747–751.
76. Wishart, D.S., Tzur, D., Knox, C., Eisner, R., Guo, A.C., Young, N., Cheng, D., Jewell, K., Arndt, D., Sawhney, S., Fung, C., Nikolai,

L., Lewis, M., Coutouly, M.-A., Forsythe, I., Tang, P., Shrivastava, S., Jeroncic, K., Stothard, P., Amegbey, G., Block, D., Hau, D.D., Wagner, J., Miniaci, J., Clements, M., Gebremedhin, M., Guo, N., Zhang, Y., Duggan, G.E., MacInnis, G.D., Weljie, A.M., Dowlatabadi, R., Bamforth, F., Clive, D., Greiner, R., Li, L., Marrie, T., Sykes, B.D., Vogel, H.J., and Querengesser, L. (2007) HMDB: the human metabolome database. *Nucleic Acids Res.* **35**, D521–D526.

77. Cui, Q., Lewis, I.A., Hegeman, A.D., Anderson, M.E., Li, J., Schulte, C.F., Westler, W.M., Eghbalnia, H.R., Sussman, M.R., and Markley, J.R. (2008) Metabolite identification via the Madison metabolomics consortium database. *Nat. Biotechnol.* **26**(2), 162–164.

78. Kind, T. and Fiehn, O. (2007) Seven golden rules for heuristic filtering of molecular formulas obtained by accurate mass spectrometry. *BMC Bioinformatics.* **27**(8), 105.

79. Hegeman, A.D., Schulte, C.F., Cui, Q., Lewis, I.A., Huttlin, E.L., Eghbalnia, H., Harms, A.C., Ulrich, E.L., Markley, J.L., and Sussman, M.R. (2007) Stable isotope assisted assignment of elemental compositions for metabolomics. *Anal. Chem.* **79**(18), 6912–6921.

80. Weckwerth, W., Wenzel, K., and Fiehn, O. (2004) Process for the integrated extraction, identification, and quantification of metabolites, proteins, and RNA to reveal their co-regulation in biochemical networks. *Proteomics.* **4**, 78–83.

81. Frey, I.M., Rubio-Aliaga, I., Siewert, A., Sailer, D., Drobyshev, A., Beckers, J., de Angelis, M.H., Aubert, J., Hen, A.B., Fiehn, O., Eichinger, H.M., and Daniel, H. (2007) Profiling at mRNA, protein, and metabolite levels reveals alterations in renal amino acid handling and glutathione metabolism in kidney tiddue of Pept2-/- mice. *Physiol. Genomics.* **28**, 301–310.

82. Trauger, S.A., Kalizak, E., Kalisiak, J., Morita, H., Weinberg, M.V., Menon, A.L., Poole, F.L. II, Adams, M.W.W., and Siuzdak, G. (2008) Correlating the transcriptome, proteome, and metabolome in the environmental adaptation of a hyperthermophile. *J. Proteome Res.* **7**, 1027–1035.

83. Schauer, N., Semel, Y., Roessner, U., Gur, A., Balbo, I., Carrari, F., Pleban, T., Perez-Melis, A., Bruedigam, C., Kopka, J., Willmitzer, L., Zamir, D., and Fernie, A.R. (2006) Comprehensive metabolic profiling and phenotyping of interspecific introgression lines for tomato improvement. *Nat. Biotech.* **24**, 447–454.

84. Lu, Y., Savage, L.J., Ajjawi, I., Imre, K.M., Yoder, D.W., Benning, C., DellaPenna, D., Ohlrogge, J.B., Osteryoung, K.W., Weber, A.P., Wilkerson, C.G., and Last, R.L. (2008) New connections across pathways and cellular processes: industrialized mutant screening reveals novel associations between diverse phenotypes in *Arabidopsis. Plant Physiol.* **146**, 1482–1500.

85. Wienkoop, S., Morgenthal, K., Wolschin, F., Scholz, M., Selbig, J., and Weckwerth, W. (2008) Integration of metabolomic and proteomic phenotypes – analysis of data-covariance dissects starch and RFO metabolism from low and high temperature response in *Arabidopsis thaliana. Mol. Cell Proteomics.* 7, 1725–1736.

Index

Note: The letters '*f*', '*n*' and '*t*' following locators refer to figure, note number and table respectively

D

E

F

G

H

I

J

K

L

M

MIX
Papier aus verantwortungsvollen Quellen
Paper from responsible sources
FSC® C105338

If you have any concerns about our products,
you can contact us on
ProductSafety@springernature.com

In case Publisher is established outside the EU,
the EU authorized representative is:
Springer Nature Customer Service Center GmbH
Europaplatz 3, 69115 Heidelberg, Germany

Printed by Libri Plureos GmbH
in Hamburg, Germany